CLIFFS KEYNOTE REVIEWS

Algebra

by

KAJ L. NIELSEN

and

CHARLOTTE M. GEMMEL

CLIFF'S NOTES, INC. • LINCOLN, NEBRASKA 68501

ISBN 0-8220-1760-1

© Copyright 1969

BY CLIFF'S NOTES, INC.

All rights reserved. No part of this book may be reproduced or utilized in any form or by any means, electronic or mechanical, including photocopying or recording, or by any information storage and retrieval system, without permission in writing from the publisher.

L. C. Catalogue Card Number: 69-18677

Printed in the United States of America

CONTENTS

1	Sets	Page 1
2	Relations between Sets	7
3	Operations on Sets	13
4	Axiomatic Systems	19
5	The Natural Numbers	25
6	Mathematical Induction: Problems	31
7	The Real Numbers	37
8	The Complex Numbers	43
9	Algebraic Expressions: Polynomials	49
10	Algebraic Expressions: Fractional and Radical	55
11	Relations and Functions	61
12	Equations	67
13	The Linear Equation	73
14	Matrices and Determinants	79
15	Systems of Linear Equations	85
16	The Quadratic Equation	91
17	Systems of Quadratic Equations	97
18	The Conics	103
19	Equations of Higher Degree	109
20	Inequalities	115
21	Exponential and Logarithmic Functions	121
22	Sequences, Series, Limits	127
23	Permutations and Combinations	133
24	Probability	139
25	Algebra of Ordered Pairs, Vectors	145
	Final Examination	151
	Appendices	159
	Tables	166
	Dictionary-Index	175
	Graph Paper	181

TO THE STUDENT

This KEYNOTE is a flexible study aid designed to help you REVIEW YOUR COURSE QUICKLY and USE YOUR TIME TO STUDY ONLY MATERIAL YOU DON'T KNOW.

FOR GENERAL REVIEW

Take the SELF-TEST on the first page of any topic and turn the page to check your answers.

Read the SOLUTIONS of any questions you answered incorrectly.

If you are satisfied with your understanding of the material, move on to another topic.

If you feel that you need further review, read the column of BASIC FACTS. Then, to check your comprehension of the material, do the EXERCISES immediately following the group of problems which gave you difficulty. The solutions to the EXERCISES appear on the final page of each topic.

For a more detailed discussion of the material, read the column of ADDITIONAL INFORMATION.

FOR QUICK REVIEW

Read the column of BASIC FACTS for a rapid review of the essentials of a topic and then take the SELF-TEST if you want to test your understanding of the material.

ADDITIONAL HELP FOR EXAMS

Review terms in the DICTIONARY-INDEX.
Test yourself by taking the sample FINAL EXAM.

1

SETS

SELF-TEST

DIRECTIONS: Write your answers in the numbered regions to the right. To check your answers, turn the page. Study the solutions to the problems you missed, and do the exercises following any problem or group of problems in which you had errors.

1. Describe the set of counting numbers less than six by the roster method.

2. Describe the set of counting numbers less than six by the set-builder method.

3. Write a mathematical symbol for the set of integers between 8 and 9.

4. Write mathematical symbols for four half-lines in Figure 1-1.

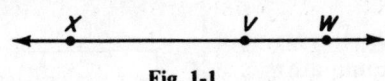

Fig. 1-1

5. Is the set of real numbers in the interval $[-6, 0]$ finite or infinite?

6. Is the point set $\{P\}$ finite or infinite?

7. Is the following statement true or false?

    The set of integers is dense.

8. Is the following statement true or false?

    The interval $(-4, -3]$ is an infinite set.

9. What sort of geometric object is indicated by $\overset{\circ}{\overline{RQ}}$?

10. Locate and reproduce the representation of $\overleftrightarrow{DB}$ in Figure 1-2.

11. Locate and reproduce the representation of $\angle ABC$ in Figure 1-2.

12. Locate and reproduce the representation of D in Figure 1-2.

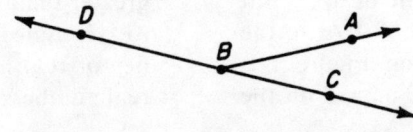

Fig. 1-2

SOLUTIONS

1. The set of counting numbers consists of integers greater than zero. In the roster, or tabulation, method, this set is written $\{1, 2, 3, 4, 5\}$.

2. The set of counting numbers less than six can be described in several different set-builder ways. Three possibilities are given.

$$\{x \text{ (integer)} \mid 0 < x < 6\}$$
$$\{x \text{ (counting number)} \mid x \leq 5\}$$
$$\{x \text{ (counting number)} \mid x < 6\}$$

EXERCISES I

Describe each of the given sets by the roster method and the set-builder method.

i. {fractions whose numerators are seven and whose denominators are multiples of three greater than six and less than twenty-one}

ii. the set consisting of the three smallest integers larger than eleven

iii. {integers less than -4 and greater than or equal to -9}

---

3. Since there is no integer between 8 and 9, {integers between 8 and 9} is the empty set, ϕ or $\{\ \}$.

4.

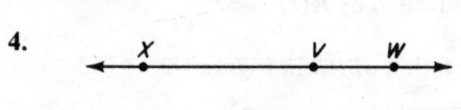

Fig. 1-1

The half-lines, $\overrightarrow{XW}$, $\overrightarrow{VW}$, $\overrightarrow{WX}$, and $\overrightarrow{VX}$ are represented on this figure. Of course, there are also half-lines to the left of X and to the right of W. But since no points to the left of X or to the right of W are named on Figure 1-1, these half-lines cannot be named on the basis of the figure.

EXERCISES II

Write each of the following expressions in mathematical symbols.

i. the set which contains no elements

ii. a is a member of the set B

iii. the open interval of real numbers from -4 to 4

iv. the half-open, half-closed interval of real numbers from 1 to 2

v. the universal set

---

5. $[-6, 0]$ is an infinite set because the real numbers are dense. That is, between any two real numbers there is another real number.

6. The point set $\{P\}$ is a finite set containing one element.

EXERCISES III

Are the following sets finite or infinite?

i. $\overleftrightarrow{PQ}$

ii. {real numbers}

iii. {letters of the English alphabet}

iv. {grains of sand on the earth}

v. $\{x \mid x \text{ is a counting number}\}$

vi. $\{2, 4, 6, \ldots, 22, 24, 26\}$

vii. $\{2, 4, 6, \ldots\}$

viii. $\overline{AB}$

---

7. The set of integers is not dense because there are no integers between two consecutive integers.

8. $(-4, -3]$ contains all real numbers greater than -4 and less than or equal to -3. Since there are an infinite number of real numbers between any two real numbers, this interval is an infinite set.

EXERCISES IV

Are the following statements true or false?

i. Any point set which extends indefinitely in two directions is a line.

ii. Two rays with a common endpoint form an angle.

iii. The interval [2, 90) is an infinite set.

iv. A ray is a point set which extends indefinitely in two directions.

v. A line segment extends indefinitely in no direction.

vi. $\overset{\circ}{A}\overset{\circ}{B}$ is a point set.

vii. $\overset{\circ}{A}\overset{\circ}{B}$ is an infinite set.

9. '$\overset{\circ}{RQ}$' names a line segment which contains the point Q and all points between R and Q. $\overset{\circ}{RQ}$ does not contain the point R.

EXERCISES V

What sort of geometric object is indicated by each symbol?

i. $\overrightarrow{QR}$ iii. $\overline{TR}$ v. $\overset{\circ}{XZ}$
ii. $\overleftrightarrow{ST}$ iv. $\overrightarrow{VW}$ vi. $\angle ABC$

10. The line DB contains D, B, and all points to the left and right of D and B.

11. The angle ABC contains $\overrightarrow{BA}$ and $\overrightarrow{BC}$.

12. $\{D\}$ is a point set with one member.

Solutions to the exercises appear on page 6.

ANSWERS

1. $\{1, 2, 3, 4, 5\}$

2. $\{x \text{ (counting number)} \mid x < 6\}$

3. ϕ

4. $\overrightarrow{VW}, \overrightarrow{VX}, \overrightarrow{WX}, \overrightarrow{XW}$

5. infinite

6. finite

7. false

8. true

9. line segment open at one end

10.

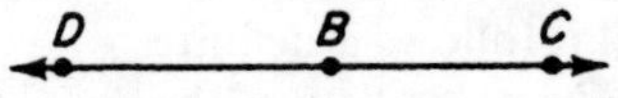

```
D       B       C
●───────●───────●
```

11.

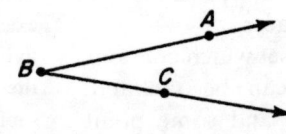

12. D
    ●

BASIC FACTS

A **set** is any collection of objects. The objects which constitute a set are called its **elements** or its **members**. Capital letters ($A, B, C, \ldots$) are usually used to denote sets. Lower-case letters ($a, b, c, \ldots$) are usually used to denote their elements.

The symbol, $\in$, indicates that an object is a member of a set. $a \in B$ is read "a is an element of B." $a \notin B$ is read "a is not an element of B."

A **variable** is a symbol used to represent various elements of a given set. The given set is called the **replacement set** or the **domain** of the variable.

The **universal set** is that set which, in a particular discussion, contains all the elements under consideration. The universal set is denoted U or I.

The **null set**, or the **empty set**, is the set which contains no elements. The null set is denoted ϕ or $\{\ \}$.

A **finite set** is any set which contains a finite (countable) number of elements.

A finite set is often denoted by listing its elements between braces. If the set is large, however, and if it follows a definite pattern, it may be given by listing the first few elements, three dots, and the last element, all between braces.

An **infinite set** is any set which contains more elements than can be counted. Some sets of numbers and some point sets are infinite sets.

An infinite set can be given by listing a few of its elements and writing three dots before or after them to indicate that an infinite number of elements have not been mentioned.

The **set of real numbers** is an infinite set. It has no greatest member, it has no least member, and it is dense—between any two members there are more members than can be counted.

ADDITIONAL INFORMATION

There are three commonly-employed methods of describing a set in terms of the brace, $\{\ \}$, which is read, "the set whose elements are." The **roster, or tabulation, method** consists in listing each element of the set within the brace. In such a list, no element should be mentioned more than once. If practical or theoretical reasons necessitate omitting mention of some elements, three dots are used to indicate this. The **descriptive-phrase method** consists in placing within the brace a phrase which clearly describes the elements of the set. This method is advantageous when a set contains a large number of elements or when all the elements cannot be named. The **rule, or set-builder, method** consists in stating one or more properties possessed by those and only those objects which are elements of the set. A description given according to the set-builder method is written in the form $\{x \mid p(x)\}$, where the vertical bar is read "such that." $\{x \mid p(x)\}$ is read, "the set of all x such that x has the property indicated by $p(x)$."

A is the set of all positive integers less than five	
tabulation method	$A = \{1, 2, 3, 4\}$
descriptive-phrase method	$A = \{\text{positive integers less than five}\}$
set-builder method	$A = \{x \text{ (integer)} \mid 0 < x < 5\}$

It is important to note that a set is determined by its elements and not by the method according to which it is described. If two different descriptions indicate precisely the same elements, they refer to the same set. The set specified by $\{a, e, i, o, u\}$, for example, is identical with the set specified by {the vowels of the English alphabet}.

A set may be **finite** even though it contains an extremely large number of elements. It would be a long and difficult job, for example, to name and count the members of the set of persons living in the city of New York at the present time; but, in principle, the members of this set could be enumerated, and therefore the set is finite. The members of an **infinite set**, on the other hand, can not be counted. No list could mention all of them because no matter how many are mentioned, there are always more.

The **set of counting numbers** is an example of an infinite set which could be described,

$$A = \{1, 2, 3, 4, 5, 6, 7, 8, 9, 10, 11, 12, 13, 14, \ldots\}.$$

It seems fairly obvious that no matter how far this process of listing counting numbers is carried, there will always be counting numbers which are not mentioned on the list. The three dots after '14' indicate this.

The **set of integers**, $\{\ldots, -2, -1, 0, 1, 2, \ldots\}$, is another example of an infinite set. The three dots at the beginning of the roster

Since there are an infinite number of real numbers between any two real numbers, **intervals of real numbers** are also infinite sets.

A **closed interval** is a set of real numbers, x, such that $a \leq x \leq b$ where $a < b$. A closed interval is written, $[a, b]$.

An **open interval** is a set of real numbers, x, such that $a < x < b$ where $a < b$. An open interval is written, (a, b).

A **half-open, half-closed interval** is a set of real numbers, x, such that $a < x \leq b$ where $a < b$. This interval is written, $(a, b]$.

A **half-closed, half-open interval** is a set of real numbers, x, such that $a \leq x < b$ where $a < b$. This interval is written, $[a, b)$.

A **point set** is any set consisting of points of Euclidean space. Some simple point sets are illustrated in Figure 1-3.

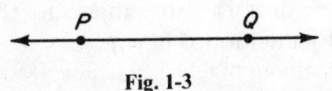

Fig. 1-3

A **line** is a set of points along a straight path determined by any two points of Euclidean space. A line extends indefinitely in both directions and constitutes an infinite set. The line in Figure 1-3 is denoted, $\overleftrightarrow{PQ}$.

A **line segment** is a set of points between two points (**endpoints**) on a line. If the endpoints are included, the segment is denoted $\overline{PQ}$; if they are not included, it is denoted $\overset{\circ\circ}{PQ}$. Since there are an infinite number of points between any two points in Euclidean space, a line segment is an infinite set.

Algebra is concerned mainly with sets of numbers and sets of expressions which represent numbers. Sets of points are used as a pictorial representation of sets of numbers.

description indicate that the set has no least member, and the three dots at the end indicate that the set has no greatest member.

The **set of real numbers** also has no least member and no greatest member. In addition, it has the property that between any two real numbers there is an infinite number of real numbers. This fact can be illustrated by considering the real numbers 0 and 1. Half-way between these numbers is the real number $\frac{1}{2}$; and half-way between 0 and $\frac{1}{2}$ is the real number $\frac{1}{4}$. Half-way between 0 and $\frac{1}{4}$ is the real number $\frac{1}{8}$; and half-way between 0 and $\frac{1}{8}$ is the real number $\frac{1}{16}$. Since this process can be continued indefinitely, there are more numbers than can be counted between 0 and 1.

In addition to the methods of description mentioned above, **point sets** can be represented by pictures of geometric objects. Figure 1-4 contains pictures of several point sets. The figure as a whole represents a **plane**, a set of points which extends indefinitely in all directions in a flat region of space.

The line $\overleftrightarrow{AB}$, which extends indefinitely in two directions, divides the plane into **half-planes**. Each half-plane contains all the points on one side of the line, but neither contains the line itself. The point C divides the line $\overleftrightarrow{AB}$ into two **half-lines**. Each half-line contains all the points on one side of the point C on the line AB, but neither includes the point C. The half-line above C is denoted, $\overset{\circ\to}{CB}$ or $\overset{\circ\to}{BC}$; and the half-line below C is denoted, $\overset{\circ\to}{AC}$ or $\overset{\circ\to}{CA}$.

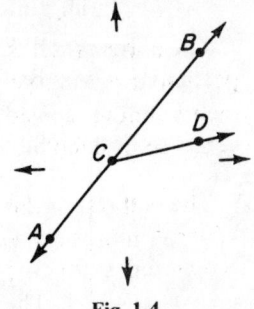

Fig. 1-4

The geometric object which includes the point C and all the points above C is the **ray**, $\overrightarrow{CB}$. A ray is a set of points on a line including an endpoint and all the points on one side of the endpoint. Thus, $\overrightarrow{AC}$ (or $\overrightarrow{CA}$) and $\overrightarrow{CD}$ are also rays. If two rays have a common endpoint, they form an **angle**. The rays $\overrightarrow{CA}$ and $\overrightarrow{CD}$ form $\angle ACD$ (read, "angle ACD"). **Line segments** represented on Figure 1-4 include $\overline{AB}$, $\overset{\circ\circ}{AB}$, $\overline{DC}$, $\overline{CB}$, and $\overline{BC}$. $\overset{\circ\circ}{AB}$, the symbol for an open-ended line segment, is usually read, "the set of all points between A and B"; and $\overline{AB}$, the symbol for a line segment including endpoints, is usually read "the set of all points from A to B."

The plane, the half-plane, the line, the half-line, the ray, the angle, and the line segment are all **infinite point sets**. All points sets, however, are not infinite. The set consisting of the point C and the set consisting of the points A, B, and C are examples of **finite point sets.**

Point sets are sometimes denoted by the capital Greek letters Γ (gamma) and Δ (delta) rather than by capital English letters. The point set consisting of the points A, B, and C, for example, may be given by, $\Gamma = \{A, B, C\}$.

SOLUTIONS TO EXERCISES

I

i. {fractions whose numerators are 7 and whose denominators are multiples of 3 greater than 6 and less than 21}

Roster Method: $\{\frac{7}{9}, \frac{7}{12}, \frac{7}{15}, \frac{7}{18}\}$

Set-builder Method:
$$\{y \mid y = \tfrac{7}{3x}, x \text{ is an integer}, 2 < x < 7\}$$

ii. the set consisting of the three smallest integers larger than eleven

Roster Method: $\{12, 13, 14\}$

Set-Builder Method:
$$\{x \text{ (integer)} \mid 11 < x < 15\}$$

iii. {integers less than -4 and greater than or equal to -9}

Roster Method:
$$\{-9, -8, -7, -6, -5\}$$

Set-Builder Method:
$$\{x \text{ (integer)} \mid -9 \leq x < -4\}$$

II

i. The empty set is denoted ϕ or $\{\ \}$.

ii. $a \in B$ is read, "a is a member of the set B."

iii. $(-4, 4)$ is the interval of real numbers between -4 and 4. The real numbers -4 and 4 are not members of this set.

iv. $(1, 2]$ is the interval containing all real numbers x such that $1 < x \leq 2$.

v. The universal set is denoted U.

III

i. $\overleftrightarrow{PQ}$ is an infinite set.

ii. The set of real numbers is an infinite set.

iii. The set of letters of the English alphabet is a finite set.

iv. The set of grains of sand on the earth is a very large, but finite, set. Even though it would be practically impossible to count the members of this set, the process of counting them would, in principle, come to an end.

v. The set of counting numbers is an infinite set.

vi. $\{2, 4, 6, \ldots, 26\}$ is a finite set. The elements not mentioned can be ascertained by following the pattern indicated by the first three elements and ending with the element which precedes 26. Caution must be exercised when using the three dots in this way, because it is possible for different patterns to be represented by a series of three elements.

vii. $\{2, 4, 6, \ldots\}$ is an infinite set. The three dots with no element following them indicate that the series continues indefinitely in the manner shown by the first three elements. (This use of the three dots is also subject to the difficulty mentioned in **vi**.)

viii. $\overline{AB}$ is an infinite set.

IV

i. false (An angle is a counterexample.)
ii. true   **v.** true
iii. true   **vi.** true
iv. false   **vii.** true

V

	Symbol	Object Denoted
i.	$\overrightarrow{QR}$	ray
ii.	$\overleftrightarrow{ST}$	line
iii.	$\overline{TR}$	line segment
iv.	$\overset{\circ}{\overrightarrow{VW}}$	half-line
v.	$\overset{\circ}{XZ}$	line segment open at X
vi.	$\angle ABC$	angle

2 RELATIONS BETWEEN SETS
SELF-TEST

DIRECTIONS: Write your answers in the numbered regions to the right. To check your answers, turn the page. Study the solutions to the problems you missed, and do the exercises following any problem or group of problems in which you had errors.

Use Figure 2-1 to determine the truth or falsity of the statements in **1–4**.

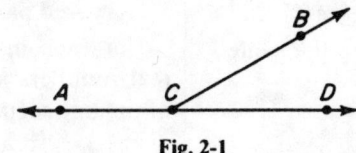

Fig. 2-1

1. $\{A\} \subseteq \angle BCD$
2. $\overleftrightarrow{AC} = \overleftrightarrow{CD}$
3. $\overline{AC} = \overline{CD}$
4. $\overline{AC} \subset \overline{CD}$

5. The set {Chris, Bill, Cathy} is
   a. equivalent to the set {1, 3, 10}
   b. equal to the set {3, 10, 1}
   c. identical to the set {4, 3, 2, 1}
   d. none of these

6. Represent [0, 2) as a line segment.

7. List all of the subsets of {3, 6, 9}.

8. Which of the subsets in problem 7 are proper subsets?

9. How many subsets has the set $\{\alpha, \beta, \gamma, \delta\}$?

10. In a math class, 85% of the students passed the final exam. Is this a proper subset of the math class?

11. If $A = (5, 9]$ and $B = (5, 9)$, is $B \subset A$?

12. If $R = \{3, 2, 1\}$ and $T = \{4, 5\}$, form the set $R \times T$ using the tabulation method.

13. What is the cardinal number of the set formed in **12**?

14. Can a one-to-one correspondence be established between the set of real numbers and the set of grains of sand on the earth?

15. Graph the set of real numbers.

SOLUTIONS

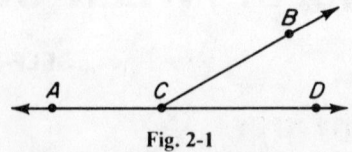

Fig. 2-1

1. $\{A\} \subseteq \angle BCD$ FALSE
   On the basis of Figure 2-1, $\angle BCD$ consists of the rays $\vec{CB}$ and $\vec{CD}$. The point A falls outside both rays.

2. $\overleftrightarrow{AC} = \overleftrightarrow{CD}$ TRUE
   On the basis of Figure 2-1, '$\overleftrightarrow{AC}$' and '$\overleftrightarrow{CD}$' are two ways of naming the same line.

3. $\overline{AC} = \overline{CD}$ FALSE
   The line segments $\overline{AC}$ and $\overline{CD}$ are not equal point sets because they have different elements.

4. $\overline{AC} \subset \overline{CD}$ FALSE
   $\overline{AC}$ is not a proper subset of $\overline{CD}$.

EXERCISES I

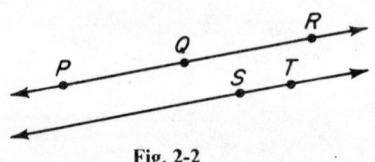

Fig. 2-2

On the basis of Figure 2-2 determine the truth or falsity of the following statements.

  i. $\overline{QR} \subset \overline{PR}$ iii. $\overline{PQ} \cong \overline{ST}$
  ii. $\overleftrightarrow{PQ} \perp \overleftrightarrow{ST}$ iv. $\{P, Q, R\} \not\subset \overset{\frown}{PR}$

5. The set {Chris, Bill, Cathy} is equivalent to the set {1, 3, 10} because the two sets have the same cardinal number.

6. The line segment representing the interval [0, 2) contains the point representing 0, but does not contain the point representing 2.

EXERCISES II

i. Which of the following sets are identical?
   A = {David, Jane, Mary, Susan}
   B = {1, 2, 3, 5}
   C = {Jane, Susan, Mary, David}
   D = {2, 4, 3, 1}

ii. Is the set of letters in the word *committee* equivalent to the set {2, 4, 6, 8, 10, 12}?

iii. Is the ordered pair (3, 8) equal to the ordered pair (8, 3)?

For each figure, write the interval of real numbers symbolized and the name of the line segment representing it.

iv. 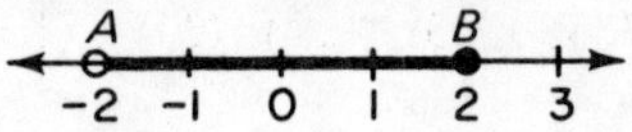
Fig. 2-3

v. 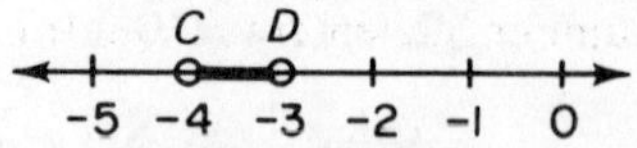
Fig. 2-4

vi. 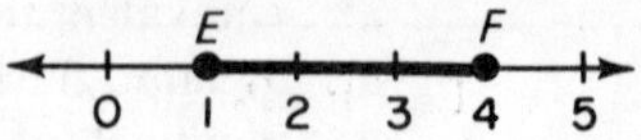
Fig. 2-5

7. Since {3, 6, 9} has three elements, $2^n = 2^3$ = eight subsets. These include the empty set (which is a subset of every set), the set itself, {3}, {6}, {9}, {3, 6}, {3, 9}, and {6, 9}.

8. Proper subsets are subsets which are not equal to the set itself. Of the eight subsets of {3, 6, 9}, only {3, 6, 9} is not a proper subset.

9. $\{\alpha, \beta, \gamma, \delta\}$ has 2^4 = 16 subsets.

10. Let A = {math students}.
    Let B = {85% of math students}.
    Then $B \subseteq A$, but $B \neq A$. Therefore, $B \subset A$.

11. If A = (5, 9] and B = (5, 9), then $B \subset A$. (5, 9] contains 9 and all real numbers between 5 and 9. (5, 9) contains all real numbers between 5 and 9, but not 5 and 9. Therefore, (5, 9) is a proper subset of (5, 9].

Relations Between Sets 9

EXERCISES III

A is the real number interval $[-10, 10]$
B is the real number interval $(-10, 10)$
C is the real number interval $(-15, 10]$

Are the following statements true or false?

i. $A \subseteq B$ vi. $10 \in B$
ii. $B \subset C$ vii. $-10 \in C$
iii. $B \subseteq B$ viii. $-10 \in A$
iv. $C \supset A$ ix. A and C are disjoint.
v. A is finite.   x. $(-15, 10) \subset C$

---

12. $R \times T$ is the set of all ordered pairs (r, t) such that $r \in R$ and $t \in T$. Since $R = \{3, 2, 1\}$ and $T = \{4, 5\}$, $R \times T = \{(3, 4), (3, 5), (2, 4), (2, 5), (1, 4), (1, 5)\}$.

13. The cardinal number of a set is the number of elements it contains.

    $n(R \times T) = 6$

EXERCISES IV

If possible, establish a one-one correspondence between each pair of sets.

i. $\{2, 3, 6\}$ and $\{1, 5, 9\}$

ii. {natural numbers}, {positive even integers}

iii. {positive odd integers}, {positive multiples of 3}

Graph each of the following sets.

iv. {negative real numbers}   vii. $\{0\}$
v. $\{2, 4, 6, 8\}$ viii. ϕ
vi. $\{(-1, 2), (3, 4), (5, -6)\}$ ix. $U \times U$

---

14. No, because the set of real numbers is an infinite set and the set of grains of sand on the earth is a finite set.

15. The graph of the set of real numbers includes all the points on the number line.

Solutions to the exercises appear on page 12.

ANSWERS

1. false		1
2. true		2
3. false		3
4. false		4
5. a		5
6. 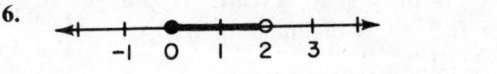		6 
7. ϕ, $\{3\}$, $\{6\}$, $\{9\}$, $\{3, 6\}$, $\{3, 9\}$, $\{6, 9\}$, $\{3, 6, 9\}$		7
8. all except $\{3, 6, 9\}$		8
9. 16		9
10. yes		10
11. yes		11
12. $\{(3, 4), (3, 5), (2, 4), (2, 5), (1, 4), (1, 5)\}$		12
13. 6		13
14. no		14
15. 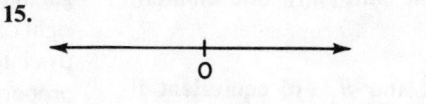		15

BASIC FACTS

Two sets, A and B, are **equal** if and only if every element of A is an element of B and every element of B is an element of A. $A = B$ is read, "A is equal to B" or "A is identical to B."

If A and B are not equal sets, a slash may be drawn through the equality symbol to indicate this.

In general, a slash drawn through the symbol for a relation indicates the negation of the relation.

A set, A, is a **subset** of a set, B, if and only if every element of A is an element of B. $A \subseteq B$, or $B \supseteq A$, is read, "A is a subset of B," "A is included in B," or "B includes A."

A set, A, is a **proper subset** of a set, B, if and only if $A \subseteq B$ and $A \neq B$. $A \subset B$, or $B \supset A$, is read, "A is a proper subset of B," "A is properly included in B," or "B properly includes A."

Two sets, A and B, are **disjoint** if and only if no element of A is an element of B. That is, if and only if the two sets have no elements in common.

Two sets, A and B, are **overlapping** if and only if at least one element of A is an element of B.

Two sets, A and B, are in **one-to-one correspondence** if and only if there exists a pairing of the elements of A with the elements of B such that each element of A corresponds to one and only one element of B, and each element of B corresponds to one and only one element of A.

Two sets, A and B, are **equivalent** if and only if a one-to-one correspondence can be established between the elements of the two sets. That is, if and only if A and B have the same number of ele-

ADDITIONAL INFORMATION

The relations between sets can be represented pictorially by **Venn diagrams.** The universal set (U) is represented by a rectangular region and any subset of U is represented by a simple closed region within the rectangle. The points in the interior of these regions represent the elements of the set. Consider the sets A and B. Figure 2-7 represents the case in which A and B are **equal**, or **identical**. And since A and B are equal, every element of A is an element of B (A is a **subset** of B) and every element of B is an element of A (B is a subset of A). Moreover, if A and B are equal, a **one-to-one correspondence** can be established between their members, and the two sets are **equivalent**. Figure 2-8, in which A and B are not equal, represents the case in which A is a **proper subset** of B. Figure 2-9 represents the case in which B is a proper subset of A. Figure 2-10 represents **disjoint** sets. These sets are not equal, but they could be equivalent; that is, they could have the same number of elements. Figure 2-11 represents nonequal **overlapping sets.** Figure 2-7 represents overlapping sets which are equal.

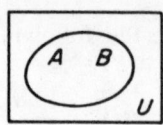

Fig. 2-7

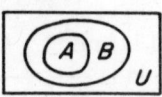

Fig. 2-8

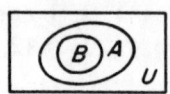

Fig. 2-9

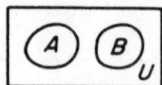

Fig. 2-10

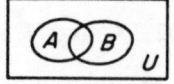

Fig. 2-11

The **subset relation,** $A \subseteq B$, may be written for all x, $x \in A \rightarrow x \in B$. From this definition, Theorems 2-1 and 2-2 follow.

THEOREM 2-1: Every set is a subset of itself.
THEOREM 2-2: The null set (ϕ) is a subset of every set.
THEOREM 2-3: Every set of n elements has exactly 2^n subsets.

To draw a **number line**, draw a picture of a line, select an arbitrary point on that line to represent zero, and choose an arbitrary unit say ⊢────────⊣ = one unit. Using this unit and beginning with 1, mark off points for the positive integers to the right of zero; and, beginning with -1, mark off points for the negative integers to the left of zero. Fractions may be marked off in proportional parts of units. The number associated with each point is called the **coordinate** of the point. (See Figure 2-12.)

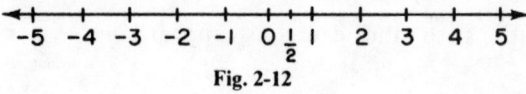

Fig. 2-12

ments. $A \equiv B$ is read, "A is equivalent to B."

If two sets are equivalent, they have the same **cardinal number.** $n(A) = 5$ is read, "the cardinal number of the set A is five" or "the set A contains five elements."

The cardinal numbers of infinite sets are called **transfinite numbers.**

There exists a one-to-one correspondence between the real numbers and the points on a line in Euclidean space. A pictorial representation of this one-to-one correspondence is called the **real number line.**

Two number lines at right angles to each other constitute the axes of a **Cartesian (rectangular) coordinate system.** The point of intersection is the **origin.** The horizontal line is the X-**axis**; the vertical line is the Y-**axis.**

Any point in the Cartesian plane (the plane determined by these axes) can be specified by an ordered pair of real numbers called the coordinates of the point.

The **Cartesian set,** or Cartesian product, of $A \times B$ is the set of all ordered pairs (a, b) such that a is an element of A and b is an element of B. In symbols, $A \times B = \{(a,b) \mid a \in A$ and $b \in B\}$. If both A and B are finite sets, an array of points (a lattice) can be used to illustrate their Cartesian product.

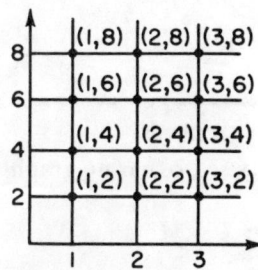

Fig. 2-6: The Cartesian set of $A \times B$ when $A = \{1,2,3\}$ and $B = \{2,4,6,8\}$.

Any subset of the real numbers may be represented (graphed) as the corresponding subset of the number line. **Intervals of real numbers,** which are subsets of the set of real numbers, may be represented on the number line by line segments, which are subsets of a line. An open interval is represented by a line segment which is open at both ends. A closed interval is represented by a line segment which contains both end points. A half-open, half-closed interval is represented by a line segment which is open at the left end. And a half-closed, half-open interval is represented by a line segment which is open at the right end.

An **ordered pair** is a set containing two elements and having the property that the left (first) element and the right (second) element can be distinguished from each other by virtue of the order in which they appear. An ordered pair is written in parentheses instead of the brace, and is thus distinguished from other sets containing two elements. The context in which (a, b) appears should indicate whether an ordered pair or an open interval is meant.

Any subset of $U \times U$ can be represented (graphed) on the Cartesian plane. The **horizontal coordinate (abscissa)** of any point P is the directed perpendicular distance, x, from YY' to P. It is positive when P is to the right of YY', and negative when P is to the left of YY'. The **vertical coordinate (ordinate)** of P is the directed perpendicular distance, y, from XX' to P. It is positive when P is above XX', and negative when P is below XX'.

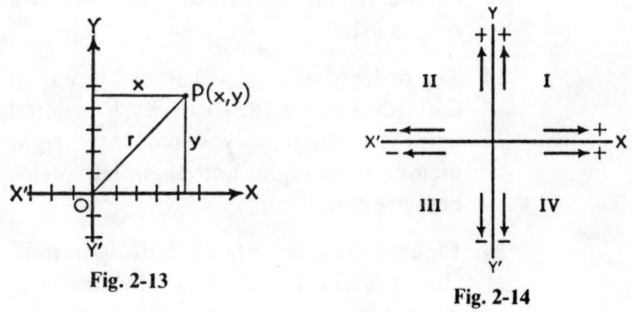

Fig. 2-13   Fig. 2-14

The line OP in Figure 2-13 is called the **radius vector,** r. If P coincides with O, then r is zero; otherwise it is a positive quantity.

The coordinate axes divide the plane into four **quadrants** and these are numbered counterclockwise. Figure 2-14 indicates the signs of x and y in each quadrant.

The **Cartesian set of** $U \times U$ (read, "U cross U"), where U is the set of real numbers, is the set of all ordered pairs of real numbers.

There is a one-to-one correspondence between the points in the Cartesian plane and the Cartesian set of $U \times U$.

SOLUTIONS TO EXERCISES

I

Refer to Figure 2-2.

i. $\overleftrightarrow{QR} \subseteq \overleftrightarrow{PR}$ TRUE

ii. $\overleftrightarrow{PQ} \perp \overleftrightarrow{ST}$ FALSE

iii. $\overleftrightarrow{PQ} \cong \overleftrightarrow{ST}$ TRUE
All lines in Euclidean space are congruent.

iv. $\{P, Q, R\} \not\subseteq \overset{\circ}{PR}$ TRUE
The segment $\overset{\circ}{PR}$ does not include the points P and R.

II

i. {David, Jane, Mary, Susan}
= {Jane, Susan, Mary, David}
Two sets are identical (equal) if they have the same elements.

ii. $\{2, 4, 6, 8, 10, 12\} \equiv \{c, o, m, i, t, e\}$
Two sets are equivalent if they have the same number of elements, and no element is repeated when the members of a set are listed.

iii. The ordered pair (3, 8) is not equal to the ordered pair (8, 3) because ordered pairs are equal only when their right members are equal and their left members are equal.

iv. Figure 2-3 represents the half-open half-closed interval $(-2, 2]$ as the line segment $\overset{\smile}{AB}$.

v. Figure 2-4 represents the open interval $(-4, -3)$ as the line segment $\overset{\circ}{CD}$.

vi. Figure 2-5 represents the closed interval $[1, 4]$ as the line segment $\overline{EF}$.

III

$A = [-10, 10], B = (-10, 10), C = (-15, 10]$

i. $A \subseteq B$ FALSE   iii. $B \subseteq B$ TRUE
ii. $B \subset C$ TRUE   iv. $C \supset A$ TRUE

v. A is finite. FALSE   vii. $-10 \in C$ TRUE
vi. $10 \in B$ FALSE   viii. $-10 \in A$ TRUE
ix. A and C are disjoint.   FALSE
x. $(-15, 10) \subset C$ TRUE

IV

i. $\{2 \quad 3 \quad 6\}$
   $\updownarrow \quad \updownarrow \quad \updownarrow$
$\{1 \quad 5 \quad 9\}$ is one of the several possible pairings.

ii. $\{1 \quad 2 \quad 3 \quad 4 \ldots n \ldots\}$
   $\updownarrow \quad \updownarrow \quad \updownarrow \quad \updownarrow \quad \updownarrow$
$\{2 \quad 4 \quad 6 \quad 8 \ldots 2n \ldots\}$, where n is a positive integer.

iii. $\{1 \quad 3 \quad 5 \ldots 2n - 1 \ldots\}$
    $\updownarrow \quad \updownarrow \quad \updownarrow \quad\quad \updownarrow$
$\{3 \quad 6 \quad 9 \ldots \quad 3n \ldots\}$, where n is a positive integer.

iv. Fig. 2-15: {Negative Real Numbers}

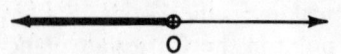

v. Fig. 2-16: {2, 4, 6, 8}

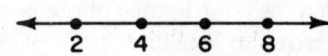

vi. Fig. 2-17: $\{(-1, 2), (3, 4), (5, -6)\}$

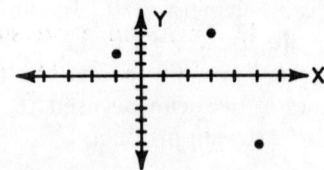

vii. Fig. 2-18: {0}

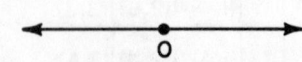

viii. The empty set, ϕ, has no graph.

ix. Fig. 2-19: $U \times U$

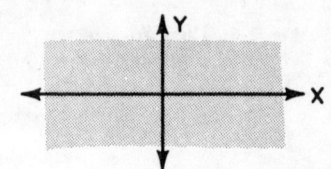

3 OPERATIONS ON SETS

SELF-TEST

DIRECTIONS: Write your answers in the numbered regions to the right. To check your answers, turn the page. Study the solutions to the problems you missed, and do the exercises following any problem or group of problems in which you had errors.

List the elements in the sets in **1-5** if $U = \{a,b,c,d,e,f,g,h,i,j\}$, $L = \{b,c,d,f,g,h,j\}$, $M = \{a,e,i\}$, and $N = \{a,b,c,d,e\}$.

1. $(M \cup N) \cap L$
2. $N' \cap (L \cap M)$
3. $N' \cap (L \cup M)$
4. $(L \cup M) \cup N$
5. $(N' \cap L)' \cap N$

6. If $A = \{x \text{ (real)} \mid x > 0\}$ and $B = \{x \text{ (real)} \mid x < 0\}$, does $A \cup B = \{\text{real numbers}\}$?

7. Express U' without using the symbol for complementation.

8. If $A \subseteq U$, express $A \cup \phi'$ without using the symbols for union and complementation.

9. Express $[1,4] \cup [2,5]$ without using $\cup$ or $\cap$.

10. Express $(1,3) \cap [2,4]$ without using $\cup$ or $\cap$.

11. Draw a Venn diagram to represent $(A \cup B) \subseteq B$.

12. Write an expression involving A, B, and C which exactly describes each of the eight disjoint regions in Figure 3-1.

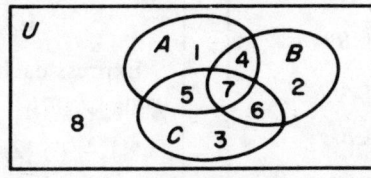

Fig. 3-1

13. Use a Venn diagram to test the truth of the following statement.

$$P \cap (P \cup Q) = P$$

14. Is the relation given in **13** a law?

13

14 Operations on Sets

SOLUTIONS

In **1-5**, find the sets within parentheses first.

$U = \{a, b, c, d, e, f, g, h, i, j\}$
$L = \{b, c, d, f, g, h, j\}$
$M = \{a, e, i\}$
$N = \{a, b, c, d, e\}$

1. $(M \cup N) \cap L = \{a,b,c,d,e,i\} \cap \{b,c,d,f,g,h,j\} = \{b,c,d\}$

2. $N' \cap (L \cap M) = \{f,g,h,i,j\} \cap \phi = \phi$

3. $N' \cap (L \cup M) = \{f,g,h,i,j\} \cap \{a,b,c,d,e,f,g,h,i,j\} = \{f,g,h,i,j\}$

4. $(L \cup M) \cup N = \{a, b, c, d, e, f, g, h, i, j\} \cup \{a, b, c, d, e\} = \{a, b, c, d, e, f, g, h, i, j\}$

5. $(N' \cap L)' \cap N = \{f, g, h, i,\}' \cap N$
   $= \{a, b, c, d, e, j\} \cap \{a, b, c, d, e\}$
   $= \{a, b, c, d, e\}$

6. $A = \{x \text{ (real)} \mid x > 0\}$
   $B = \{x \text{ (real)} \mid x < 0\}$
   $A \cup B$ contains all real numbers except 0.

EXERCISES I

Let $U = \{1,2,3,4,5,6,7,8,9,10\}$, $A = \{1,3,5,7,9\}$, and $B = \{3,6,9\}$.

i. List the elements in the set $A \cup B$.

ii. List the elements in the set $A \cap B$.

iii. Is $A \cap B$ a subset of $A \cup B$?

iv. List the elements in the set A'.

v. List the elements in the set B'.

vi. List the elements in the set $(A \cap B)'$.

vii. List the elements in the set $A' \cap B'$.

viii. Is $(A \cap B)' = A' \cap B'$?

7. Since U contains all elements in the universe under consideration, and since the complement of a set contains all elements in the universe not contained in the set, the complement of U is ϕ.

8. Since $\phi' = U, A \cup \phi' = A \cup U$.
   Since $A \subseteq U, A \cup U = U$.

EXERCISES II

Let $A \subseteq U$. Express each of the following subsets of U without using $'$, $\cup$, or $\cap$.

i. ϕ' iv. $A \cup U'$

ii. $A \cap U'$ v. $(A')'$

iii. $A \cap \phi$ vi. $A \cap A'$

9 and **10** can be done graphically by shading each set on the number line. The intersection of two sets will be the set of points which is shaded twice. The union of two sets will be the set containing all the shaded points.

9.

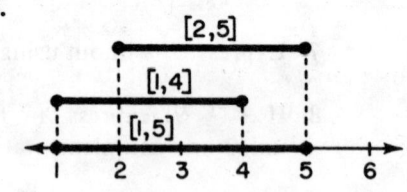

Fig. 3-2: $[1,4] \cup [2,5] = [1,5]$

10.

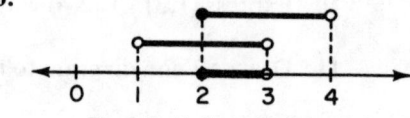

Fig. 3-3: $(1,3) \cap [2,4) = [2,3)$

EXERCISES III

Express each of the following without using $\cup$ or $\cap$.

i. $[1,4] \cap [2,5]$ iv. $[2,6] \cup [3,4]$

ii. $[1,2] \cup [2,3]$ v. $[2,6] \cap [3,4]$

iii. $[1,2] \cap [2,3]$ vi. $(1,3) \cup [3,4)$

11. $(A \cup B) \subseteq B$

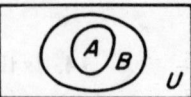

Fig. 3-4

The Venn diagram (Figure 3-4) shows that if $A \subseteq B$ then $A \cup B = B$. Therefore $(A \cup B) \subseteq B$ since every set is a subset of itself.

12. The expressions given in the answer column are not the only expressions which may be used to describe these regions in Figure 3-1. Check other answers by means of Venn diagrams.

EXERCISES IV

In **i–iv**, draw a Venn diagram to illustrate the given sets.

i. $B \subset A$ iii. $A \cap B \neq \phi$
ii. $A \cap B = \phi$ iv. $A \cap B = B$ and $A \subseteq B$

13.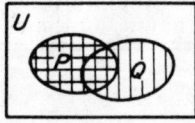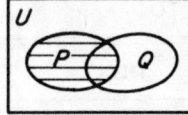

Fig. 3-5   Fig. 3-6

P is horizontally shaded in both Venn diagrams. In Figure 3-5, $P \cup Q$ is vertically shaded; and, therefore, $P \cap (P \cup Q)$ is cross-hatched. Since the cross-hatched region in Figure 3-5 is the same as the horizontally-shaded region in Figure 3-6, the diagrams indicate that '$P \cap (P \cup Q) = P$' is true.

14. $P \cap (P \cup Q) = P$

Let $x \in [P \cap (P \cup Q)]$.
Then $x \in P$ and $x \in P \cup Q$. Hence $P \cap (P \cup Q) \subseteq P$.
Let $x \in P$.
Then $x \in P$ and $x \in P \cup Q$, or $x \in [P \cap (P \cup Q)]$. Hence $P \subseteq [P \cap (P \cup Q)]$. Therefore $P \cap (P \cup Q) = P$.

EXERCISES V

Test each of the following relationships by Venn diagrams. SUGGESTION: Use overlapping circles.

i. $P' \cup Q' = P \cap Q'$
ii. $P \cup (Q \cup R) = (P \cup Q) \cup R$
iii. $(P \cap Q) \cup (P \cap R) = P \cap (Q \cup R)$

Solutions to the exercises are on page 18.

ANSWERS

1. $\{b, c, d\}$	1
2. ϕ	2
3. $\{f, g, h, i, j\}$	3
4. $\{a, b, c, d, e, f, g, h, i, j\}$	4
5. $\{a, b, c, d, e\}$	5
6. no	6
7. ϕ	7
8. U	8
9. [1, 5]	9
10. [2, 3]	10
11. 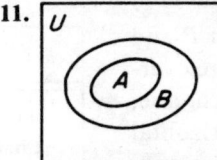	11
12. 1: $A \cap (B \cup C)'$, 2: $B \cap (A \cup C)'$, 3: $C \cap (A \cup B)'$, 4: $(A \cap B) \cap C'$, 5: $(A \cap C) \cap B'$, 6: $(B \cap C) \cap A'$, 7: $(A \cap B) \cap C$, 8: $[(A \cup B) \cup C]'$	12
13. true	13
14. yes	14

BASIC FACTS

Sets can be operated on in various ways to form new sets.

The basic operations on sets are complementation, difference, intersection, and union. Difference is complementation relative to a subset of U. Since $U \subseteq U$, the laws governing difference hold for complementation.

The **complement** of a set P relative to a universal set U is the set of all elements of U which are not elements of P. The complement of P relative to U is written P' (read, "P complement"). In symbols, $P' = \{x \mid x \notin P\}$.

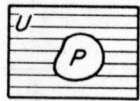

The Venn diagram for P' is shown in Figure 3-7. P' is the set of points indicated by horizontal lines.

Fig. 3-7: P'

The complement of a set Q relative to a set P which is a subset of U is the set of all elements of P which are not elements of Q. The complement of Q relative to P is written $P - Q$ (read, "the **difference** of P and Q"). In symbols, $P - Q = \{x \mid x \in P \text{ and } x \notin Q\}$.

If P and Q are non-empty sets, $P - Q$ is non-empty when $Q \subset P$, when P and Q are disjoint, or when P and Q are overlapping. (See Figure 3-8.) In each case, $P - Q$ is the set indicated by horizontal lines. If $P \subseteq Q$, $P - Q$ is the empty set, ϕ.

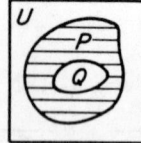 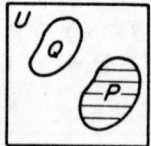 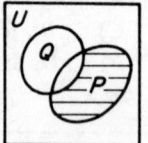

Fig. 3-8: $P - Q$

The **intersection** of two sets, P and Q, is the set of all elements which belong to both P and Q. The intersection of P and Q is written $P \cap Q$ (read, "P intersect Q"). In symbols,

$P \cap Q = \{x \mid x \in P \text{ and } x \in Q\}.$

ADDITIONAL INFORMATION

The Laws of Operations on Sets

The **principle of duality** states that in any true statement involving only union, intersection, or complementation, if $\cup$ and $\cap$ are interchanged and U and ϕ are interchanged, a true statement results. The principle of duality cannot be used with a statement involving difference. However, if a difference, $A - B$, is replaced with its equivalent, $A \cap B'$, the principle can be applied.

LAWS GOVERNING COMPLEMENTATION			
COMPLEMENTATION RELATIVE TO U		DIFFERENCE
LAW	DUAL	
$(A')' = A$ $A \cup A' = U$ $(A \cup B)' = A' \cap B'$ $U' = \phi$	$A \cap A' = \phi$ $(A \cap B)' = A' \cup B'$ $\phi' = U$	$A - B = A \cap B'$ $A - \phi = A$ $\phi - A = \phi$ $A - A = \phi$ $(A - B) - C = A - (B \cup C)$ $A - (B - C) = (A - B) \cup$ $\quad (A \cap C)$ $A \cup (B - C) = (A \cup B) -$ $\quad (C \cup A)$ $A \cap (B - C) = (A \cap B) -$ $\quad (A \cap C)$

LAWS GOVERNING UNION AND INTERSECTION	
LAW	DUAL
$A \cup A = A$ Commutative Law: $A \cup B = B \cup A$ Associative Law: $A \cup (B \cup C) = (A \cup B) \cup C$ Distributive Law: $A \cup (B \cap C) = (A \cup B) \cap$ $\quad (A \cup C)$ $A \cup U = U$ $A \cup \phi = A$	$A \cap A = A$ $A \cap B = B \cap A$ $A \cap (B \cap C) = (A \cap B) \cap C$ $A \cap (B \cup C) = (A \cap B) \cup$ $\quad (A \cap C)$ $A \cap \phi = \phi$ $A \cap U = A$

Testing the Laws

The laws governing operations on sets can be tested by means of Venn diagrams, or, if the sets are point sets, by sketches of geometric objects. Such diagrammatic tests, however, are only indications of the validity of a law. They are not proofs because one diagram cannot take all possible cases into consideration.

If P and Q are non-empty sets, $P \cap Q$ is non-empty when $P \subseteq Q$, when $Q \subseteq P$, or when P and Q overlap. (See Figure 3-9.) In each case $P \cap Q$ is the set indicated by horizontal lines. If P and Q are disjoint, $P \cap Q = \phi$.

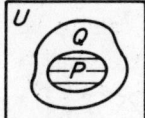 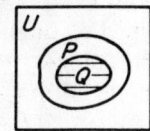 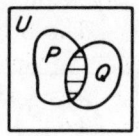

Fig. 3-9: $P \cap Q$

The **union** of two sets, P and Q, is the set of all elements which belong to P, to Q, or to both P and Q. The union of P and Q is written $P \cup Q$ (read, "P union Q"). In symbols,

$$P \cup Q = \{x \mid x \in P \text{ or } x \in Q\}.$$

If either P or Q are non-empty, $P \cup Q$ is non-empty. (See Figure 3-10.)

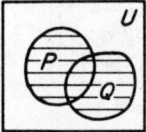 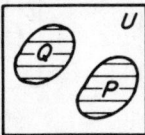 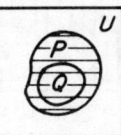

Fig. 3-10: $P \cup Q$

Some familiar geometric objects are formed by operations on point sets.

An **angle** is the union of two rays such that the intersection of those rays is the endpoint of both.

A **polygon** is the union of line segments such that each segment intersects two others and each such intersection is an endpoint of exactly two of the segments.

The result of an operation on sets of real numbers can be represented on the number line; and the result of an operation on sets of ordered pairs of real numbers can be represented on the Cartesian plane.

An open sentence, $p(x)$, is an expression which contains a variable x and which is neither true nor false. It becomes true or false when replacements are made for x from the domain. The set of replacements for which $p(x)$ is true is called its truth set.

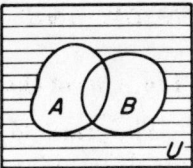

Fig. 3-11

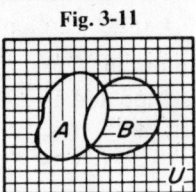

Fig. 3-12

Consider, for example, the law $(A \cup B)' = A' \cap B'$. Recalling that two sets are equal if they contain precisely the same elements, draw different Venn diagrams for the left and right members of the equality, and compare the resulting sets. The unmarked region in Figure 3-11 represents $A \cup B$, and the horizontally shaded region represents $(A \cup B)'$. The horizontally shaded region in Figure 3-12 represents A', the vertically shaded region represents B', and the crosshatched region represents $A' \cap B'$. Since the crosshatched region representing $A' \cap B'$ is precisely the same region as the horizontally shaded region representing $(A \cup B)'$, $A' \cap B' = (A \cup B)'$ is the case in which A and B are overlapping sets.

Proving the Laws

A proof of a law which holds for all sets may be accomplished by (1) verifying the relationship for each possible case of an element x of the universe U, or by (2) employing the definitions of the operations to show that both members of the equality have the same meaning for any element x of U.

Operations on Specific Sets

Even though diagrammatic techniques are insufficient for establishing a law of set operations, they can be used to determine the set resulting from an operation on specific sets. The result of an operation on sets of real numbers can be determined by using the real number line; the result of an operation on sets of ordered pairs of real numbers can be determined on the Cartesian plane; and the result of an operation on point sets can be determined by sketching the geometric objects. Most other sets are best represented by Venn diagrams.

Truth Sets

Given the **open sentence**, $p(x)$, there exists a set P which consists of all elements of U for which the open sentence is true. $P = \{x \mid p(x)\}$ is the truth set of $p(x)$. There also exists a set P' which consists of all elements of U for which the open sentence is false. $P' = \{x \mid \sim p(x)\}$ is the truth set of $\sim p(x)$. $\sim$ is read "not" or "the negation of."

Given two open sentences, $p(x)$ and $q(x)$, the set $P \cap Q$ is the set of all elements for which $p(x)$ and $q(x)$ are both true. $P \cap Q = \{x \mid p(x) \wedge q(x)\}$ is the truth set of $p(x) \wedge q(x)$. $\wedge$ is read "and" or "the conjunction of." The set $P \cup Q$ is the set of all elements for which $p(x)$ is true or $q(x)$ is true or both. $P \cup Q = \{x \mid p(x) \vee q(x)\}$ is the truth set of $p(x) \vee q(x)$. $\vee$ is read "or" or "the disjunction of."

SOLUTIONS TO EXERCISES

I

$U = \{1,2,3,4,5,6,7,8,9,10\}$
$A = \{1,3,5,7,9\}$
$B = \{3,6,9\}$

i. $A \cup B$ consists of all elements in A or in B. Since no element is repeated in listing the elements of a set, $A \cup B = \{1,3,5,6,7,9\}$.

ii. Since $A \cap B$ consists of elements which A and B have in common, $A \cap B = \{3,9\}$.

iii. Since every element of $A \cap B$ is in $A \cup B$, $A \cap B \subseteq A \cup B$.

iv. Since A' consists of all elements in U not in A, $A' = \{2,4,6,8,10\}$.

v. $B' = \{1,2,4,5,7,8,10\}$

vi. To obtain $(A \cap B)'$, find $A \cap B = \{3,9\}$ and take its complement. $(A \cap B)' = \{1,2,4,5,6,7,8,10\}$

vii. $A' = \{2,4,6,8,10\}$
$B' = \{1,2,4,5,7,8,10\}$
$A' \cap B' = \{2,4,8,10\}$

viii. A' and B' are not equal sets because they do not have the same elements.

II

i. $\phi' = U$
ii. $A \cap U' = \phi$
iii. $A \cap \phi' = A$
iv. $A \cup U' = A$
v. $(A')' = A$
vi. $A \cap A' = \phi$

III

i.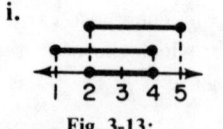
Fig. 3-13:
$[1,4] \cap [2,5] = [2,4]$

ii.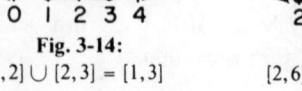
Fig. 3-14:
$[1,2] \cup [2,3] = [1,3]$

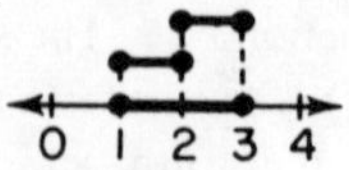

iii.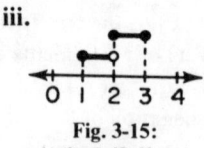
Fig. 3-15:
$[1,2] \cap [2,3] = \phi$

iv.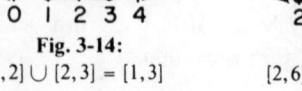
Fig. 3-16:
$[2,6] \cup [3,4] = [2,6]$

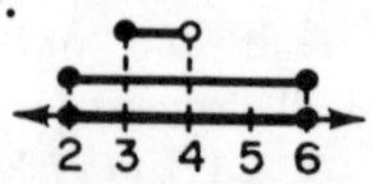

v.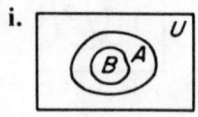
Fig. 3-17:
$[2,6] \cap [3,4] = [3,4]$

vi.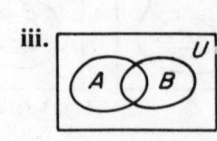
Fig. 3-18:
$(1,3) \cup [3,4] = (1,4)$

IV

i.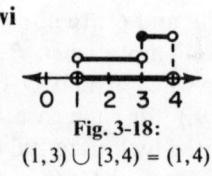
Fig. 3-19

ii.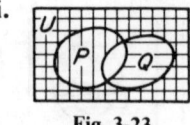
Fig. 3-20

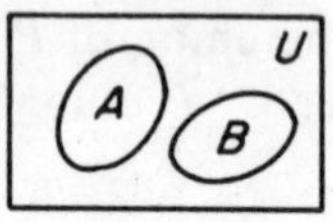

iii.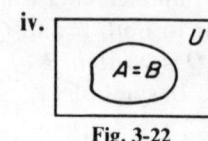
Fig. 3-21

iv.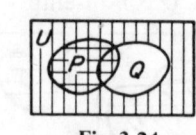
Fig. 3-22

V

i.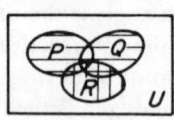
Fig. 3-23   Fig. 3-24

Since the shaded region in Figure 3-23 is not the same as the cross-hatched region in Figure 3-24, '$P' \cup Q' = P \cap Q'$' is false.

ii.
Fig. 3-25   Fig. 3-26

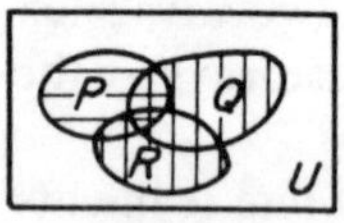

Since the shaded regions in Figures 3-25 and 3-26 are the same, the Venn diagrams indicate that $P \cup (Q \cup R) = (P \cup Q) \cup R$.

iii.
Fig. 3-27   Fig. 3-28

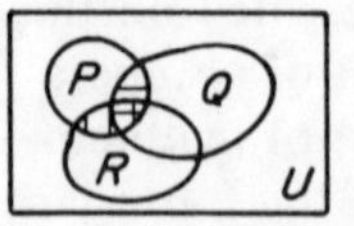

Since the crosshatched region in Figure 3-27 is the same as the shaded region in Figure 3-28, the Venn diagrams indicate that $(P \cap Q) \cup (P \cap R) = P \cap (Q \cup R)$.

4 AXIOMATIC SYSTEMS
SELF-TEST

DIRECTIONS: Write your answers in the numbered regions to the right. To check your answers, turn the page. Study the solutions to the problems you missed, and do the exercises following any problem or group of problems in which you had errors.

1. Which of the relations listed below could not be classified as dyadic (two-place) relations for the universe specified?
   a. the between relation, $U = \{$points on a line$\}$
   b. the perpendicular relation, $U = \{$lines in a plane$\}$
   c. the sum relation, $U = \{$real numbers$\}$
   d. the father relation, $U = \{$people$\}$
   e. the converse relation, $U = \{$implications$\}$

2. Consider the dyadic relation '—is a factor of—' in the universe of positive integers. Is this relation reflexive? Is it symmetric? Is it transitive?

3. Is the relation in **2** an equivalence relation?

4. $\{(1,1), (2,2), (8,8)\}$ defines a relation. Is this an equivalence relation? If not, what property or properties are not satisfied.

5. Let S be the set $\{a, b, c\}$. Define, by listing ordered pairs, a relation in S which is symmetric but not transitive.

6. Let R be the set of real numbers, and let $*$ be the operation of averaging any two elements a and b of R.
   a. Evaluate $a * b$ for $(3,7)$, $(4,14)$ and $(2,17)$.
   b. Is $*$ an operation on the set of whole numbers?

7. $a * b = 2ab$ is an operation on the set R of real numbers.
   a. Is $*$ commutative on this set?
   b. Is $*$ associative on this set?
   c. What is the identity element of $*$ in the set of real numbers?
   d. Find the inverse of any element a, where a is a real number.

8. Bertrand Russell once remarked that mathematics may be defined as the subject in which we never know what we are talking about, nor whether what we are saying is true. Was Russell referring to pure or applied mathematics?

1 _____

2 _____

3 _____

4 _____

5 _____

6 a _____
  b _____

7 a _____
  b _____
  c _____
  d _____

8 _____

19

SOLUTIONS

1. If a relation is dyadic, a meaningful statement results when the relation is used to associate two elements of the universe.
   a. 'Point A is between point B' is meaningless. Therefore, the between relation is not dyadic. The between relation is a triadic (three-place) relation of the form '— is between — and —.'
   b. '$\overleftrightarrow{AB}$ is perpendicular to $\overleftrightarrow{CD}$' is meaningful. Therefore, the perpendicular relation is dyadic.
   c. '— is the sum of — and —' is a triadic relation.
   d. '— is the father of —' is a dyadic relation.
   e. '— is the converse of —' is a dyadic relation.

2. A relation is reflexive in a given universe if $a\,R\,a$ holds for every element of that universe; it is symmetric if $a\,R\,b \rightarrow b\,R\,a$ holds; and it is transitive if $(a\,R\,b$ and $b\,R\,c) \rightarrow a\,R\,c$ holds. (a, b, and c are elements of U.)

   To determine whether or not these properties hold, substitute specific elements of the universe for a, b, and c; and substitute the given relation for R. If the properties do not hold in the specific case, they will not hold in the general case. If they do hold in the specific case tested, they must then be tested to determine whether or not they hold in the general case.

   To determine the properties of '— is a factor of —' in the set of positive integers, let $a = 2$, $b = 4$, and $c = 16$. Then,

   $a\,R\,a$ holds because 2 is a factor of 2

   $a\,R\,b \rightarrow b\,R\,a$ does not hold because 'If 2 is a factor of 4, then 4 is a factor of 2' is false. Therefore, the operation is not reflexive in the universe specified.

   $(a\,R\,b$ and $b\,R\,c) \rightarrow a\,R\,c$ holds because if 2 is a factor of 4 and 4 is a factor of 16 then 2 is a factor of 16.

   Now the reflexive and transitive properties must be considered for the general case. The relation '— is a factor of —' is reflexive for all elements of U because every positive integer is a factor of itself. The relation is transitive in U because if a is a factor of b, '$b\,R\,c$' can be written in the form '$a\,R\,c$,' which is to say that a is a factor of c.

EXERCISES I

Classify the following dyadic relations as reflexive, symmetric, transitive, or any combination of these.
   i. '— is congruent to —,' $U = \{$triangles in a plane$\}$
   ii. '— is parallel to —,' $U = \{$lines in a plane$\}$
   iii. '— contradicts —,' $U = \{$propositions$\}$
   iv. '— is the mother —,' $U = \{$people$\}$

---

3. The relation discussed in 2 is not an equivalence relation because it is not symmetric.

EXERCISES II

Which of the relations in Exercises I are equivalence relations?

---

4. $\{(1, 1), (2, 2), (8, 8)\}$ describes an equivalence relation. Every element of each ordered pair is related to itself (reflexive property), and the symmetric and transitive properties are trivially satisfied.

EXERCISES III

Each of the following sets of ordered pairs describes a relation. Which of these relations are equivalence relations? For those which are not equivalence relations, state the unsatisfied property or properties.
   i. $\{(1,1), (3,3), (1,3)\}$
   ii. $\{(3,3), (4,4), (3,4), (4,3)\}$
   iii. $\{(1,1), (3,1), (1,3)\}$
   iv. $\{(5,5), (4,4), (3,3), (4,5), (3,5), (5,4), (5,3), (4,3)\}$

5. $S = \{a, b, c\}$
There are various relations in S which are symmetric but not transitive. $\{(a,b), (b,a)\}$ is an example. Check other answers by the method discussed in **2**.

EXERCISES IV

Let S be the set $\{a, b, c\}$. Define, by listing ordered pairs, a relation in S which satisfies the given conditions.

i. transitive but not symmetric
ii. reflexive but neither transitive nor symmetric

---

6. **a.** $3 * 7 = \frac{3+7}{2} = 5; \; 4 * 14 = 9;$
   $2 * 17 = 9\frac{1}{2}$

   b. $*$ is not an operation on the set of whole numbers; for example, $9\frac{1}{2}$ is not a whole number.

EXERCISES V

Let I be the set of integers, and let $\circ$ be a rule defined for integers as $a \circ b = a - 3b$.

i. Evaluate $a \circ b$ for $(3, 12)$, $(-8, -3)$ and $(7, 2)$.
ii. Does $\circ$ yield a unique result when performed on any two elements of I?
iii. Is the result of $\circ$ always an element of I?
iv. Is $\circ$ an operation on the set I?
v. If $\circ$ is an operation, is it unary or binary?

---

7. $a * b = 2ab$
   a, b. $*$ is commutative and associative on R.
   c. $a * x = 2ax$. If x is the identity element, $a * x = a$. Hence $a = 2ax$ and $x = \frac{1}{2}$.
   d. If x is the inverse element for a, $a * x = \frac{1}{2}, \frac{1}{2} = 2ax$ and $x = \frac{1}{4a}$.

ANSWERS

1. a and c

2. reflexive and transitive

3. no

4. yes

5. $\{(a,b), (b,a)\}$

6. **a.** $5, 9, 9\frac{1}{2}$
   b. no

7. **a.** yes
   b. yes
   c. $\frac{1}{2}$
   d. $\frac{1}{4a}$

8. pure

8. Pure mathematics consists of axiomatic systems in which all non-primitive terms are defined in terms of undefined terms, and all theorems are deduced from postulates containing undefined terms. Because the primitive terms are undefined, we can never know what we are talking about in pure mathematics. Since the theorems are established solely on the basis of logic, they possess only formal (as opposed to factual) truth. Therefore, we never know whether or not what we say is (factually) true.

Solutions to the exercises appear on page 24.

Axiomatic Systems

BASIC FACTS

An **axiomatic system** is any logical system having primitive terms and postulates. Defined terms and theorems are formulated on the basis of the primitive terms and the postulates.

The **primitive terms** are undefined technical terms including a set of elements, relations which can hold among the elements, and operations which can be performed on the elements.

All non-primitive technical terms used in the system are **defined terms**. These terms must be described by means of the primitive terms. Defined terms may be introduced whenever it is convenient to do so.

The **postulates** (axioms, assumptions, or primitive statements) are a set of unproven statements containing only primitive terms. The postulate set of an axiomatic system must be **consistent**; that is, it must not imply contradictory statements.

All non-primitive statements of the system are **theorems**. These statements must be logically deduced (proved) from the postulates.

A **relation** is an association among two or more things.

Relations are classified according to the number of elements associated. A dyadic relation associates two elements; a triadic, three; a tetratic, four; and so on.

$(x, y) \in R$, or $x R y$ is read "(x, y) is an ordered pair belonging to the dyadic relation R."

A relation in a set S determines a set of ordered n-tuples which is a subset of a cross-product of S. If the relation is dyadic, it determines a set of ordered pairs which is a subset of $S \times S$; if the relation is triadic, it determines a set of ordered triples which is a subset of $S \times (S \times S)$; if the relation is tetratic, it determines a set of ordered four-tuples which is a subset of $(S \times S) \times (S \times S)$; and so on.

ADDITIONAL INFORMATION

An **axiomatic system** is a formal system based ultimately on **primitive terms** which are devoid of definite meaning. However, an axiomatic system may be given an interpretation by assigning terms of definite meaning to the primitive terms of the system such that all the postulates become true statements when these assignments are substituted for the primitive terms. The system resulting from such an interpretation is called a **model** of the axiomatic system. An uninterpreted axiomatic system is part of pure mathematics and an interpretation of an axiomatic system is part of applied mathematics.

Models of axiomatic systems can be used to determine the **consistency** of the postulate set of the system. If the model is concrete—if it is such that the primitive terms are interpreted as objects, relations, and operations from the real world—and if no contradictory statements are implied by the interpreted postulates, the postulate set is said to be absolutely consistent because contradictions in the real world are impossible. If the model is ideal—if it is derived from another postulate set—and if no contradictory statements are implied by the interpreted postulates, the original postulate set is said to be relatively consistent because its consistency is relative to the consistency of the postulate set in terms of which it is interpreted.

Many of the relations encountered in mathematics are **dyadic relations**. The table below gives a few examples.

DYADIC RELATIONS	
Set Theory	Algebra
'— is an element of —'	'— is equal to —'
'— is identical to —'	'— is less than —'
'— is a subset of —'	'— is greater than —'
'— and — are disjoint'	'— is a divisor of —'
'— and — are equivalent'	'— is a factor of —'
'— and — overlap'	'— is the cube of —'

Some commonly encountered **triadic relations** are '— is between — and —,' '— is the sum of — and —,' and '— is the product of — and —.'

Since an **equivalence relation** is any dyadic relation which is reflexive, symmetric, and transitive, it is necessary to test for these three properties in order to determine whether or not a given dyadic relation is an equivalence relation.

Groups, rings, fields, and ordered fields are special types of axiomatic systems. The set of real numbers forms an ordered field under the operations of addition and multiplication. The set of real numbers and the set of complex numbers form a field under the operations of addition and multiplication.

A dyadic relation, R, is **reflexive** in a set S if, for each element, a, of S, $a\,R\,a$.

A dyadic relation, R, is **symmetric** in a set S if $a\,R\,b$ implies $b\,R\,a$; where a and b are elements of S.

A dyadic relation, R, is **transitive** in a set S if ($a\,R\,b$ and $b\,R\,c$) implies $a\,R\,c$, where a, b, and c are elements of S.

A dyadic relation which is reflexive, symmetric, and transitive is called an **equivalence relation.**

An **operation** on a set S is a **rule** which assigns a unique element of S to each element of S or some cross-product of S.

Operations are classified according to the number of elements on which the operation is performed. A **unary operation** assigns a unique element of S to each element of S; that is, the operation is performed on one element of S. A **binary operation** assigns a unique element of S to each element of $S \times S$; that is, the operation is performed on two elements of S. A **tertiary operation** assigns a unique element of S to each ordered triple of $S \times (S \times S)$; and so on.

A binary operation, $*$, on a set S is **commutative** if for all elements, a and b, of S, $a * b = b * a$.

A binary operation, $*$, on a set S is **associative** if for all elements, a, b, and c, of S, $(a * b) * c = a * (b * c)$.

A binary operation, $*$, on a set S is **distributive** over a binary operation, $\circ$, if for all elements, a, b, and c, of S, $a * (b \circ c) = (a * b) \circ (a * c)$.

A binary operation, $*$, on a set S has **cancellation** if for all elements, a, b, and c, of S, $(a * b = a * c)$ implies $b = c$.

A binary operation, $*$, on a set S has an **identity element**, i, in S if for all elements, a, of S, $a * i = a$ and $i * a = a$.

A binary operation, $*$, on a set S has **inverse elements**, a', in S if for each element, a, of S, there exists an a' such that $a * a' = i$ and $a' * a = i$.

An **operation** on a set, S, is a rule which assigns a unique element of S to each element of S or $S \times \cdots \times S$. From this definition, it is evident that a rule is an operation on a set S only if S is **closed** under that rule. That is, only if the result of applying the rule on any element of S or a cross-product of S is an element of S. Sometimes this condition is omitted from the definition of an operation on a set. When this is the case, every rule which assigns an element of S to at least one element of S or $S \times \cdots \times S$ is an operation, but a set need not be closed under its operations. In this text, the term 'operation on S' implies that the set S is closed under the operation. The rule for addition, for example, is an operation on the set, $S = \{0, 1, 2, 3, \ldots\}$ because to each ordered pair (e_1, e_2) of $S \times S$ the rule assigns an element $e_3 \in S$ such that $e_3 = e_1 + e_2$. The rule for subtraction, however, is not an operation on S because it does not assign an element of S to every ordered pair of $S \times S$. Subtraction assigns to $(5, 7)$ the element -2, but $-2 \notin S$.

From the definition of an operation it also follows that the result of performing the operation must be unique; that is, there is one and only one result each time the operation is performed. Finding a common factor is not an operation on the set, $S = \{0, 1, 2, 3, \cdots\}$ because the result of applying the rule is not a unique element of S.

Most of the operations in logic and mathematics are unary or binary operations. Squaring, negation, and complementation are **unary operations.** Addition, multiplication, conjunction, disjunction, union, and intersection are **binary operations.**

To determine whether an operation, $*$, on a set S is commutative, associative, distributive over $\circ$, or has cancellation, use the following technique.

1. Evaluate each member of the defining equality for specific elements of S. If the results are not equal, then operation S does not have the property. If the results are equal, proceed to 2.
2. Evaluate each member of the defining equality using x and y as representatives for any two values of S. If the results are equal, the operation on S has the property. If the results are not equal the operation on S does not have the property.

SOLUTIONS TO EXERCISES

I

i. The relation '— is congruent to —' is reflexive, symmetric, and transitive in the set of triangles in a Euclidean plane.

ii. '— is parallel to —' is symmetric and transitive in the set of lines in a plane. It is not reflexive because parallel lines are disjoint point sets.

iii. '— contradicts —' is symmetric and transitive in the set of propositions. It is not reflexive.

iv. '— is the mother of —' is neither reflexive, symmetric, nor transitive in the set of people.

II

'— is congruent to —' is the only relation in Exercises I which is reflexive, symmetric, and transitive in the universe specified. Therefore, '— is congruent to —' is the only equivalence relation in Exercises I.

III

i. $\{(1, 1), (3, 3), (1, 3)\}$ does not describe an equivalence relation because the symmetric property does not hold. $(1, 3)$ is in the set but $(3, 1)$ is not.

ii. $\{(3, 3), (4, 4), (3, 4), (4, 3)\}$ describes an equivalence relation.

iii. $\{(1, 1), (3, 1), (1, 3)\}$ does not describe an equivalence relation because the reflexive property ($a\,R\,a$) does not hold when $a = 3$.

iv. $\{(5, 5), (4, 4), (3, 3), (4, 5), (3, 5), (5, 4), (5, 3), (4, 3)\}$ does not describe an equivalence relation because neither the symmetric property nor the transitive property holds. The symmetric property does not hold because $(4, 3)$ is in the set but $(3, 4)$ is not. The transitive property does not hold because $(3, 5)$ and $(5, 4)$ are in the set but $(3, 4)$ is not.

IV

$S = \{a, b, c\}$

i. $\{(a,a), (b,b), (a,b)\}$ is a relation in S which is transitive but not symmetric. (Other answers are possible.)

ii. $\{(a,a), (b,b), (c,c), (c,b), (b,a)\}$ is a relation in S which is reflexive but neither symmetric nor transitive. (Other answers are possible.)

V

i. If $\circ$ is defined for the set of integers (I) as $a \circ b = a - 3b$, then
$$3 \circ 12 = 3 - 3(12) = 3 - 36 = -33$$
$$-8 \circ - 3 = 1$$
$$7 \circ 2 = 1$$

ii. There is never more than one result when the rule is evaluated on elements of I.

iii. From arithmetic we know that multiplying and subtracting integers always yields an integer.

iv. $\circ$ is an operation on I.

v. $\circ$ is a binary operation—it is performed on two elements.

5

THE NATURAL NUMBERS

SELF-TEST

DIRECTIONS: Write your answers in the numbered regions to the right. To check your answers, turn the page. Study the solutions to the problems you missed, and do the exercises following any problem or group or problems in which you had errors.

1. Can it be proved that five is a natural number by using the first four of Peano's axioms?

2. If n' is the successor of n, what is the successor of $(3 \cdot 4')'$?

3. Compute $6 \cdot 2 + 4$ by using Peano's axioms, the definitions of the numerals for the natural numbers, and the definitions of addition and multiplication.

4. If the nth term of a sequence is $2n(n + 1)$, what is the 8th term?

5. If the nth term of a sequence is $(n - 3)^{n+1}$, what is the $(k + 1)$st term?

6. What is a possible nth term for the following series?

    $$1 + 5 + 8^2 + 11^3 + 14^4 + \ldots$$

7. Is the formula $2^{n-1} < 3n - 2$ true for $n = 1$, $n = 2$, and $n = 3$?

8. Suppose a formula is correct for $n = 1, 2$ and 3. Does the establishment of this fact verify the formula for all natural numbers?

9. Can it be proved by mathematical induction that every even number greater than four is the sum of two odd prime numbers? (A prime number has no factors but itself and 1, and is not equal to 1. A composite number is neither 1 nor prime.)

10. Let $P = \{\text{prime numbers}\}$, $C = \{\text{composite numbers}\}$, $A = \{1\}$, $E = \{\text{even numbers}\}$, $O = \{\text{odd numbers}\}$, and $N = \{\text{natural numbers}\}$. Are the following statements true or false?

    a. $P \cap C = N$ d. $E \subseteq C$
    b. $P \cup (C \cup A) = N$ e. $O \cap A = A$
    c. $E \cup O = N$

SOLUTIONS

1. Let n' be the successor of n. By P_1 and P_2, 1, $1'$, $(1')'$, $((1')')'$, and $(((1')')')'$ are natural numbers. By P_3 and P_4, 1, $1'$, $(1')'$, $((1')')'$, and $(((1')')')'$ are all different natural numbers. By definition, $(((1')')')' = 5$.

2. $(3 \cdot 4')' = (3 \cdot 5)' = (15)' = 16$
   $16' = 17$

EXERCISES I

State the successor of each of the given natural numbers.

- **i.** $1'$
- **ii.** $8 + 2'$
- **iii.** $3 \cdot 6'$
- **iv.** $48' + 27'$
- **v.** $(4 + 7')'$

---

3. $6 \cdot 2 + 4 = 6 \cdot 1' + 4$ Def. of 2
   $= 6 \cdot 1 + 6 + 4$ Def. of mult.
   $= 6 + 6 + 4$ Def. of mult.
   $= 6 + ((((1')')')' + 4$ Def. of 6
   $= (((((6')')')')' + 4$ Def. of add.
   $= 12 + 4$ Def. of 12
   $= 12 + ((1')')'$ Def. of 4
   $= (((12')')')'$ Def. of add.
   $= 16$ Def. of 16

EXERCISES II

Compute the following by using Peano's axioms and the definitions for the numerical symbols and addition and multiplication.

- **i.** $6 + 1$
- **ii.** $4 + 2$
- **iii.** $3 \cdot 2$
- **iv.** $5 \cdot 2$
- **v.** $5 \cdot 2 + 3 \cdot 2$

---

4. nth term: $2n(n + 1)$
   8th term: $2 \cdot 8(8 + 1) = 16(9) = 144$

5. nth term: $(n - 3)^{n+1}$
   $(k + 1)$st term: $(k + 1 - 3)^{k+1+1}$
   $= (k - 2)^{k+2}$

EXERCISES III

i. If the nth term of a sequence is $3n$, write
- **a.** the 6th term
- **b.** the kth term
- **c.** the $(k + 1)$st term
- **d.** the $(k + 3)$rd term

ii. If the nth term of a sequence is $n(n + 1)$, write
- **a.** the 3rd term
- **b.** the kth term
- **c.** the $(k + 1)$st term
- **d.** the $(k + 3)$rd term

iii. If the nth term of a sequence is $2n - 1$, write
- **a.** the 8th term
- **b.** the kth term
- **c.** the $(k + 1)$st term
- **d.** the $(k + 3)$rd term

---

6. $1 + 5 + 8^2 + 11^3 + 14^4 + \ldots$

To determine the nth term, first consider the exponents. The third, fourth, and fifth terms have exponents of the form $n - 1$. The first term, then, has the exponent 0 and is equal to 1. The second term is of the form $3n - 1$, and its exponent must be 1. The third, fourth, and fifth terms are also of the form $3n - 1$. Therefore, a general term is $(3n - 1)^{n-1}$.

EXERCISES IV

Write a kth term and a $(k + 1)$st term.

- **i.** $2 + 4 + 6 + 8 + \cdots$
- **ii.** $1^2 + 2^4 + 3^6 + 4^8 + \cdots$
- **iii.** $1 + 8 + 12^2 + 16^3 + \cdots$

---

7. $2^{n-1} < 3n - 2$
   $n = 1$: $2^0 < 3 - 2, 1 < 1$ FALSE
   $n = 2$: $2^{2-1} < 6 - 2, 2 < 4$ TRUE
   $n = 3$: $2^{3-1} < 9 - 2, 4 < 7$ TRUE

8. A formula's being correct for $n = 1$, $n = 2$, and $n = 3$ does not satisfy the second condition for the Axiom of Mathematical Induction.

EXERCISES V

Determine whether the given formula is correct for $n = 1$, $n = 2$, and $n = 3$.

i. $1 + 3 + 5 + 7 + \cdots + (2n - 1) = n^2$
ii. $2 + 4 + 6 + \cdots + 2n = n(n + 1) + (n - 1)$
iii. $2^n > 2(n + 1)$

9. The statement that every even number greater than four is the sum of two odd prime numbers is known as *Goldbach's Conjecture*. The idea was advanced by Goldbach in 1742, but it has never been proved or disproved. It cannot be proved by mathematical induction (at least at the present time) because there is no known formula for the prime numbers.

EXERCISES VI

The following is a list of true statements. For each statement determine whether proof by mathematical induction would be possible.

i. The square of every integer leaves a remainder of 0, 1 or 4 when divided by 5.
ii. The square of every even integer is divisible by 4.
iii. The number of regions into which n straight lines divide a plane can never exceed 2^n.
iv. $\frac{a}{b} + \frac{c}{d} = \frac{(ad + bc)}{bd}$ if $b \neq 0$ and $d \neq 0$ and a, b, c, d are integers.
v. $\sqrt{5}$ is irrational.
vi. If $x = 3$, then $x^2 = 9$.

10. a. $P \cap C = N$ FALSE
    b. $P \cup (C \cup A) = N$ TRUE
    c. $E \cup O = N$ FALSE : 0 is an even number
    d. $E \subseteq C$ FALSE
    e. $O \cap A = A$ TRUE

Solutions to the exercises are on page 30.

ANSWERS

1. yes
2. 17
3. 16
4. 144
5. $(k - 2)^{k+2}$
6. $(3n - 1)^{n-1}$
7. false for $n = 1$
   true for $n = 2$ and $n = 3$
8. no
9. no
10. a. false
    b. true
    c. false
    d. false
    e. true

BASIC FACTS

The **natural numbers** constitute the infinite set $\{1, 2, 3, \ldots\}$. These are the numbers used in counting, and are often referred to as **counting numbers.**

There are three essentials in counting: something to count with, the concept of "comes after," and a starting point. Guiseppi **Peano** (1858–1932) developed a set of **axioms for the natural numbers** taking these three essentials, respectively, as the undefined terms "natural number," "successor," and "1." The following axioms, advanced by Peano, state the properties which these undefined terms possess and form a basis for the logical development of the system of natural numbers.

P_1: 1 is a natural number.

P_2: Every natural number n has a successor n' which is a natural number.

P_3: 1 is not the successor of any natural number.

P_4: No two natural numbers have the same successor.

P_5: **Axiom of Mathematical Induction**
If S is a set of natural numbers which has the properties
$$1 \in S$$
$$K \in S \to K' \in S$$
then all natural numbers belong to S.

Starting with this set of axioms, all the terms and relations necessary for ordinary arithmetic and algebra may be defined.

The **numerals** (the symbols for numbers) are defined as follows. By P_1, 1 is a natural number. By P_2, 1 must have a successor 1'. The successor of 1 is defined as 2. Then, by P_2, 2 is a natural number and it, in turn, has a successor 2' which is defined as 3. This process can be continued indefinitely since, by P_3, it is impossible to return to 1; and, by P_4, it is impossible to return to any other number previously defined.

The operation of **addition** is defined as:

ADDITIONAL INFORMATION

The Axiom of Mathematical Induction

It is necessary to include axiom P_5 in the set of **Peano's axioms** because without it it is impossible to state that *all* natural numbers are in the set S. By using P_1 to P_4, it is possible to prove that any given natural number is in the set S, but this is not equivalent to stating that all natural numbers are in the set. The latter is the effect of the Axiom of Mathematical Induction.

Axiom P_5——the Axiom of Mathematical Induction——may be understood intuitively in terms of the following analogy. Consider an infinitely tall ladder with its base on the ground. To make certain that we can climb to the top of this ladder, it is not sufficient to climb onto the first rung, and then onto the second rung, and so on; because at any given rung we have actually climbed only a finite number of rungs. However, we would be certain that we could climb as many rungs as we wish if we could assure ourselves of two things:

1. We can get our foot on a low rung.
2. We can climb from any rung to the next rung.

Clearly, if we can do these two things, we can climb the ladder as far as we please.

Condition 1 of the analogy corresponds to the first statement of P_5 and condition 2 corresponds to the second statement of P_5. If we let S be the set of all rungs to be climbed, then P_5 becomes

If S is the set of all rungs of the ladder and S is such that
1. the first rung is in S
2. if any rung is in S, then its successor (the next rung of the ladder) is in S

then all rungs are in S.

Proof by Mathematical Induction

The Axiom of Mathematical Induction provides a method of proof which is extremely useful in proving theorems which assert that certain propositions are true for *all* natural numbers. This method is called mathematical induction; and since integers, rational numbers, real numbers, and complex numbers can be written in terms of operations on natural numbers, the method is useful in all areas of number theory.

To take an example, even though $\frac{1}{2}$ is a rational number, it is possible to use mathematical induction to show that the sum of a sequence involving this and other rational numbers is given by a certain formula. The only qualification is that in the general term, n, must represent a natural number. Thus, we could use mathematical induction to show that $\frac{1}{1 \cdot 2} + \frac{1}{2 \cdot 3} + \frac{1}{3 \cdot 4} + \cdots + \frac{1}{n(n+1)} = \frac{n}{n+1}$ where n is a natural number.

$$n + 1 = n'$$
$$n + k' = (n + k)'.$$

This defines the sum of any two numbers. The operation of **multiplication** is defined in terms of addition.

$$n \cdot 1 = n$$
$$n \cdot k' = n \cdot k + n$$

This defines the product of any two natural numbers.

These axioms and definitions imply the following **properties of the natural numbers**.

The set of natural numbers is **closed** under addition and multiplication. That is, the sum and product of any two natural numbers are natural numbers.

Addition and multiplication are **commutative**. That is, if a and b are any two natural numbers, $a + b = b + a$ and $ab = ba$.

Addition and multiplication are **associative**. That is, if a, b and c are any three natural numbers, $a + (b + c) = (a + b) + c$ and $a(bc) = (ab)c$.

Multiplication is **distributive** over addition. That is, if a, b and c are any three natural numbers, $a(b + c) = ab + ac$.

1 is the **identity for multiplication** since, for every natural number a, $1 \cdot a = a$ and $a \cdot 1 = a$

The Venn diagram (Figure 5-1) shows that {prime numbers}, {composite numbers}, and {1} are disjoint **subsets of the natural numbers**. The union of these three sets is the set of natural numbers. The union of {positive even numbers $(2n)$} and {positive odd numbers $(2n - 1)$} is also the set of natural numbers.

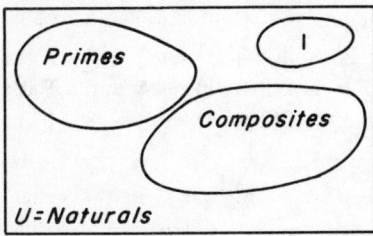

Fig. 5-1

In a **proof by mathematical induction**, the axiom P5 is applied in the following way.

Let S be the set of natural numbers for which the statement is to be proved true. Denote the statement by p_n, and show that
1. $1 \in S$
   (Show that p_n is true for $n = 1$.)
2. $k \in S \rightarrow k + 1 \in S$
   (Assume that p_n is true for $n = k$, and use this to prove that p_n is true for $n = k + 1$.)

Then since S satisfies P5, S contains all natural numbers.

In order to illustrate this technique, we will prove that the sum of the first n even natural numbers is $n(n + 1)$ where n is a natural number. The first even natural number is $2 \cdot 1$, the second is $2 \cdot 2$, and the nth is $2n$.

for $n = 1$, $2 = 1(1 + 1)$
for $n = 2$, $2 + 4 = 2(2 + 1)$
for $n = 3$, $2 + 4 + 6 = 3(3 + 1)$
for $n = 4$, $2 + 4 + 6 + 8 = 4(4 + 1)$ and so on

The theorem can be verified for these values of n, but the "and so on" does not prove the theorem for other values of n. Hence a proof by mathematical induction is appropriate.

PROOF: Let S be the set of natural numbers for which p_n is true, where p_n is the statement $2 + 4 + 6 + \cdots + 2n = n(n + 1)$.

1. Show p_n is true for $n = 1$.
   $2 = 1(1 + 1) \leftrightarrow 2 = 2$. Therefore $1 \in S$.
2. Assume p_n is true for $n = k$ and prove p_n is true for $n = k + 1$.
   HYPOTHESIS: $2 + 4 + 6 + \cdots + 2k = k(k + 1)$
   CONCLUSION: $2 + 4 + 6 + \cdots + 2k + 2(k + 1)$
   $= (k + 1)(k + 1 + 1)$
   $= (k + 1)(k + 2)$
   $= k^2 + 3k + 2$

The conclusion above is obtained by substituting $n = k + 1$ in p_n. In order to prove that *if* the formula p_n is true for $n = k$, *then* this conclusion is true, we must obtain the conclusion from the hypothesis. This can be done by adding the $(k + 1)$st term to both members of the equation constituting the hypothesis and simplifying. Then the second condition ($k \in S \rightarrow k + 1 \in S$) is fulfilled. It follows that all natural numbers are elements of S; or, which is the same thing, that p_n is true for all natural numbers.

SOLUTIONS TO EXERCISES

I

First find the number, then find its successor. If there are parentheses, work from the inside out.

i. $(1')' = 2' = 3$
ii. $(8 + 2')' = (8 + 3)' = 11' = 12$
iii. $(3 \cdot 6')' = (3 \cdot 7)' = 21' = 22$
iv. $(48' + 27')' = (49 + 28)' = 77' = 78$
v. $((4 + 7')')' = ((4 + 8)')' = (12')' = 13' = 14$

II

If addition and multiplication occur in the same problem, multiply first.

i. $6 + 1 = 6'$ Def. of add.
         $= 7$ Def. of 7

ii. $4 + 2 = 4 + 1'$ Def. of 2
         $= (4 + 1)'$ Def. of add.
         $= (4')'$ Def. of add.
         $= 5' = 6$ Def. of 5 and 6

iii. $3 \cdot 2 = 3 \cdot 1'$ Def. of 2
         $= 3 \cdot 1 + 3$ Def. of mult.
         $= 3 + 3$ Def. of mult.
         $= 3 + 2'$ Def. of 3
         $= 3 + (1')'$ Def. of 2
         $= (3 + 1')'$ Def. of add.
         $= ((3 + 1)')'$ Def. of add.
         $= ((3')')'$ Def. of add.
         $= ((4)')'$ Def. of 4
         $= 5' = 6$ Def. of 5 and 6

iv. $5 \cdot 2 = 5 \cdot 1'$ Def. of 2
         $= 5 \cdot 1 + 5$ Def. of mult.
         $= 5 + 5$ Def. of mult.
         $= 5 + 4'$ Def. of 5
         $= 5 + (((1')')')'$ Def. of 4, 3, 2
         $= (5 + ((1')')')'$ Def. of add.
         $= ((5 + (1')')')'$ Def. of add.
         $= (((5 + 1')')')'$ Def. of add.
         $= ((((5 + 1)')')')'$ Def. of add.
         $= ((((5')')')')'$ Def. of add.
         $= 10$ Def. of 6, 7, 8, 9, 10

v. Use the results of **iii** and **iv** and combine. The sum is 16.

III

i.
 a. $n = 6, 3(6) = 18$
 b. $n = k, 3(k) = 3k$
 c. $n = k + 1, 3(k + 1) = 3k + 3$
 d. $n = k + 3, 3(k + 3) = 3k + 9$

ii.
 a. $n = 3, 3(3 + 1) = 3(4) = 12$
 b. $n = k, k(k + 1) = k^2 + k$
 c. $n = k + 1, (k + 1)(k + 1 + 1) = (k + 1)(k + 2) = k^2 + 3k + 2$
 d. $n = k + 3, (k + 3)(k + 3 + 1) = (k + 3)(k + 4) = k^2 + 7k + 12$

iii.
 a. $n = 8, 2(8) - 1 = 16 - 1 = 15$
 b. $n = k, 2(k) - 1 = 2k - 1$
 c. $n = k + 1, 2(k + 1) - 1 = 2k + 2 - 1 = 2k + 1$
 d. $n = k + 3, 2(k + 3) - 1 = 2k + 6 - 1 = 2k + 5$

IV

i. $2 + 4 + 6 + 8 + \cdots + 2k + 2(k + 1)$
ii. $1^2 + 2^4 + 3^6 + 4^8 + \cdots + k^{2k} + (k + 1)^{2(k+1)}$
iii. $1 + 8 + 12^2 + 16^3 + \cdots + (4k)^{k-1} + 4(k + 1)^k$

V

i. $1 + 3 + 5 + 7 + \cdots + (2n - 1) = n^2$
$n = 1: 1 = 1^2 \leftrightarrow 1 = 1$ TRUE
$n = 2: 1 + 3 = 2^2 \leftrightarrow 4 = 4$ TRUE
$n = 3: 1 + 3 + 5 = 3^2 \leftrightarrow 9 = 9$ TRUE

ii. $2 + 4 + 6 + \cdots + 2n = n(n + 1) + (n - 1)$
$n = 1: 2 = 1(2) + 0 \leftrightarrow 2 = 2$ TRUE
$n = 2: 2 + 4 = 2(3) + 1 \leftrightarrow 6 = 7$ FALSE
$n = 3: 2 + 4 + 6 = 3(4) + 2 \leftrightarrow 12 = 14$ FALSE

iii. $2^n > 2(n + 1)$
$n = 1: 2^1 > 2(1 + 1) \leftrightarrow 2 > 4$ FALSE
$n = 2: 2^2 > 2(2 + 1) \leftrightarrow 4 > 6$ FALSE
$n = 3: 2^3 > 2(3 + 1) \leftrightarrow 8 > 8$ FALSE

VI

i. yes   **iii.** yes   **v.** no
ii. yes   **iv.** no   **vi.** no

6 MATHEMATICAL INDUCTION: PROBLEMS

SELF-TEST

DIRECTIONS: Write your answers in the numbered regions to the right. To check your answers, turn the page. Study the solutions to the problems you missed, and do the exercises if you had several errors.

1. Prove the following statement by mathematical induction.

   $1 + 3 + 5 + \cdots + (2n - 1) = n^2$, for any natural number n

2. Find the value of the 80th term of the series in **1**.

3. Use **1** to find the sum of the first twenty odd natural numbers.

4. Prove the following statement by mathematical induction.

   $$\frac{1}{1 \cdot 2} + \frac{1}{2 \cdot 3} + \frac{1}{3 \cdot 4} + \cdots + \frac{1}{n(n + 1)} = \frac{n}{n + 1}$$

   for any natural number n

5. Find the value of the fifty-ninth term of the series given in **4**.

6. Find the sum of the first one thousand terms of the series given in **4**.

7. Will the sum of the terms of the series in **4** ever be greater than or equal to one?

8. Prove the following statement by mathematical induction.

   For any natural number n, $4^n - 1$ is divisible by 3.

32 Mathematical Induction: Problems

SOLUTIONS

1. p_n: $1 + 3 + 5 + \cdots + (2n - 1) = n^2$
   Let S be the set of all natural numbers for which p_n is true.
   Show that p_n is true for $n = 1$: $1 = 1^2 \leftrightarrow 1 = 1$, $\therefore 1 \in S$
   Assume that p_n is true for $n = k$ and use this to prove that p_n is true for $n = k + 1$:

$1 + 3 + 5 + \cdots + (2k - 1) = k^2$	Hypothesis
$1 + 3 + 5 + \cdots + (2k - 1) + (2k + 1) = k^2 + (2k + 1)$	Add the $(k + 1)$st term, $(2k + 1)$, to both members of the equation.
$\quad = k^2 + 2k + 1$	Remove the parentheses.
$\quad = (k + 1)^2$	Factor.

   Since the result of operating on the hypothesis is precisely the conclusion obtained by substituting $k + 1$ for n in p_n, the second part of the Axiom of Mathematical Induction ($k \in S \to k + 1 \in S$) is satisfied. It follows that all natural numbers are elements of S and that p_n is true for all natural numbers n.

2. Substitute $n = 80$ in the general term.

   $2(80) - 1 = 160 - 1 = 159$

3. Substitute $n = 20$ in the right member of p_n.

   $20^2 = 400$

4. p_n: $\frac{1}{1 \cdot 2} + \frac{1}{2 \cdot 3} + \frac{1}{3 \cdot 4} + \cdots + \frac{1}{n(n + 1)} = \frac{n}{n + 1}$

   Let S be the set of all natural numbers for which p_n is true.

   Show that p_n is true for $n = 1$: $\frac{1}{1 \cdot 2} = \frac{1}{1 + 1} \leftrightarrow \frac{1}{2} = \frac{1}{2}$, $\therefore 1 \in S$

   Assume that p_n is true for $n = k$ and use this to prove that p_n is true for $n = k + 1$:

$\frac{1}{1 \cdot 2} + \frac{1}{2 \cdot 3} + \cdots + \frac{1}{k(k + 1)} = \frac{k}{k + 1}$	Hypothesis
$\frac{1}{1 \cdot 2} + \frac{1}{2 \cdot 3} + \cdots + \frac{1}{k(k + 1)} + \frac{1}{(k + 1)(k + 2)}$	Add the $(k + 1)$st term to both members of the equation.
$\quad = \frac{k}{k + 1} + \frac{1}{(k + 1)(k + 2)}$	
$\quad = \frac{k(k + 2)}{(k + 1)(k + 2)} + \frac{1}{(k + 1)(k + 2)}$	$(k + 1)(k + 2)$ is a common denominator.
$\quad = \frac{k^2 + 2k + 1}{(k + 1)(k + 2)}$	Combine and remove parentheses.
$\quad = \frac{(k + 1)(k + 1)}{(k + 1)(k + 2)} = \frac{k + 1}{k + 2}$	Factor the numerator.

   Since $\frac{k + 1}{k + 2}$ is also the result obtained by substituting $k + 1$ for n in p_n, $k \in S \to k + 1 \in S$ and S contains all natural numbers.

5. Substitute $n = 59$ in the nth term.

   $\frac{1}{59(60)} = \frac{1}{3540}$

6. Substitute $n = 1000$ in the right member of p_n.

   $\frac{1000}{1000 + 1} = \frac{1000}{1001}$

7. Since n is a natural number the numerator of $\frac{n}{n + 1}$ is always one less than the denominator.

   Therefore, the value of the fraction is always less than 1.

8. p_n: $4^n - 1$ is divisible by 3 for any natural number n.
   Let S be the set of natural numbers for which p_n is true.
   Show p_n is true for $n = 1$: $4^1 - 1 = 4 - 1 = 3$, an integer divisible by 3, $\therefore 1 \in S$
   Assume that p_n is true for $n = k$ and use this to prove that p_n is true for $n = k + 1$.

(Continued on page 33, bottom.)

EXERCISES

i. a. Prove by mathematical induction that $1^2 + 2^2 + 3^2 + \cdots + n^2 = \dfrac{n(n+1)(2n+1)}{6}$ for any natural number n.

  b. Find the value of the 60th term of this series.

  c. Find the sum of the squares of the first sixty natural numbers.

ii. Prove by mathematical induction that for all integers $x \neq 1$, $x^n - 1$ is divisible by $x - 1$ for any natural number n.

iii. Prove by mathematical induction that $2^n > 2(n + 1)$, where n is any natural number greater than 3.

iv. Prove by mathematical induction that $(ab)^n = a^n b^n$, where n is any natural number.

  a. Prove by mathematical induction that $0 + 1^2 + 2^2 + \cdots + (n-1)^2 = \dfrac{n(n-1)(2n-1)}{6}$.

  b. Is the sum of the first n terms of this series equal to the sum of the first $n - 1$ terms of the series in **ia**?

ANSWERS

1. The statement satisfies both conditions of the Axiom of Mathematical Induction. The proof is on page 32.

2. 159

3. 400

4. The statement satisfies both conditions of the Axiom of Mathematical Induction. The proof is on page 32.

5. $\dfrac{1}{3540}$

6. $\dfrac{1000}{1001}$

7. no

8. The statement satisfies both conditions of the Axiom of Mathematical Induction. The proof is on page 32.

$4^k - 1$ is divisible by 3	Hypothesis
$\dfrac{4^k - 1}{3} = x$, x an integer	If a number is divisible by 3, its quotient is an integer.
$4^k - 1 = 3x$	Multiply both members by 3.
$4^1(4^k - 1) = 4(3x)$	Multiply both members by 4 since we need a 4^{k+1} in the conclusion.
$4^{k+1} - 4 = 12x$	Simplify.
$4^{k+1} - 1 = 12x + 3$	Add 3 to both members. This makes the left member identical to the conclusion obtained by substituting $n = k + 1$ in p_n.
$\dfrac{4^{k+1} - 1}{3} = 4x + 1$	Divide both members by 3. Since the right member is divisible by 3, so is the left member.

Hence, if $k \in S$, then $k + 1 \in S$. Since S satisfies the Axiom of Mathematical Induction, S contains all natural numbers and the statement p_n is true for all natural numbers.

Solutions to the exercises appear on pages 34, 35, and 36.

SOLUTIONS TO EXERCISES

i. a. p_n: $1^2 + 2^2 + 3^2 + \cdots + n^2 = \dfrac{n(n + 1)(2n + 1)}{6}$

Let S be the set of all natural numbers for which p_n is true.

Show that p_n is true when $n = 1$: $1^2 = \dfrac{1(1 + 1)(2 + 1)}{6} \leftrightarrow 1 = \dfrac{1(2)(3)}{6} \leftrightarrow 1 = 1$, $\therefore 1 \in S$

Assume that p_n is true for $n = k$ and use this to prove that p_n is true for $n = k + 1$.

HYPOTHESIS: $1^2 + 2^2 + 3^2 + \cdots + k^2 = \dfrac{k(k + 1)(2k + 1)}{6}$ Substitute $n = k$ in p_n

CONCLUSION: $1^2 + 2^2 + 3^2 + \cdots + k^2 + (k + 1)^2$ Substitute $n = k + 1$ in k_n

$= \dfrac{(k + 1)(k + 1 + 1)(2(k + 1) + 1)}{6} = \dfrac{(k + 1)(k + 2)(2k + 3)}{6}$

PROOF OF THE CONCLUSION:

$1^2 + 2^2 + 3^2 + \cdots + k^2 = \dfrac{k(k + 1)(2k + 1)}{6}$ Hypothesis

$1^2 + 2^2 + 3^2 + \cdots + k^2 + (k + 1)^2 = \dfrac{k(k + 1)(2k + 1)}{6} + (k + 1)^2$ Add the $(k + 1)$st term $(k + 1)^2$ to both members of the equation.

$= \dfrac{k(k + 1)(2k + 1)}{6} + \dfrac{6(k + 1)^2}{6}$ Change $(k + 1)^2$ to a fraction with a denominator of 6.

$= \dfrac{k(k + 1)(2k + 1) + 6(k + 1)^2}{6}$ Combine fractions.

$= \dfrac{(k + 1)(k(2k + 1) + 6(k + 1))}{6}$ Factor out $(k + 1)$.

$= \dfrac{(k + 1)(2k^2 + k + 6k + 6)}{6}$ Multiply.

$= \dfrac{(k + 1)(2k^2 + 7k + 6)}{6}$ Combine.

$= \dfrac{(k + 1)(k + 2)(2k + 3)}{6}$ Factor.

Since the result of operating on the hypothesis is the same as the result of substituting $n = k + 1$ in p_n, $k \in S \rightarrow k + 1 \in S$.

Since S satisfies the Axiom of Mathematical Induction, S contains all natural numbers and the statement p_n is true for all natural numbers.

b. Substitute $n = 60$ in the nth term.

$60^2 = 3600$

c. Substitute $n = 60$ in the right member of p_n.

$\dfrac{60(61)(121)}{6} = 10(61)(121) = 73{,}810$

ii. p_n: for all integers $x \neq 1$, $x^n - 1$ is divisible by $x - 1$ for any natural number n. Let S be the set of natural numbers for which p_n is true. Show p_n is true for $n = 1$: $x^1 - 1 = x - 1$ which is divisible by $x - 1$, $\therefore 1 \in S$. Assume that p_n is true for $n = k$ and use this to prove that p_n is true for $n = k + 1$.

$x^k - 1$ is divisible by $x - 1$	Hypothesis
$\dfrac{x^k - 1}{x - 1} = y$, y an integer	If a number is divisible by $x - 1$, x an integer $\neq 1$, its quotient is an integer.
$x^k - 1 = y(x - 1)$	Multiply both members by $x - 1$.
$x^1(x^k - 1) = x^1 y(x - 1)$	Multiply both members by x.
$x^{k+1} = xy(x - 1) + x$	Simplify and add x to both members.
$x^{k+1} - 1 = xy(x - 1) + (x - 1)$	Add -1 to both members.
$\dfrac{x^{k+1} - 1}{x - 1} = xy + 1$	Divide both members by $x - 1$. Since the right member is divisible by $x - 1$, so is the left member.

Hence, if $k \in S$, then $k + 1 \in S$. Since S satisfies the Axiom of Mathematical Induction, S contains all natural numbers and the statement p_n is true for all natural numbers.

iii. p_n: $2^n > 2(n + 1)$, where n is any natural number greater than 3.

COMMENT: We do not wish to prove this true for all natural numbers. In fact, the statement is not true for $n = 1, 2, 3$. Hence, we begin by verifying the statement for $n = 4$ rather than for $n = 1$.

Let S be the set of natural numbers for which p_n is true.
Show p_n true for $n = 4$: $2^4 > 2(4 + 1) \leftrightarrow 16 > 10$, $\therefore 4 \in S$
Assume that p_n is true for $n = k$ and use this to prove that p_n is true for $n = k + 1$.

HYPOTHESIS: $2^k > 2(k + 1)$
CONCLUSION: $2^{k+1} > 2(k + 1 + 1)$
$\qquad\qquad\quad > 2(k + 2)$

PROOF OF THE CONCLUSION:

$2^k > 2(k + 1)$	Hypothesis
$2(2^k) > 2(2(k + 1))$	Multiply both members by 2 since we need 2^{k+1} in the left member of the conclusion.
$2^{k+1} > 4k + 4$	Simplify.
$2^{k+1} > 2(2k + 2)$	Factor out 2 since we need this in the conclusion.
$2(2k + 2) > 2(k + 2)$	Since k is a natural number, $2k + 2 > k + 2$, hence $2(2k + 2) > 2(k + 2)$.
$2^{k+1} > 2(k + 2)$	Transitive Property for $>$

Since S satisfies both the conditions of Axiom of Mathematical Induction for $n > 3$, S contains all natural numbers greater than 3 and the statement is true for all natural numbers greater than 3.

iv. p_n: $(ab)^n = a^n b^n$ for all natural numbers.
Let S be the set of natural numbers for which the statement is true.
Show p_n true for $n = 1$: $(ab)^1 = a^1 b^1 \leftrightarrow ab = ab$, $\therefore 1 \in S$
Assume that p_n is true for $n = k$ and use this to prove that p_n is true for $n = k + 1$.

HYPOTHESIS: $(ab)^k = a^k b^k$
CONCLUSION: $(ab)^{k+1} = a^{k+1} b^{k+1}$
PROOF OF THE CONCLUSION:
$(ab)^k = a^k b^k$ Hypothesis
$(ab)^1 (ab)^k = (ab)^1 a^k b^k$ Multiply both members by ab since we need ab^{k+1} in the left member.
$(ab)^{k+1} = a^1 b^1 a^k b^k$ Simplify.
$(ab)^{k+1} = a^{k+1} b^{k+1}$ Multiply.

Since S satisfies both conditions of the Axiom of Mathematical Induction, S contains all natural numbers and the statement p_n is true for all natural numbers.

v. a. p_n: $0 + 1^2 + 2^2 + \cdots + (n-1)^2 = \dfrac{n(n-1)(2n-1)}{6}$

Let S be the set of natural numbers for which p_n is true.
Show p_n true for $n = 1$: $0^2 = \dfrac{1(1-1)(2(1)-1)}{6} \leftrightarrow 0 = 0$ $\therefore 1 \in S$.
Assume that p_n is true for $n = k$ and use this to prove that p_n is true for $n = k + 1$.
HYPOTHESIS: $0 + 1^2 + 2^2 + \cdots + (k-1)^2 = \dfrac{k(k-1)(2k-1)}{6}$

CONCLUSION: $0 + 1^2 + 2^2 + \cdots + (k-1)^2 + k^2 = \dfrac{(k+1)(k)(2k+1)}{6}$

PROOF OF CONCLUSION:

$0 + 1^2 + 2^2 + \cdots + (k-1)^2 = \dfrac{k(k-1)(2k-1)}{6}$ Hypothesis.

$0 + 1^2 + 2^2 + \cdots + (k-1)^2 + k^2 = \dfrac{k(k-1)(2k-1)}{6} + k^2$ Add the $(k+1)$st term, k^2, to both members.

$= \dfrac{k(k-1)(2k-1) + 6k^2}{6}$ Change k^2 to a fraction with a denominator of 6 and combine fractions.

$= \dfrac{k((k-1)(2k-1) + 6k)}{6}$ Factor out k.

$= \dfrac{k(2k^2 + 3k + 1)}{6}$ Multiply.

$= \dfrac{k(k+1)(2k+1)}{6}$ Factor.

Since the result of operating on the hypothesis is the same as the result of substituting $n = k + 1$ in p_n, $k \in S \to k + 1 \in S$.

Since S satisfies the Axiom of Mathematical Induction, S contains all natural numbers and the statement p_n is true for all natural numbers.

b. Yes.

7 THE REAL NUMBERS

SELF-TEST

DIRECTIONS: Write your answers in the numbered regions to the right. To check your answers, turn the page. Study the solutions to the problems you missed, and do the exercises following any problem or group of problems in which you had errors.

1. In **a–d**, answer true or false.
   a. {prime numbers} $\subseteq$ {real numbers}
   b. {natural numbers} $\subseteq$ {irrational numbers}
   c. {rational numbers} $\cap$ {irrational numbers} = {real numbers}
   d. {natural numbers} $\not\subseteq$ {real numbers}

2. Each statement is an application of one of the following properties: closure, associativity, commutativity, distributivity, identity, or inverse. For each statement, give the appropriate property.
   a. $7 + (3 \times 2) = 7 + (2 \times 3)$
   b. $3(-4) + 3(7) = 3(-4 + 7)$
   c. $\sqrt{2} + \frac{1}{7}$ is a real number
   d. $-9 + 0 = 9$
   e. $7 + (8 - 3) = (8 - 3) + 7$
   f. $8 + 3(\frac{1}{3}) = 8 + 1$

3. Perform the indicated operations.
   a. $-(-5) + (-8) - 2(6)$
   b. $(-8)(3)(-2)(-1)$
   c. $(-3)(-5) + (8)(-6)$
   d. $|-7| - |2| + |-6| \cdot |-2|$
   e. $-4 - (-6 + 3)$
   f. $2^3 \times 2^2 \times 2^4$
   g. $3^7 \div 3^5$
   h. $\sqrt[3]{27^2}$
   i. $9^{\frac{3}{2}} + (3 \cdot 4)^2$

4. Find the prime factorization of 430.

5. Find the greatest common divisor (GCD) of 68 and 1476.

6. Are 54 and 77 relatively prime?

7. Express the rational number $\frac{8}{10}$ as a repeating decimal.

8. Express $0.\overline{859}$ as the quotient of two integers.

37

SOLUTIONS

1. a. True   c. False
   b. False   d. False

2. a. Commutativity
   b. Distributivity
   c. Closure
   e. Additive identity
   f. Multiplicative inverse

   These properties are explained in the Basic Facts.

EXERCISES I

i. State the multiplicative inverse of 5. Of -3. Of 0.
ii. State the additive inverse of -10. Of 0. Of 8.
iii. If a and b are irrational, is $a - b$ always irrational?
iv. Find y if $4(y - 7) = 0$.
v. Are the real numbers commutative under division?
vi. Is $\frac{-2}{3}$ equal to $-\frac{2}{3}$?

---

3. When an expression involves several operations, perform all the operations within parentheses first. Then perform all multiplications and divisions. After this, perform all additions and subtractions.

   a. $-(-5) + (-8) - 2(6)$
      $= 5 - 8 - 12 = 5 - 20 = -15$
   b. $(-8)(3)(-2)(-1) = -48$
   c. $(-3)(-5) + (8)(-6)$
      $= 15 + (-48) = 15 - 48 = -33$
   d. $|-7| - |2| + |-6| \cdot |-2|$
      $= 7 - 2 + 6 \cdot 2$
      $= 7 - 2 + 12 = 17$
   e. $-4 - (-6 + 3) = -4 - (-3)$
      $= -4 + 3 = -1$
   f. $2^3 \times 2^2 \times 2^4 = 8 \times 4 \times 16 = 512$
   g. $3^7 \div 3^5 = 3^{7-5} = 3^2 = 9$
   h. $\sqrt[3]{27^2} = (\sqrt[3]{27})^2 = 3^2 = 9$
   i. $9^{\frac{3}{2}} + (3 \cdot 4)^2 = (\sqrt{9})^3 + 12^2$
      $= 3^3 + 12^2$
      $= 27 + 144 = 171$

EXERCISES II

Compute each expression.

i. $-4 + 3 - 7(2 + 8)$
ii. $-4[3 - 6(2 - 4)]$
iii. $|-22| + |-2|(-|-14|)$
iv. $4^4 \times 4^{-5} \times 4$
v. $(-2)^3 \div (-2)^6$
vi. $\sqrt[4]{16^2} - \sqrt[3]{27}$
vii. $(-4)(-\frac{1}{2})(8) - (7)(0)(-3)$
viii. $[(4 + 2)3]^2$
ix. $\sqrt[3]{(5)^3}$
x. $7(8 - 3)^3$

---

4. To find the prime factors, divide successively by primes starting with the smallest prime divisor and continuing until the quotient is a prime.

   $\begin{array}{r|l} 2 & 430 \\ 5 & 215 \\ & 43 \end{array}$ $430 = 2 \times 5 \times 43$

EXERCISES III

In i–iii, find the prime factorization of the given number.

i. 819   ii. 5280   iii. 2850

---

5. $\left.\begin{array}{l} 68 = 2^2 \times 17 \\ 1476 = 2^2 \times 3^2 \times 41 \end{array}\right\}$ Prime factorizations

   GCD $= 2^2 = 4$ Product of numbers which are in both factorizations

---

EXERCISES IV

In i–iii, find the GCD of the given numbers.

i. 123 and 530   iii. 2057 and 11,339
ii. 7469 and 2464

---

6. $\begin{array}{r|l} 2 & 54 \\ 3 & 27 \\ 3 & 9 \\ & 3 \end{array}$ $\begin{array}{r|l} 7 & 77 \\ & 11 \end{array}$

   GCD $= 1$, $\therefore$ Relative primes

7. Divide the denominator into the numerator, and continue dividing until a sequence of digits repeats itself.

$$\frac{8}{10} = .8\overline{0}$$

EXERCISES V

Express each of the following rational numbers as a repeating decimal.

i. $\frac{1}{7}$
ii. $\frac{7}{5}$
iii. $\sqrt{6.25}$
iv. $\frac{7}{12}$
v. $43\frac{2}{3}$

---

8. $x = 0.\overline{859}$ (1)
    $1000x = 859.\overline{859}$ (2)
    $999x = 859$ Subtract (1) from (2)
    $x = 859/999$

EXERCISES VI

Express each of the following decimal expansions as the quotient of two integers.

i. $3.\overline{6}$
ii. $7.\overline{36}$
iii. $6.2\overline{45}$
iv. $1.46\overline{72}$
v. $0.0006\overline{742}$
vi. $675.38\overline{574892}$

Solutions to the exercises appear on page 42.

ANSWERS

1. a. true
   b. false
   c. false
   d. false

2. a. commutativity
   b. distributivity
   c. closure
   d. identity
   e. commutativity
   f. inverse

3. a. -15
   b. -48
   c. -33
   d. 17
   e. -1
   f. 512
   g. 9
   h. 9
   i. 171

4. $2 \times 5 \times 43$

5. 4

6. yes

7. $.8\overline{0}$

8. $\frac{859}{999}$

BASIC FACTS

Many mathematical sentences have empty solution sets if {natural numbers} is the universe. '$3 + x = 2$,' '$6x = 5$,' and '$x^2 = 2$' are some examples. To obtain non-empty solution sets for equations of this type, the set of natural numbers must be extended.

The set whose elements include solutions to the equations like those mentioned above is the set of **real numbers**. The Venn diagram (Figure 7-1) illustrates some important **subsets of {real numbers}**.

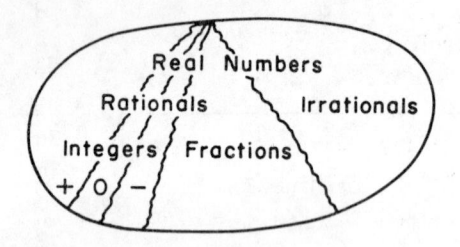

Fig. 7-1

The set of **integers** contains all natural numbers, negatives of natural numbers, and zero.

Extending the set of numbers from the naturals to the integers enables us to solve *all* equations of the form $a + x = b$ where a and b are integers. The solution is the integer $x = b - a$.

The set of **rational numbers** contains all decimal numbers of the form $\frac{a}{b}$, where a and b are integers, $b \neq 0$.

Extending the set of numbers from the integers to the rationals enables us to solve all equations of the form $ax = b$, where a and b are rational numbers, $a \neq 0$. The solution is the rational number $x = \frac{b}{a}$.

The set of **irrational numbers** contains all decimal numbers not expressible in the form $\frac{a}{b}$, where a and b are integers, $b \neq 0$.

ADDITIONAL INFORMATION

Two important properties which hold for the elements of the set of **integers** are divisibility and factorization. An integer $b \neq 0$ is said to divide an integer a if and only if there exists an integer x such that $a = bx$. Then a is said to be divisible by b, and b is said to be a **factor** or a **divisor** of a. For example, 3 divides 27 since there exists an integer 9 such that $3 \times 9 = 27$, while 3 does not divide 25 since there is no integer x such that $3 \times x = 25$.

The set of **positive integers** includes three disjoint subsets the union of which is equal to the whole set. These are the unit 1, the prime integers, and the composite integers. A positive integer not equal to 1 is a **prime** if and only if its only positive divisors are itself and 1. A positive integer is **composite** if it is neither prime nor the unit 1. Two positive integers are **relatively prime** if and only if there is no positive integer except 1 which divides both of them.

Every composite integer can be expressed as a product of primes in exactly one way, ignoring the order in which the factors are written. Consider some of the various ways of factoring 36. $36 = 2 \times 2 \times 3 \times 3 = 2^2 \times 3^2$ is in terms of the prime numbers 2 and 3. All the other factorizations of 36 could be put into this form by factoring their factors. But $2 \times 2 \times 3 \times 3$ cannot be factored further because the factors are all prime numbers.

The **common divisors** of two integers a and b are any integers which divide both a and b. The greatest of these common divisors is called the **greatest common divisor** (GCD). To find the GCD of two integers, express each integer by its prime factorization; consider those primes which are factors of both given integers; take the product of these with each raised to the highest power that is a factor of both given integers.

The set of **rational numbers** is identical with the set of **repeating decimals**. That is, every rational number may be expressed as a repeating decimal and every repeating decimal is an expression for a rational number. A rational number may be expressed as a repeating decimal by carrying out the ordinary process of division. For example, $\frac{8}{7} = 1.142857142857\ldots = 1.\overline{142857}$. The bar indicates that the numbers below it repeat indefinitely.

To write a repeating decimal in the form $\frac{a}{b}$, multiply the repeating decimal twice by different powers of 10 to obtain two expressions in which only the repeating digits follow the decimal point. When one of these expressions is subtracted from the other, the decimals disappear. Then division by the difference between the powers of 10 yields a number of the form $\frac{a}{b}$ where a and b are integers. See the solution to problem 8 for an example.

The set of **irrational numbers** is identical to the set of non-repeating, non-terminating decimals. π, e, $\sqrt{2}$, $\sqrt{3}$, and $\sqrt{5}$ are examples of irrational numbers. Three dots written after the first few digits of the decimal approximations of these numbers indicate that the decimals do not terminate.

Extending the set of numbers to include the irrationals enables us to solve equations of the form $x^2 = a$, where a is not a perfect square. The solutions are the irrational numbers $x = \pm \sqrt{a}$.

The set of **real numbers** is the union of the set of rational numbers and the set of irrational numbers.

Addition and multiplication of real numbers satisfy the following properties. (a, b, and c represent any real number.)

- **Closure Properties**
  $a + b$ is a unique real number.
  $a \times b$ is a unique real number.
- **Associative Properties**
  $(a + b) + c = a + (b + c)$
  $(a \times b) \times c = a \times (b \times c)$
- **Commutative Properties**
  $a + b = b + a$
  $a \times b = b \times a$
- **Identity Properties**
  Zero is the identity element for addition because, for all a, $a + 0 = a$.
  One is the identity element for multiplication because, for all a, $a \times 1 = a$.
- **Inverse Properties**
  $-a$ is the additive inverse (or negative) of a because, for all a, $a + (-a) = 0$.
  $\frac{1}{a}$ is the multiplicative inverse (or reciprocal) of a because, for all a except 0, $a \times \frac{1}{a} = 1$.
- **Distributive Property**
  $a \times (b + c) = (a \times b) + (a \times c)$
- **Trichotomy Property**
  For each a, exactly one of the following is true: a is positive, a is zero, a is negative.
- **Closure Property for Positive Reals**
  If a and b are positive, then $a + b$ and $a \times b$ are positive.

Since the set of real numbers has the first six properties, it is a **field** under $+$ and $\times$. Since it has the last two properties, it is an **ordered field**.

All properties of the **real numbers** can be derived from the field and order properties. Some derived properties are given below. Each letter represents any real number, unless otherwise stated.

			Examples
$\lvert a \rvert$	DEF	The absolute value of a is given by $\lvert a \rvert = \begin{cases} a \text{ if } a \text{ is positive} \\ 0 \text{ if } a \text{ is zero} \\ -a \text{ if } a \text{ is negative.} \end{cases}$	$\lvert 4 \rvert = 4$ $\lvert -8 \rvert = 8$
$a - b$	DEF	The **difference** of two real numbers is given by $a - b = a + (-b)$.	$-3 - 7$ $= -3 + (-7)$ $= -10$ $6 - (-3) = 6 + 3$
	THM	$a(b - c) = ab - ac$	$2(5 - 3)$ $= 2 \cdot 5 - 2 \cdot 3$ $= 4$
ab	THMS	The **product** of a positive number and a negative number is a negative number.	$8(-3) = -24$
		The product of two negative numbers is a positive number.	$-5(-3) = 15$
		$a(0) = 0$	$5(0) = 0$
		$-(-a) = a$	$-(-12) = 12$
		$-(a + b) = -a - b$	$-(-3 + 6)$ $= 3 - 6 = -3$
		$(-a)(-b) = ab$	$(-4)(-7) = (4)(7)$
$\frac{a}{b}$	DEF	The **quotient** of two real numbers is given by $\frac{a}{b} = a \times \frac{1}{b}$.	$\frac{-16}{4} = -16 \times \frac{1}{4}$ $5 \div 2 = 5 \times \frac{1}{2}$ $= 2\frac{1}{2}$
	THMS	$\frac{-a}{b} = \frac{a}{-b} = -\frac{a}{b}, b \neq 0$	$\frac{-2}{3} = \frac{2}{-3} = -\frac{2}{3}$
		$\frac{-a}{-b} = \frac{a}{b}, b \neq 0$	$\frac{-8}{-5} = \frac{8}{5}$
a^n	DEFS	$a^n = \underbrace{a \cdot a \cdots a}_{n \text{ factors}}, n$ natural	$2^4 = 2 \cdot 2 \cdot 2 \cdot 2$
		$a^0 = 1, a \neq 0$	$9^0 = 1$
		$a^{-n} = \frac{1}{a^n}, a \neq 0$	$4^{-3} = \frac{1}{4^3}$
	THMS	For a^m, a^n, b^n real and m, n rational,	
		$a^m \cdot a^n = a^{m+n}$	$2^5 \cdot 2^2 = 2^{5+2} = 2^7$
		$\frac{a^m}{a^n} = a^{m-n}, a \neq 0$	$\frac{2^5}{2^3} = 2^{5-3} = 2^2$

(Continued on page 42.)

42 *The Real Numbers*

(Continued from page 41.)

$\sqrt[n]{a}$ DEF

$(a^m)^n = a^{mn}$ $(2^3)^4 = 2^{12}$
$(ab)^n = a^n b^n$ $(2x)^3 = 2^3 x^3$
$\left(\dfrac{a}{b}\right)^n = \dfrac{a^n}{b^n}, b \neq 0$ $\left(\dfrac{2}{3}\right)^4 = \dfrac{2^4}{3^4}$

$\sqrt[n]{a} = a^{\frac{1}{n}}$, n natural, $a^{\frac{1}{n}}$ real
$\sqrt[2]{a}$ is written $\sqrt{a}$

$\sqrt[3]{8} = 8^{\frac{1}{3}} = 2$
since $2^3 = 8$

THMS

For positive real radicals and m, n natural,

$\sqrt[n]{a} \cdot \sqrt[n]{b} = \sqrt[n]{ab}$ $\sqrt{3} \cdot \sqrt{2} = \sqrt{6}$

$\dfrac{\sqrt[n]{a}}{\sqrt[n]{b}} = \sqrt[n]{\dfrac{a}{b}}, b \neq 0$ $\dfrac{\sqrt[3]{3}}{\sqrt[3]{7}} = \sqrt[3]{\dfrac{3}{7}}$

$\sqrt[n]{a^m} = (\sqrt[n]{a})^m = a^{\frac{m}{n}}$ $\sqrt[3]{8^2} = (\sqrt[3]{8})^2 = 8^{\frac{2}{3}}$

SOLUTIONS TO EXERCISES

I

i. The multiplicative inverse of 5 is $\frac{1}{5}$. Of -3 is $-\frac{1}{3}$. Zero has no multiplicative inverse.

ii. The additive inverse of -10 is 10. Of 0 is 0. Of 8 is -8.

iii. If a and b are irrational, $a - b$ is not always irrational. Example:
$$(2 + \sqrt{2}) - \sqrt{2} = 2,$$
a rational number.

iv. $4(y - 7) = 0 \rightarrow y - 7 = 0 \rightarrow y = 7$

v. The real numbers are not commutative under division.
Example: $4 \div 8 \neq 8 \div 4$

vi. $\dfrac{-2}{3} = -\dfrac{2}{3}$

II

i. $-4 + 3 - 7(2 + 8)$
$= -4 + 3 - 7(10)$
$= -4 + 3 - 70 = -71$

ii. $-4[3 - 6(2 - 4)] = -4[3 - 6(-2)]$
$= -4[3 + 12]$
$= -4[15] = -60$

iii. $|-22| + |-2|(-|-14|)$
$= 22 + 2(-14)$
$= 22 - 28 = -6$

iv. $4^4 \times 4^{-5} \times 4 = 4^0 = 1$

v. $(-2)^3 \div (-2)^6 = (-2)^{-3} = \dfrac{1}{(-2)^3}$
$= \dfrac{1}{-8} = -\dfrac{1}{8}$

vi. $\sqrt[4]{16^2} - \sqrt[3]{27} = 4 - 3 = 1$

vii. $(-4)\left(-\dfrac{1}{2}\right)(8) - (7)(0)(-3) = 16$

viii. $[(4 + 2)3]^2 = [6 \cdot 3]^2 = 18^2 = 324$

ix. $\sqrt[3]{(5)^3} = 5$

x. $7(8 - 3)^3 = 7(5)^3 = 7(125) = 875$

III

i. $3 | \underline{819}$
$3 | \underline{273}$
$7 | \underline{91}$
$13 \quad\quad 819 = 3^2 \times 7 \times 13$

ii. $5280 = 2^5 \times 3 \times 5 \times 11$

iii. $2850 = 2 \times 3 \times 5^2 \times 19$

IV

i. $123 = 3 \times 41$
$530 = 2 \times 5 \times 53$, GCD $= 1$

ii. $7469 = 7 \times 1067$
$2464 = 2^5 \times 7 \times 11$, GCD $= 7$

iii. $2057 = 11^2 \times 17$
$11{,}339 = 17 \times 23 \times 29$, GCD $= 17$

V

i. $\dfrac{1}{7} = .\overline{142857}$ iii. $\sqrt{6.25} = 2.5\overline{0}$

ii. $\dfrac{7}{5} = 1.4\overline{0}$ iv. $\dfrac{7}{12} = .58\overline{3}$

v. $43\dfrac{2}{3} = 43.\overline{6}$

VI

i. $x = 3.\overline{6}$ $\quad$ (1)
$10x = 36.\overline{6}$ $\quad$ (2)
$9x = 33$ $\quad\quad$ (2) $-$ (1)
$x = \dfrac{33}{9}$

ii. $7.\overline{36} = \dfrac{729}{99}$ iii. $6.2\overline{45} = \dfrac{6183}{990}$

iv. $1.46\overline{72} = \dfrac{14526}{9900}$

v. $0.0006\overline{742} = \dfrac{6736}{9990000}$

vi. $675.38\overline{574892} = \dfrac{67538507354}{99999900}$

8 THE COMPLEX NUMBERS

SELF-TEST

DIRECTIONS: Write your answers in the numbered regions to the right. To check your answers, turn the page. Study the solutions to the problems you missed, and do the exercises following any problem or group of problems in which you had errors.

In 1–3, find the values of x and y.

1. $3 + 2i - 4 - 6i = x + yi$

2. $2(\cos 30° + i \sin 30°) = x + yi$

3. $3(\cos 18° + i \sin 18°) \times 2(\cos 72° + i \sin 72°) = x + yi$

4. Change $-3 + 3i$ to its trigonometric form.

5. z is the complex number $x + yi$, and $\bar{z}$ is its conjugate. Find the product of z and $\bar{z}$.

In 6 and 7, determine the quadrant of the Argand diagram in which the given complex number is located.

6. $3 - 5i$

7. $-a - bi$ if $a > 0$ and $b < 0$

In 8–10, determine the modulus and the amplitude of the specified complex number.

8. $3 + \sqrt{3}\,i$

9. the complex number z, if $z = \dfrac{6 \text{ cis } 60°}{2 \text{ cis } 30°}$

10. $z = \dfrac{5 + i}{3 - 2i}$

1
$x =$ _____
$y =$ _____

2
$x =$ _____
$y =$ _____

3
$x =$ _____
$y =$ _____

4 _____

5 _____

6 _____

7 _____

8 _____

9 _____

10 _____

43

44 The Complex Numbers

SOLUTIONS

1. Collect real and imaginary parts.
   $3 + 2i - 4 - 6i$
   $= (3 - 4) + (2 - 6)i = -1 - 4i$

2. $2(\cos 30° + i \sin 30°) = 2\left(\frac{\sqrt{3}}{2} + \frac{1}{2}i\right)$
   $= \sqrt{3} + i$

3. $(3)(2)[\cos(18° + 72°) + i \sin(18° + 72°)]$
   $= 6(\cos 90° + i \sin 90°)$
   $= 6(0 + i) = 6i$

EXERCISES I

In i–vii, find the values of x and y.

i. $2 + 3i^3 - i^2 + 7i + 1 = x + yi$

ii. $(3 + 2i)(-2i + 4) = x + yi$

iii. $(2 - 3i)(3 + i) - (2 + i)(i - 3) = x + yi$

iv. $(4 + i) \div (i - 2) = x + yi$

v. $4 + \sqrt{-8} + 3 - \sqrt{-2} = x + yi$

vi. $(2 + i)(3 - 2i) \div (-2i + 2) = x + yi$

vii. $\dfrac{12(\cos 142° + i \sin 142°)}{6(\cos 97° + i \sin 97°)} = x + yi$

---

4. $r = \sqrt{(-3)^2 + 3^2} = 3\sqrt{2}$
   $\tan \theta = \dfrac{3}{-3} = -1$ and $\theta = 135°$
   $-3 + 3i = 3\sqrt{2}(\cos 135° + i \sin 135°)$

EXERCISES II

i. Find the conjugate of the complex number obtained by simplifying $(-1 - i) \div (1 - i)$.

ii. Find the cube roots of 1.

iii. Find $(\sqrt{3} + i)^6$.

iv. Change $8(\cos 45° + i \sin 45°)$ to rectangular form.

---

5. Let $z = x + yi$, then $\overline{z} = x - yi$ and $z\overline{z} = x^2 + y^2$.

EXERCISES III

In i–iv, z is a complex number of the form $x + yi$, and $\overline{z}$ is its conjugate.

i. Under what condition is $\overline{z} = z$?

ii. Find the value of the difference of z and $\overline{z}$.

iii. Find the conjugate of $\overline{z}$.

iv. Find the conjugate of the sum of z and $\overline{z}$.

In 6 and 7, use the fact that the Argand diagram of $x + yi$ is the point $P(x, y)$.

6. Since $x + yi = 3 - 5i$, $P(x, y) = P(3, -5)$ which is in quadrant IV.

7. If $x + yi = -a - bi$ and $a > 0$ and $b < 0$, then $x < 0$ and $y > 0$ and $P(x, y)$ is in quadrant II.

EXERCISES IV

In i–iii, determine the quadrant of the Argand diagram in which the given complex number is located.

i. $-5 - 2i$

ii. $7 + 3i$

iii. $2 - 3i$

iv. $-3 + 2i$

v. the conjugate of $5 - 2i$

vi. the number $x + yi$ if $x + yi = (2 - 3i)(4 + i)$

In 8–10, use the formulas
$r = \sqrt{x^2 + y^2}$ and $\theta = \arctan \dfrac{y}{x}$
and the location of $P(x, y)$.

8. $x + yi = 3 + \sqrt{3}\,i$
$r = \sqrt{9 + 3} = \sqrt{12} = 2\sqrt{3}$
$\theta = \arctan \dfrac{\sqrt{3}}{3} = 30°$

9. $z = \dfrac{6 \text{ cis } 60°}{2 \text{ cis } 30°} = 3 \text{ cis } 30°,\ r = 3,\ \theta = 30°$

10. $z = \dfrac{5 + i}{3 - 2i} = 1 + i,\ r = \sqrt{2},\ \theta = 45°$

EXERCISES V

In i–vii, determine the modulus and the amplitude of the specified complex number.

i. $5 + 5i$

ii. $x + yi$ if $x = -1$ and $y = -1$

iii. the complex number 3

iv. $z = (4 \text{ cis } 37°)(8 \text{ cis } 12°)$

v. $z = (2 + 3i)(2 - 3i)$

vi. $z = (-2 - 3i)(-3 - 2i)$

vii. $z = \sqrt{4} \text{ cis } 30°$

Solutions to the exercises appear on page 48.

ANSWERS

1. $x = -1$
   $y = -4$

2. $x = \sqrt{3}$
   $y = 1$

3. $x = 0$
   $y = 6$

4. $3\sqrt{2} \text{ cis } 135°$

5. $x^2 + y^2$

6. IV

7. II

8. $2\sqrt{3},\ 30°$

9. $3,\ 30°$

10. $\sqrt{2},\ 45°$

---

1 x = _____
   y = _____

2 x = _____
   y = _____

3 x = _____
   y = _____

4 _____

5 _____

6 _____

7 _____

8 _____

9 _____

10 _____

46 The Complex Numbers

BASIC FACTS

The **imaginary number**, i, has the interesting property that any power of it can be reduced to a number of first degree which is either itself, its negative, 1, or -1.

$i = i$ $\qquad i^3 = -i$ $\qquad i^5 = i$
$i^2 = -1$ $\qquad i^4 = 1$ $\qquad$ etc.

A number of the form $a + bi$ is called a **complex number**. The number a is the real part, and the number bi is the imaginary part. If $a = 0$, then $a + bi = bi$ is a **pure imaginary number**. If $b = 0$, then $a + bi = a$ is a real number.

The complex numbers are sometimes written as **ordered pairs** of real numbers, and graphed on a rectangular coordinate system. Several complex numbers are graphed in Figure 8-1.

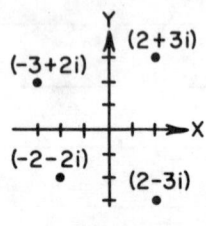

Fig. 8-1

Complex numbers which differ only in the signs of their imaginary parts are called **conjugate complex numbers**.

Two complex numbers are said to be **equal** if and only if their real parts are equal and their imaginary parts are equal. That is, $a + bi = x + yi$ if and only if $a = x$ and $b = y$.

Operations on Complex Numbers

To **add** or **subtract** two complex numbers, add or subtract the real and imaginary parts separately.

$(a + bi) + (c + di)$
$\qquad = (a + c) + (b + d)i$
$(a + bi) - (c + di)$
$\qquad = (a - c) + (b - d)i$

ADDITIONAL INFORMATION

Addition and multiplication of complex numbers is **closed**. That is, the sum and product of two complex numbers is always a complex number.

addition: $\qquad (a + bi) + (c + di) = (a + c) + (b + d)i$
multiplication: $(a + bi)(c + di) = (ac - bd) + (bc + ad)i$

Addition and multiplication of complex numbers is **commutative**. That is, the sum or product of two complex numbers, z_1 and z_2, is the same regardless of the order in which the terms are added or multiplied.

addition: $\qquad z_1 + z_2 = z_2 + z_1$
multiplication: $z_1 z_2 = z_2 z_1$

Addition and multiplication of complex numbers is **associative**. That is, the sum or product of three complex numbers is the same regardless of the manner in which they are grouped.

addition: $\qquad z_1 + (z_2 + z_3) = (z_1 + z_2) + z_3$
multiplication: $z_1 (z_2 z_3) = (z_1 z_2) z_3$

Multiplication is **distributive** over addition.

$z_1 (z_2 + z_3) = z_1 z_2 + z_1 z_3$
$(z_2 + z_3) z_1 = z_2 z_1 + z_3 z_1$

The complex numbers have been defined as **ordered pairs** of real numbers (a, b). This is simply another notation for a complex number, and it is understood that $(a, b) = a + bi$. This notation leads naturally to the representation of a complex number as a point with coordinates (a, b) in a rectangular coordinate system. The plane on which they are graphed is called the **complex plane**, and the figure is called an **Argand diagram**.

The sum or difference of two complex numbers can be obtained geometrically. Consider the operation $(a + bi) + (c + di)$. Locate the two points $P_1(a, b)$ and $P_2(c, d)$ in the complex plane and complete the parallelogram with sides OP_1 and OP_2. The endpoint of the diagonal of this parallelogram is the sum of the two complex numbers, $(a + c)$ and $(b + d)i$. To find the difference, $(a + bi) - (c + di)$, locate the points $P_1(a, b)$ and $P_2(-c, -d)$ and complete the parallelogram with the sides OP_1 and OP_2. The endpoint of the diagonal is the point representing the difference.

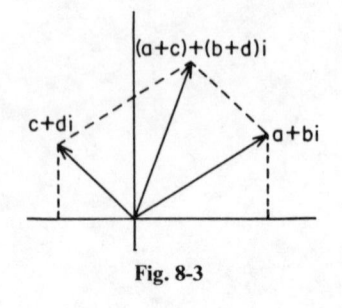

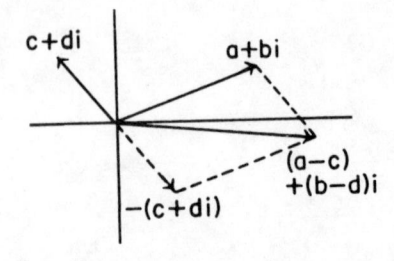

Fig. 8-3 $\qquad\qquad$ **Fig. 8-4**

To find the product of two complex numbers, multiply them together as two binomials and substitute -1 for i^2.

$$(a + bi)(c + di)$$
$$= ac + adi + bci + bdi^2$$
$$= (ac - bd) + (ad + bc)i$$

To express the quotient of two complex numbers as a single complex number, multiply both numerator and denominator by the conjugate of the denominator.

$$\frac{a + bi}{c + di} = \frac{(a + bi)(c - di)}{(c + di)(c - di)}$$
$$= \frac{ac + bd}{c^2 + d^2} + \frac{bc - ad}{c^2 + d^2} i$$

The Trigonometric Form of a Complex Number

From Figure 8–2, it can be seen that $a + bi = r(\cos \theta + i \sin \theta)$ where

$$a = r \cos \theta \qquad b = r \sin \theta$$
$$r = \sqrt{a^2 + b^2} \qquad \tan \theta = \frac{b}{a}.$$

The quantity, r, is called the **modulus** and is always positive; the angle, θ, is called the **argument** or **amplitude**.

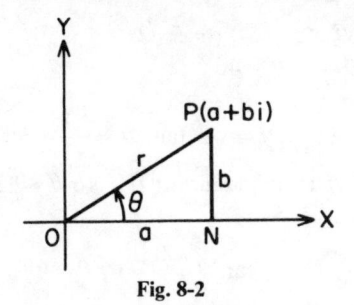

Fig. 8-2

De Moivre's Theorem states that if n is any positive integer, then

$$[r(\cos \theta + i \sin \theta)]^n$$
$$= r^n (\cos n\theta + i \sin n\theta).$$

The nth root of a complex number, $z = \sqrt[n]{a + bi}$, is

$$\sqrt[n]{r} \left[\cos \left(\frac{\theta}{n} + \frac{2k\pi}{n} \right) + i \sin \left(\frac{\theta}{n} + \frac{2k\pi}{n} \right) \right]$$

where $k = 0, 1, 2, 3, \ldots, n - 1$.

If the symbols r and θ are defined as shown in Figure 8-2, the trigonometric form of a complex number follows from the definitions of the trigonometric functions and the theorem of Pythagoras.* From the theorem of Phythagoras, it can be seen that r, the hypotenuse of triangle OPN, is equal to the sum of the squares of a and b, the legs of that triangle. From the definitions of the trigonometric functions, we have

$$\cos \theta = \frac{a}{r} \quad \text{and} \quad a = r \cos \theta \qquad \sin \theta = \frac{b}{r} \quad \text{and} \quad b = r \sin \theta.$$

Substituting for a and b in $a+bi$, $a+bi = r \cos \theta + ir \sin \theta = r(\cos \theta + i \sin \theta)$.

The trigonometric form can be abbreviated r cis θ, which arises from c (the cosine), i (the imaginary number), and s (the sine). Since the trigonometric functions are periodic, the complete statement of a complex number in trigonometric form is

$$x + yi = r[\cos (\theta + 2n\pi) + i \sin (\theta + 2n\pi)] = r \text{ cis } (\theta + 2n\pi)$$

where n is an integer. This is also called the **polar form**.

If two complex numbers are given in polar form, their **product** can be obtained by simply multiplying their moduli and adding their amplitudes. Thus if $z_1 = r_1$ cis θ_1 and $z_2 = r_2$ cis θ_2, then the product $z_1 z_2 = r_1 r_2$ cis $(\theta_1 + \theta_2)$. This fact can be demonstrated by considering the trigonometric addition formulas.

The following formula is for the **division** of two complex numbers in trigonometric form.

$$\frac{r_1 (\cos \theta_1 + i \sin \theta_1)}{r_2 (\cos \theta_2 + i \sin \theta_2)} = \frac{r_1}{r_2} [\cos (\theta_1 - \theta_2) + i \sin (\theta_1 - \theta_2)]$$

This formula can be derived in a manner similar to that used in deriving the product formula. First multiply the numerator and denominator by $\cos \theta_2 - i \sin \theta_2$, and note that the denominator becomes $r_2 (\cos^2 \theta_2 + \sin^2 \theta_2)$. Apply the addition formulas to the resulting numerator.

The polar form is used to find the roots of a complex number. For example, to find the cube roots of $4 \sqrt{3} + 4i$ write the complex number in its polar form, 8 cis $\left(\frac{\pi}{6} + 2k\pi \right)$. Then use the formula $(a + bi)^{\frac{1}{n}} = \sqrt[n]{r}$ cis $\left(\frac{\theta}{n} + \frac{2k}{n} \right)$, $k = 0, 1, \ldots, n - 1$ to obtain the cube roots.

$$(4 \sqrt{3} + 4i) = 2 \text{ cis } \left(\frac{\pi}{18} + \frac{2k\pi}{3} \right), k = 0, 1, 2$$

In terms of degrees we have

$$z_1 = 2 \text{ cis } 10°$$
$$z_2 = 2 \text{ cis } 130°$$
$$z_3 = 2 \text{ cis } 250°$$

If we locate these complex numbers on an Argand diagram, we note that the three points fall on a circle of radius 2, the modulus of each root, and are separated by 120°, which is 360° divided by 3.

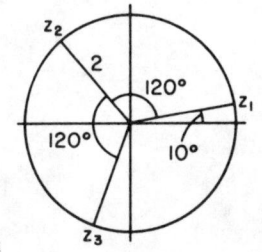

Fig. 8-5

*See Appendix A, page 159, for a brief review of trigonometric functions.

SOLUTIONS TO EXERCISES

I

i. $2 + 3i^3 - i^2 + 7i + 1$
$= 2 - 3i + 1 + 7i + 1 = 4 + 4i$

ii. $(3 + 2i)(-2i + 4)$
$= -6i - 4i^2 + 12 + 8i = 16 + 2i$

iii. $(2 - 3i)(3 + i) - (2 + i)(i - 3)$
$= 6 - 9i + 2i - 3i^2 - (2i + i^2 - 6 - 3i)$
$= 16 - 6i$

iv. $\dfrac{4 + i}{-2 + i} \cdot \dfrac{-2 - i}{-2 - i}$
$= \dfrac{-8 - 2i - 4i - i^2}{4 + i} = \dfrac{7}{5} - \dfrac{6}{5}i$

v. $4 + \sqrt{-8} + 3 - \sqrt{-2} = 4 + 2\sqrt{2}i + 3 - \sqrt{2}i$
$= 7 + \sqrt{2}i$

vi. $(2 + i)(3 - 2i) \div (-2i + 2)$
$= \dfrac{8 - i}{2 - 2i} \cdot \dfrac{2 + 2i}{2 + 2i}$
$= \dfrac{16 - 2i + 16i + 2}{4 + 4} = \dfrac{9}{4} + \dfrac{7}{4}i$

vii. $\dfrac{12(\cos 142° + i \sin 142°)}{6(\cos 97° + i \sin 97°)}$
$= \dfrac{12}{6}[\cos(142° - 97°) + i \sin(142° - 97°)]$
$= 2(\cos 45° + i \sin 45°)$
$= 2\left(\dfrac{\sqrt{2}}{2} + i\dfrac{\sqrt{2}}{2}\right) = \sqrt{2} + \sqrt{2}i$

II

i. $\dfrac{-1 - i}{1 - i} \cdot \dfrac{1 + i}{1 + i} = \dfrac{-1 - 2i + 1}{1 + 1} = -i$
The conjugate is $+i$.

ii. $1 = 1 + 0i = 1(\cos 2k\pi + i \sin 2k\pi)$
The cube roots are
$1(\cos 120°k + i \sin 120°k)$,
where $k = 0, 1, 2$.
In rectangular form,
$z_1 = 1$, $z_2 = -\dfrac{1}{2} + \dfrac{\sqrt{3}}{2}i$, $z_3 = -\dfrac{1}{2} - \dfrac{\sqrt{3}}{2}i$.

iii. $\sqrt{3} + i = 2(\cos 30° + i \sin 30°)$
Now apply De Moivre's theorem.
$(\sqrt{3} + i)^6 = 2^6[\cos 6(30°) + i \sin 6(30°)]$
$= 2^6(\cos 180° + i \sin 180°)$
$= -64$

iv. $8(\cos 45° + i \sin 45°) = 8\left(\dfrac{\sqrt{2}}{2} + \dfrac{\sqrt{2}}{2}i\right)$
$= 4\sqrt{2} + 4\sqrt{2}i$

III

i. $x + yi = x - yi$ if $y = -y$ or $y = 0$
$\therefore z = \overline{z}$ only if both z and $\overline{z}$ are real.

ii. $x + yi - (x - yi) = 2yi$

iii. $\overline{z} = x - yi$ and $\overline{\overline{z}} = x + yi = z$

iv. $z + \overline{z} = 2x$. The conjugate of $2x$ is $2x$.

IV

	POINT	QUAD-RANT		POINT	QUAD-RANT
i.	(−5, −2)	III	iv.	(−3, 2)	II
ii.	(7, 3)	I	v.	(5, 2)	I
iii.	(2, −3)	IV	vi.	(11, −10)	IV

V

i. $x + yi = 5 + 5i$, $r = 5\sqrt{2}$
$\tan \theta = 1$, $\theta = 45°$

ii. $z = -1 - i$, $r = \sqrt{2}$, $\tan \theta = \dfrac{-1}{-1} = 1$
$P(x, y)$ is in quadrant III, so $\theta = 225°$.

iii. $z = 3$, $r = 3$, $\tan \theta = \dfrac{0}{3} = 0$, $\theta = 0°$

iv. $z = (4 \text{ cis } 37°)(8 \text{ cis } 12°) = 32 \text{ cis } 49°$
$r = 32$, $\theta = 49°$

v. $z = (2 + 3i)(2 - 3i) = 4 + 9 = 13$
$r = 13$, $\theta = 0°$

vi. $z = (-2 - 3i)(-3 - 2i) = 13i$, $r = 13$
The point $P(x, y)$ is on the positive Y-axis, so $\theta = 90°$.

vii. $z = \sqrt{4} \text{ cis } 30° = 2 \text{ cis } 15°$, $r = 2$, $\theta = 15°$

9 ALGEBRAIC EXPRESSIONS: POLYNOMIALS

SELF-TEST

DIRECTIONS: Write your answers in the numbered regions to the right. To check your answers, turn the page. Study the solutions to the problems you missed, and do the exercises following any problem or group of problems in which you had errors.

1. Is the expression $\frac{1}{2}x^3 + 8x^2y - \sqrt{7}x$ a polynomial?

2. Evaluate $5x^3 + 2xy^2 - 7y^4$ for $x = 2$, $y = 3$.

3. Evaluate $(2 + 3i)x^2 + 6iy$ for $x = i$, $y = 3 - 4i$.

4. Is the expression $3x - x(7x - 4) - 3 = -7x^2 + 7x - 3$ a conditional equality or an identity?

5. Remove all grouping symbols and express as a sum of unlike terms.
$$4 - \{3x^2 + 7[x^2 + 12 + (3x + 2)] - 10x\}$$

6. Perform the indicated operations.
   a. $(4 - 3x^3 + 2y^2 - 7x^2y) + (6x^3 - 2x^2y + 7x^2 - 10)$
   b. $(2a^2 - 7ab + 5b^2 + 8) - (6a^2 - b^2 + 7)$
   c. $(5x^3y)^4$
   d. $(3x^2y^3)(-7xy^2)$
   e. $(5x + 7)(3x^2 - 2x)$
   f. $(3a - 2)(6a^2 + 4a - 7)$
   g. $(8x^2y - 4xy^2 + 12y) \div (4y)$
   h. $(10x^2 - 9x - 5) \div (5x + 3)$

7. Factor each polynomial.
   a. $3x^2 - 27y^2$
   b. $x^3 + xy^2 + xz^2 + 2x^2y + 2x^2z + 2xyz$
   c. $y^2 - 8y + 16$

SOLUTIONS

1. $\frac{1}{2}x^3 + 8x^2y - \sqrt{7}x$ is a polynomial since it contains only non-negative, integral powers of x and y and no variables in denominators.

EXERCISES I

Identify as polynomial, fractional, or irrational expressions.

i. $\frac{x^3}{2} - \sqrt{3x^2}$ ii. $\frac{1}{x} + \frac{1}{y}$

---

2. $x = 2, y = 3 \rightarrow 5x^3 + 2xy^2 - 7y^4$
  $= 5(2)^3 + 2(2)(3)^2 - 7(3)^4$
  $= 5(8) + 2(2)(9) - 7(81)$
  $= 40 + 36 - 567 = -491$

3. $x = i, y = 3 - 4i \rightarrow (2 + 3i)x^2 + 6iy$
  $= (2 + 3i)(i)^2 + 6i(3 - 4i)$
  $= (2 + 3i)(-1) + 18i - 24i^2$
  $= -2 - 3i + 18i + 24 = 22 + 15i$

4. $3x - x(7x - 4) - 3 = -7x^2 + 7x - 3$ is true for all values of x and is therefore an identity. To prove this, simplify the left member. It will be identical to the right member.

EXERCISES II

Evaluate each polynomial.

i. $-x^2y - 3xy^2 - 8$ at $x = -1, y = 3$
ii. $-x^3y - 3x - 5$ at $x = i, y = 2 + i$

Identify as a conditional equality or an identity.

iii. $5x - (-2x - 7) - 3 = 3x + 4$
iv. $x^2 - x - 6 = 0$

---

5. Remove the innermost grouping symbols first.

  $4 - \{3x^2 + 7[x^2 + 12 + (3x + 2)] - 10x\}$
  $= 4 - \{3x^2 + 7[x^2 + 12 + 3x + 2] - 10x\}$
  $= 4 - \{3x^2 + 7x^2 + 84 + 21x + 14 - 10x\}$
  $= 4 - 3x^2 - 7x^2 - 84 - 21x - 14 + 10x$
  $= -10x^2 - 11x - 94$

6. a. Combine like terms.
  $4 - 3x^3 + 2y^2 - 7x^2y$
  $\underline{-10 + 6x^3 - 2x^2y + 7x^2}$
  $-6 + 3x^3 + 2y^2 - 9x^2y + 7x^2$

  b. Change the signs of the subtrahend and combine like terms.
  $\ 2a^2 - 7ab + 5b^2 + 8$
  $- (6a^2 - b^2 + 7)$
  $= 2a^2 - 7ab + 5b^2 + 8$
  $- 6a^2 + b^2 - 7$
  $= -4a^2 - 7ab + 6b^2 + 1$

  c. Apply the laws of exponents.
  $(5x^3y)^4 = 5^4x^{12}y^4 = 625x^{12}y^4$

  d. Associate like factors and apply the laws of exponents.
  $(3x^2y^3)(-7xy^2)$
  $\ = 3(-7)(x^2 \cdot x)(y^3 \cdot y^2)$
  $\ = -21x^3y^5$

  e. $(5x + 7)(3x^2 - 2x)$
  $= (5x + 7)(3x^2) + (5x + 7)(-2x)$
  $= 15x^3 + 21x^2 - 10x^2 - 14x$
  $= 15x^3 + 11x^2 - 14x$

  f. $6a^2 + 4a - 7$
  $3a - 2$
  $\overline{18a^3 + 12a^2 - 21a}$
  $- 12a^2 - 8a + 14$
  $\overline{18a^3 - 29a + 14}$

  g. Divide each term of the dividend by the divisor.
  $(8x^2y - 4xy^2 + 12y) \div (4y)$
  $\ = 2x^2 - xy + 3$

  h.
  $2x - 3$
  $5x + 3 \overline{)10x^2 - 9x - 5}$
  $\underline{10x^2 + 6x}$
  $-15x - 5$
  $\underline{-15x - 9}$
  4

EXERCISES III

Perform the indicated operations.

i. $(-8a + 7ab - b) + (8b - ab + a)$
ii. $(3x^2 - 2xy + 11) - (9 - 5xy + x^2)$
iii. $(\frac{2}{3}a^2b^3c)^4$
iv. $(6x^7y^3)(2xyz^2)(4x^3yz^2)$
v. $(-2x)(6x^2 - 7x - 5)$
vi. $(5a - 2b)(4a + 6b)$
vii. $(8x - y)^2$
viii. $(ax^2 + bx + c)(x + a)$
ix. $(3x^2 - 2xy + 15x) \div 3x$
x. $(6x^3 + 3x^2 - 5) \div (3x + 6)$

ANSWERS

1. yes

2. -491

3. $22 + 15i$

4. identity

5. $-10x^2 - 11x - 94$

6. a. $-6 + 3x^3 + 2y^2 - 9x^2y + 7x^2$
   b. $-4a^2 - 7ab + 6b^2 + 1$
   c. $625x^{12}y^4$
   d. $-21x^3y^5$
   e. $15x^3 + 11x^2 - 14x$
   f. $18a^3 - 29a + 14$
   g. $2x^2 - xy + 3$
   h. $2x - 3$, remainder 4

7. a. Factor out a 3 and then use the relation $a^2 - b^2 = (a - b)(a + b)$, where $x^2 = a^2$ and $9y^2 = b^2$.
   $3x^2 - 27y^2 = 3(x^2 - 9y^2)$
   $ = 3(x - 3y)(x + 3y)$

   b. Factor out an x and then use the relation $a^2 + b^2 + c^2 + 2ab + 2ac + 2bc = (a + b + c)^2$, where $x = a$, $y = b$, and $z = c$.
   $x^3 + xy^2 + xz^2 + 2x^2y + 2x^2z + 2xyz$
   $= x(x^2 + y^2 + z^2 + 2xy + 2xz + 2yz)$
   $= x(x + y + z)^2$

   c. Use the relation $a^2 - 2ab + b^2 = (a - b)^2$, where $y = a$ and $4 = b$.
   $y^2 - 8y + 16 = y^2 - 2(4y) + 4^2$
   $ = (y - 4)^2$

7. a. $3(x - 3y)(x + 3y)$
   b. $x(x + y + z)^2$
   c. $(y - 4)^2$

EXERCISES IV

Factor each expression.

i. $x(a - b) + y(a - b)$
ii. $x^2 - \frac{1}{4}$
iii. $12x^2 + 4x - 21$

Solutions to the exercises appear on page 54.

Algebraic Expressions: Polynomials

BASIC FACTS

An **algebraic expression** is any expression formed by a finite number of additions, subtractions, multiplications, divisions, or extractions of roots on a finite set of numbers and a finite set of variables. The numbers and the replacement set for the variables may be real or complex. Expressions such as

$$5x^3 + \tfrac{1}{2}x - 4, \quad 2x^{\tfrac{1}{4}} - \sqrt{5}y,$$

and

$$(2 + i)x^2 - \tfrac{6i}{x}$$

are algebraic.

If the operations on the variables are restricted to addition, subtraction, and multiplication, the algebraic expression is a **polynomial**. It follows from this restriction that a polynomial contains only non-negative, integral powers of variables and no variables in denominators. In the examples above, only the first is a polynomial.

Every polynomial can be expressed as the sum of a finite number of terms, each of which is a product of numbers and variables. The numerical factor of each term is its **coefficient**; the variable factor is its **literal part**. Terms which have the same literal parts are called **similar** or **like terms**. Polynomials consisting of one, two, or three terms are called monomials, binomials, and trinomials, respectively.

The value of a polynomial depends on the values assigned to its variables. If the polynomial has real coefficients and if the values of the variables are restricted to real numbers, the polynomial represents a real number; otherwise it represents an imaginary number. Since every polynomial represents a number, the field properties of the number system hold for polynomials and their terms.

Two polynomials P and Q are equal if the value of P equals the value of Q for all values of their variables. In this case,

ADDITIONAL INFORMATION

Many mathematical expressions containing numbers and variables are nonalgebraic. Expressions such as $3 \cos x$, $\log x$, and 2^x, where x is variable, are examples.

Polynomials are just one type of algebraic expressions. (See Figure 9-1.) The set of algebraic expressions can be subdivided into two disjoint subsets: the set of rational expressions and the set of irrational expressions. **Rational expressions** are those which *can* be written without variables under radical signs. **Irrational expressions*** are those which *cannot* be written without variables under radical signs.

Rational expressions: $5 + \dfrac{1}{x^2}, \; \sqrt{3}x^3, \; \sqrt{x^2}$

(Note: $\sqrt{x^2} = x$)

Irrational expressions: $\sqrt{x}, \; \dfrac{3x + y}{\sqrt{x + y}}, \; (x^3 + 5x - 7)^{\tfrac{1}{2}}$

The set of rational expressions can be subdivided into the set of polynomial expressions and the set of fractional expressions. **Polynomial expressions** are rational expressions with no variables in denominators. **Fractional expressions*** are rational expressions with variables in denominators.

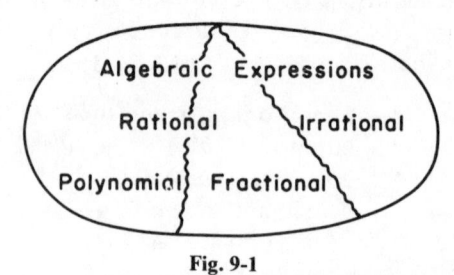

Fig. 9-1

Polynomial expressions: $2x^2 - 3y, \; \dfrac{10x - 5}{7}, \; \tfrac{2}{3}x + \sqrt{3}$

Fractional expressions: $\dfrac{1}{x}, \; \dfrac{3 + 2y}{x - 9y}, \; 4x^2 - 3x + x^{-3}$

Parentheses, ( ), brackets, [ ], and braces, { }, are **grouping symbols** used to associate parts of algebraic expressions. Grouping symbols preceded by a plus sign may be removed without altering the enclosed terms. Grouping symbols preceded by a minus sign may be removed by changing the sign of each enclosed term. Grouping symbols preceded by a factor may be removed by multiplying each enclosed term by that factor. Expressions consisting of grouping symbols within grouping symbols may be simplified by removing the innermost pair of symbols first, the remaining innermost pair next, and so on.

Addition, Subtraction, and Multiplication of Polynomials

The work for addition, subtraction, and multiplication of polynomials may be arranged vertically or horizontally. The following examples illustrate both ways.

*Fractional and irrational expressions are discussed in Chapter 10.

$P = Q$ is called an **identity**. If the value of P equals the value of Q for some, but not all, values of the variables, the polynomials are **conditionally equal** and $P = Q$ is called a **conditional equality**. The equals symbol is used for both relations. However, when there is cause for confusion, the equivalence symbol, $\equiv$, is used for the identity.

The **elementary operations** of addition, subtraction, multiplication, and division can be performed on polynomials by applying the properties of numbers. The polynomials resulting from the operations should be *simplified*, that is, expressed as a sum of unlike terms. Like terms are combined by adding their coefficients. This procedure is justified by the associative, commutative, and distributive properties. For example,

$7x^2 + 2x - 2x^2 \equiv 5x^2 + 2x$ since
$$\begin{aligned} 7x^2 + 2x - 2x^2 &= (2x + 7x^2) - 2x^2 \\ &= 2x + (7x^2 - 2x^2) \\ &= 2x + (7 - 2)x^2 \\ &= 2x + 5x^2 \end{aligned}$$

Addition of polynomials is performed by combining all like terms of the polynomials.

Subtraction of polynomials is performed by changing all signs of the subtrahend and combining like terms. This procedure is a direct consequence of the definition of subtraction of numbers, $a - b = a + (-b)$, and the property $-(c + d) = -c - d$.

Multiplication of polynomials is performed by multiplying each term of one polynomial by each term of the other and combining the like products. This procedure is based on successive applications of the distributive property.

Division of polynomials is performed by an algorithm (method) similar to that used in long division in arithmetic. If the divisor is a factor of the dividend, the quotient may be found by inspection.

Addition
$$\begin{array}{r} 6x^2 + 2x + 2 \\ 2x^2 - 3x \\ \hline 8x^2 - x + 2 \end{array}$$
$$\begin{aligned}(6x^2 + 2x + 2) + (2x^2 - 3x) \\ = 6x^2 + 2x + 2 + 2x^2 - 3x \\ = 8x^2 - x + 2\end{aligned}$$

Subtraction
$$\begin{array}{r} 6x^2 + 2x + 2 \\ \ominus 2x^2 \ominus 3x \\ \hline 4x^2 + 5x + 2 \end{array}$$
$$\begin{aligned}(6x^2 + 2x + 2) - (2x^2 - 3x) \\ = 6x^2 + 2x + 2 - 2x^2 + 3x \\ = 4x^2 + 5x + 2\end{aligned}$$

Multiplication
$$\begin{array}{r} 6x^2 + 2x + 2 \\ 2x^2 - 3x \\ \hline 12x^4 + 4x^3 + 4x^2 \\ -18x^3 - 6x^2 - 6x \\ \hline 12x^4 - 14x^3 - 2x^2 - 6x \end{array}$$
$$\begin{aligned}(6x^2 + 2x + 2)(2x^2 - 3x) \\ = (6x^2 + 2x + 2)(2x^2) \\ \quad + (6x^2 + 2x + 2)(-3x) \\ = 12x^4 + 4x^3 + 4x^2 - 18x^3 \\ \quad - 6x^2 - 6x \\ = 12x^4 - 14x^3 - 2x^2 - 6x\end{aligned}$$

Division of Polynomials is defined by the relation dividend = divisor × quotient + remainder.

To find the quotient of one polynomial divided by another, first arrange each polynomial in descending powers of a common letter. Secondly, divide the first term of the dividend by the first term of the divisor, yielding the first term of the quotient. Thirdly, multiply the whole divisor by the first term of the quotient, and subtract the product from the dividend. The remainder is a new dividend, and the second and third steps must be repeated on this dividend. Continue this process until a remainder is obtained whose first term does not contain the first term of the divisor as a factor.

$$\underbrace{4x^2 + 3x + 2}_{\text{divisor}} \overline{\smash{)}\begin{array}{l} \quad\; 4x - 5 = \text{quotient} \\ 16x^3 - 8x^2 + 5x + 7 = \text{dividend} \\ \underline{16x^3 + 12x^2 + 8x} \\ \quad\; -20x^2 - 3x + 7 \\ \quad\; \underline{-20x^2 - 15x + 10} \\ \qquad\qquad\; 12x + 17 = \text{remainder} \end{array}}$$

If polynomials P, Q, and R are related so that $P \cdot Q = R$, then P and Q are factors of R. The process of **factoring a polynomial** is usually one of trial and error. This process is facilitated by first recognizing as a factor any expression that is common to all terms. It is further facilitated by recognizing typical products whose factors should be known. Ten such products and their factors are listed below.

$ab + ac = a(b + c)$
$a^2 + 2ab + b^2 = (a + b)^2$
$x^2 + (a + b)x + ab = (x + a)(x + b)$
$a^3 + b^3 = (a + b)(a^2 - ab + b^2)$
$b^3 - b^3 = (a - b)(a^2 + ab + b^2)$
$acx^2 + (ad + bc)x + bd = (ax + b)(cx + d)$
$a^2 + b^2 + c^2 + 2ab + 2ac + 2bc = (a + b + c)^2$
$a^3 + 3a^2b + 3ab^2 + b^3 = (a + b)^3$
$a^2 - 2ab + b^2 = (a - b)^2$
$a^2 - b^2 = (a + b)(a - b)$

54 *Algebraic Expressions: Polynomials*

SOLUTIONS TO EXERCISES

I

Polynomial expressions can be written without divisions by variables and without roots of variables. Fractional expressions cannot be written without divisions by variables and cannot contain roots of variables. Irrational expressions cannot be written without roots of variables.

i. $\dfrac{x^3}{2} - \sqrt{3x^2} = \dfrac{x^3}{2} - \sqrt{3}x$ and is a polynomial expression.

ii. $\dfrac{1}{x} + \dfrac{1}{y}$ is a fractional expression.

II

i. $x = -1, y = 3 \rightarrow -x^2y - 3xy^2 - 8$
$= -(-1)^2(3) - 3(-1)(3)^2 - 8$
$= -(1)(3) - 3(-1)(9) - 8$
$= -3 + 27 - 8 = 16$

ii. $x = i, y = 2 + i \rightarrow -x^3y - 3x - 5$
$= -(i)^3(2 + i) - 3(i) - 5$
$= -(-i)(2 + i) - 3i - 5$
$= i(2 + i) - 3i - 5$
$= 2i + i^2 - 3i - 5$
$= 2i - 1 - 3i - 5$
$= -6 - i$

iii. $5x - (-2x - 7) - 3 = 3x + 4$ is a conditional equality since $5x - (-2x - 7) - 3 = 7x + 4$ and $7x + 4 = 3x + 4$ only if $x = 0$.

iv. $x^2 - x - 6 = 0$ is true only if $x = 3$ or $x = -2$. Therefore it is a conditional equality.

III

i. $(-8a + 7ab - b) + (8b - ab + a)$
$= -8a + 7ab - b + 8b - ab + a$
$= -7a + 6ab + 7b$

ii. $\begin{array}{r} 3x^2 - 2xy + 11 \\ \ominus x^2 \overset{+}{\ominus} 5xy \ominus 9 \\ \hline 2x^2 + 3xy + 2 \end{array}$

iii. $\left(\tfrac{2}{3}a^2b^3c\right)^4 = \left(\tfrac{2}{3}\right)^4 a^8 b^{12} c^4$
$\qquad = \tfrac{16}{81} a^8 b^{12} c^4$

iv. $(6x^7y^3)(2xyz^2)(4x^3yz^2)$
$= (6 \cdot 2 \cdot 4)(x^7 \cdot x \cdot x^3)(y^3 \cdot y \cdot y)(z^2 \cdot z^2)$
$= 48x^{11}y^5z^4$

v. $(-2x)(6x^2 - 7x - 5)$
$= -12x^3 + 14x^2 + 10x$

vi. $(5a - 2b)(4a + 6b)$
$= (5a - 2b)4a + (5a - 2b)6b$
$= 20a^2 - 8ab + 30ab - 12b^2$
$= 20a^2 + 22ab - 12b^2$

vii. $(8x - y)^2 = (8x - y)(8x - y)$
$\qquad\quad\; = 64x^2 - 16xy + y^2$

viii. $(ax^2 + bx + c)(x + a)$
$= (ax^2 + bx + c)x$
$\quad + (ax^2 + bx + c)a$
$= ax^3 + bx^2 + cx + a^2x^2$
$\quad + abx + ac$

ix. $(3x^2 - 2xy + 15x) \div 3x$
$= x - \tfrac{2}{3}y + 5$

x.
$$\begin{array}{r} 2x^2 - 3x + 6 \\ 3x + 6 \overline{\smash{\big)}\, 6x^3 + 3x^2 + 0x - 5} \\ \underline{6x^3 + 12x^2 } \\ -9x^2 + 0 \\ \underline{-9x^2 - 18x } \\ 18x - 5 \\ \underline{18x + 36} \\ -41 \end{array}$$

IV

i. Factor out $(a - b)$.
$x(a - b) + y(a - b) = (a - b)(x + y)$

ii. This is the difference of two perfect squares.
$x^2 - \tfrac{1}{4} = \left(x - \tfrac{1}{2}\right)\left(x + \tfrac{1}{2}\right)$

iii. Factor by trial and error.
$12x^2 + 4x - 21 = (2x + 3)(6x - 7)$

10 ALGEBRAIC EXPRESSIONS: FRACTIONAL AND RADICAL

SELF-TEST

DIRECTIONS: Write your answers in the numbered regions to the right. To check your answers, turn the page. Study the solutions to the problems you missed, and do the exercises following any problem or group of problems in which you had errors.

1. State all values of x for which the given expression is not defined.

$$\frac{2x + 7}{2x^2 - 7x + 3}$$

2. Is $\dfrac{x^2 - xy + y^2}{x - y}$ equivalent to $\dfrac{x^3 - y^3}{x^2 - y^2}$?

3. a. Simplify: $\dfrac{x^3 - 9x}{x^3 + 2x^2 - 15x}$
   b. For what values of x is this simplification invalid?

4. Simplify: $\dfrac{\sqrt{y}}{\sqrt{x} - 3}$, x and y are positive reals.

5. Perform the indicated operations. Simplify all results.
   a. $\dfrac{4x + 3}{x} + \dfrac{2x - 7}{x}$
   b. $\dfrac{x^2 - 3x + 8}{x^2} - \dfrac{2x + 5}{x}$
   c. $\dfrac{8}{2x - 3} - \dfrac{12}{3 - 2x}$
   d. $\dfrac{x^2 + 4}{x^2 - 6x - 7} + \dfrac{2x}{x - 7}$
   e. $\dfrac{3x^2 z^3}{4a} \times \dfrac{16a^2}{xz^3}$
   f. $\dfrac{15a^2 x}{-7b} \div \dfrac{10x^2}{7b^3}$
   g. $\dfrac{x^2 - 1}{3x - 3} \times \dfrac{6x + 15}{(x + 1)^2}$
   h. $\dfrac{x^2 + 4}{x^2 + x - 20} \div \dfrac{3}{x - 4}$

6. Perform the indicated operations. Assume that all variables and radicands represent positive real numbers.
   a. $6x\sqrt{4x} + 8\sqrt{x^3}$
   b. $\sqrt[5]{x^4} \cdot \sqrt[5]{x^3}$
   c. $\dfrac{x + 1}{\sqrt{1 - x^2}} \times \dfrac{x}{\sqrt{1 - x^2}}$

55

SOLUTIONS

1. $\dfrac{2x + 7}{2x^2 - 7x + 3}$ is not defined for values of x for which the denominator is zero.
$2x^2 - 7x + 3 = 0$
$(2x - 1)(x - 3) = 0$
$2x - 1 = 0 \rightarrow x = \tfrac{1}{2}$
$x - 3 = 0 \rightarrow x = 3$

2. $\dfrac{x^2 - xy + y^2}{x - y} \stackrel{?}{=} \dfrac{x^3 - y^3}{x^2 - y^2}$

Rule: $\dfrac{P}{Q} = \dfrac{R}{S}$ if $PS = QR$

$(x^2 - xy + y^2)(x^2 - y^2)$
$\stackrel{?}{=} (x - y)(x^3 - y^3)$

Multiply out both members. Since
$x^4 - x^3y + xy^3 - y^4$
$\ne x^4 - x^3y - xy^3 + y^4$,
the expressions are not equivalent.

3. a. Divide out common factors.
$\dfrac{x^3 - 9x}{x^3 + 2x^2 - 15x} = \dfrac{x(x - 3)(x + 3)}{x(x - 3)(x + 5)}$
$= \dfrac{x + 3}{x + 5}$

b. Since the original expression is not defined for $x = 0, 3,$ or -5, the simplified expression is invalid for $x = 0, 3,$ or -5.

EXERCISES I

Given: $\dfrac{2x^3 - x^2 - 28x}{x^3 - 9x^2 + 20x}$

i. For what values of x is the expression not defined?

ii. Simplify the given expression.

iii. For what values of x is answer **ii** not defined?

iv. For what values of x is the simplification invalid?

---

4. Rationalize the denominator by multiplying the numerator and denominator by $\sqrt{x} + 3$.

$\dfrac{\sqrt{y}}{\sqrt{x} - 3} \times \dfrac{\sqrt{x} + 3}{\sqrt{x} + 3} = \dfrac{\sqrt{xy} + 3\sqrt{y}}{x - 9}$

5. a. $\dfrac{4x + 3}{x} + \dfrac{2x - 7}{x}$

$= \dfrac{4x + 3 + 2x - 7}{x} = \dfrac{6x - 4}{x}$

In **b–d**, express as equivalent fractions with common denominators.

b. $\dfrac{x^2 - 3x + 8}{x^2} - \dfrac{2x + 5}{x}$

$= \dfrac{x^2 - 3x + 8}{x^2} - \dfrac{2x + 5}{x} \cdot \dfrac{x}{x}$

$= \dfrac{x^2 - 3x + 8}{x^2} - \dfrac{2x^2 + 5x}{x^2}$

$= \dfrac{x^2 - 3x + 8 - 2x^2 - 5x}{x^2}$

$= \dfrac{-x^2 - 8x + 8}{x^2}$

c. Rule: $-\dfrac{P}{Q} = \dfrac{P}{-Q}$ so $-\dfrac{12}{3 - 2x} = \dfrac{12}{2x - 3}$

$\dfrac{8}{2x - 3} - \dfrac{12}{3 - 2x} = \dfrac{8}{2x - 3} + \dfrac{12}{2x - 3}$

$= \dfrac{20}{2x - 3}$

d. $\dfrac{x^2 + 4}{x^2 - 6x - 7} + \dfrac{2x}{x - 7}$

$= \dfrac{x^2 + 4}{(x - 7)(x + 1)} + \dfrac{2x}{x - 7} \cdot \dfrac{x + 1}{x + 1}$

$= \dfrac{x^2 + 4 + 2x^2 + 2x}{(x - 7)(x + 1)}$

$= \dfrac{3x^2 + 2x + 4}{(x - 7)(x + 1)}$

In **e** and **f**, divide out common factors.

e. $\dfrac{3x^2z^3}{4a} \cdot \dfrac{16a^2}{xz^3}$

$= \dfrac{3 \cdot x \cdot x \cdot z^3 \cdot 4 \cdot 4 \cdot a \cdot a}{4 \cdot a \cdot x \cdot z^3} = 12xa$

f. $\dfrac{x^2 - 1}{3x - 3} \times \dfrac{6x + 15}{(x + 1)^2}$

$= \dfrac{(x + 1)(x - 1)}{3(x - 1)} \times \dfrac{3(2x + 5)}{(x + 1)(x + 1)}$

Algebraic Expressions: Fractional and Radical   57

$$= \frac{2x + 5}{x + 1}$$

In **g** and **h**, use the rule:

$$\frac{P}{Q} \div \frac{R}{S} = \frac{P}{Q} \times \frac{S}{R}$$

g. $\dfrac{15a^2x}{-7b} \div \dfrac{10x^2}{7b^3} = \dfrac{15a^2x}{-7b} \times \dfrac{7b^3}{10x^2}$

$= \dfrac{\cancel{5} \cdot 3 \cdot a^2 \cdot \cancel{x} \cdot \cancel{7} \cdot \cancel{b} \cdot b^2}{-1 \cdot \cancel{7} \cdot \cancel{b} \cdot 2 \cdot \cancel{5} \cdot \cancel{x} \cdot x}$

$= \dfrac{3a^2b^2}{-2x} = -\dfrac{3a^2b^2}{2x}$

h. $\dfrac{x^2 + 4}{x^2 + x - 20} \div \dfrac{3}{x - 4}$

$= \dfrac{x^2 + 4}{(x + 5)\cancel{(x - 4)}} \times \dfrac{\cancel{x - 4}}{3}$

$= \dfrac{x^2 + 4}{3(x + 5)}$

EXERCISES II

Perform the indicated operations. Simplify the results.

i. $\dfrac{3a}{x} - \dfrac{5b}{y} + \dfrac{7c}{z}$

ii. $\dfrac{a - b}{a + b} + \dfrac{7b}{a - b}$

iii. $\dfrac{y^3 + 9y}{y^2 - 9} \cdot \dfrac{y + 3}{y}$

iv. $\dfrac{8x^6y}{3z} \div \dfrac{4xy^2}{6z}$

v. $\dfrac{\dfrac{x^2 - 7x - 18}{x - 3}}{\dfrac{x + 2}{x^2 - 8x + 15}}$

---

6. a. $6x\sqrt{4x} + 8\sqrt{x^3}$
   $= 12x\sqrt{x} + 8x\sqrt{x} = 20x\sqrt{x}$

b. $\sqrt[5]{x^4} \cdot \sqrt[5]{x^3} = \sqrt[5]{x^4 \cdot x^3} = \sqrt[5]{x^7}$
   $= \sqrt[5]{x^5 \cdot x^2} = x\sqrt[5]{x^2}$

c. $\dfrac{x + 1}{\sqrt{1 - x^2}} \times \dfrac{x}{\sqrt{1 - x^2}} = \dfrac{(x + 1)x}{1 - x^2}$

   $= \dfrac{\cancel{(x + 1)}x}{(1 - x)\cancel{(1 + x)}} = \dfrac{x}{1 - x}$

Solutions to the exercises are on page 60.

ANSWERS

1. $x = \frac{1}{2}, x = 3$ 1

2. no   2

3. a. $\dfrac{x + 3}{x + 5}$ 3a

   b. $x = 0, x = 3, x = -5$ b

4. $\dfrac{\sqrt{xy} + 3\sqrt{y}}{x - 9}$ 4

5. a. $\dfrac{6x - 4}{x}$ 5a

   b. $\dfrac{-x^2 - 8x + 8}{x^2}$ b

   c. $\dfrac{3x^2 + 2x + 4}{(x - 7)(x + 1)}$ c

   d. $\dfrac{20}{2x - 3}$ d

   e. $12ax$ e

   f. $\dfrac{2x + 5}{x + 1}$ f

   g. $-\dfrac{3a^2b^2}{2x}$ g

   h. $\dfrac{x^2 + 4}{3(x + 5)}$ h

6. a. $20x\sqrt{x}$ 6a

   b. $x\sqrt[5]{x^2}$ b

   c. $\dfrac{x}{1 - x}$ c

EXERCISES III

All radicands are positive reals.

i. Rationalize the denominator of
$$\dfrac{2\sqrt{x}}{\sqrt{x - 7}}.$$

ii. Combine like terms:
$$3a\sqrt{a^3} + 2a^2\sqrt{a} - \sqrt{a^7}.$$

Algebraic Expressions: Fractional and Radical

BASIC FACTS

An algebraic expression denoting the quotient of two polynomials is called a fractional or rational expression.

A **fractional expression** represents a number for each replacement of the variable(s), excluding those replacements which make the denominator zero. Denominators with value zero are not allowed since division by zero is not defined.

Because fractional expressions represent numbers, the properties of the number system can be applied in working with these expressions.

Two fractional expressions, $\frac{P}{Q}$ and $\frac{R}{S}$, are equal or **equivalent** if the value of $\frac{P}{Q}$ equals the value of $\frac{R}{S}$ for all values of the variable(s) for which both expressions are defined. When $\frac{P}{Q}$ and $\frac{R}{S}$ are equivalent, $\frac{P}{Q} = \frac{R}{S}$ is called an **identity**.

If the value of $\frac{P}{Q}$ equals the value of $\frac{R}{S}$ for some, but not all, values of the variable(s) for which both expressions are defined, $\frac{P}{Q}$ is **conditionally equal** to $\frac{R}{S}$.

A fractional expression can be **simplified**—expressed as an equivalent fraction in lower terms—by factoring the numerator and denominator and applying the fundamental principle of fractions, namely:

$$\frac{PR}{QR} = \frac{P}{Q} \quad (Q \neq 0, R \neq 0)$$

Addition, subtraction, multiplication, and division of fractional expressions are similar to the same operations on fractional numbers. The results of these operations should always be simplified. In the following rules, P, Q, R, and S represent polynomials.

Addition and subtraction:

$$\frac{P}{Q} \pm \frac{R}{Q} = \frac{P \pm R}{Q} \quad (Q \neq 0)$$

ADDITIONAL INFORMATION

As is illustrated by the following list of examples, operations on fractional (and radical) expressions are performed in the same way as operations on fractional (and radical) numbers.

Fractional Numbers *Fractional Expressions*

Simplification

$$\frac{510}{525} = \frac{2 \cdot \cancel{3} \cdot \cancel{5} \cdot 17}{\cancel{3} \cdot \cancel{5} \cdot 5 \cdot 7} = \frac{34}{35} \qquad \frac{x^3 - x}{x^3 - 2x^2 - 3x} = \frac{\cancel{x}\cancel{(x+1)}(x-1)}{\cancel{x}\cancel{(x+1)}(x-3)}$$

Addition and Subtraction

$$\frac{5}{12} + \frac{1}{2} = \frac{5}{2 \cdot 6} + \frac{1}{2} \qquad \frac{3x}{x^2 - x - 6} + \frac{7}{x+2}$$

$$= \frac{5}{2 \cdot 6} + \frac{1}{2} \cdot \frac{6}{6} \qquad = \frac{3x}{(x-3)(x+2)} + \frac{7}{x+2}$$

$$= \frac{5}{2 \cdot 6} + \frac{6}{2 \cdot 6} \qquad = \frac{3x}{(x-3)(x+2)} + \frac{7}{x+2} \cdot \frac{x-3}{x-3}$$

$$= \frac{5+6}{2 \cdot 6} = \frac{11}{12} \qquad = \frac{3x}{(x-3)(x+2)} + \frac{7x - 21}{(x-3)(x+2)}$$

$$\qquad \qquad = \frac{3x + 7x - 21}{(x-3)(x+2)} = \frac{10x - 21}{(x-3)(x+2)}$$

The LCD is 12, but any common denominator could be used. The product of the given denominators is always a common denominator.

The common denominator that is easiest to work with is found by factoring all the denominators and using each factor the greatest number of times it occurs in any one denominator.

Multiplication

$$\frac{2}{7} \times \frac{21}{40} = \frac{\cancel{2}}{\cancel{7}} \times \frac{3 \cdot \cancel{7}}{\cancel{2} \cdot 20} \qquad \frac{x^2 - 1}{x - 5} \times \frac{x^2 - x - 20}{x^3 + x}$$

$$= \frac{3}{20} \qquad \qquad = \frac{\cancel{x^2 + 1}}{\cancel{x - 5}} \times \frac{\cancel{(x-5)}(x+4)}{x\cancel{(x^2+1)}}$$

All common factors are divided out before multiplying.

x cannot be divided out of this expression since x is a term, not a factor, of the numerator.

Division

$$\frac{3}{20} \div \frac{9}{10} = \frac{3}{20} \times \frac{10}{9} \qquad \frac{x^2 - 16}{x^2 + 7x} \div \frac{x + 4}{x^2 - x}$$

$$\qquad \qquad = \frac{x^2 - 16}{x^2 + 7x} \times \frac{x^2 - x}{x + 4}$$

From this point on the operations are the same as those for multiplication (see above).

If the expressions have unlike denominators, they can be written as equivalent expressions having common denominators as follows:

$$\frac{P}{Q} \pm \frac{R}{S} = \frac{PS}{QS} \pm \frac{RQ}{SQ} = \frac{PS \pm RQ}{QS} \quad (Q, S \neq 0)$$

Multiplication:

$$\frac{P}{Q} \times \frac{R}{S} = \frac{PR}{QS} \quad (Q, S \neq 0)$$

To insure that the product is simplified, factor the numerators and denominators and divide out common factors before finding the product.

Division:

$$\frac{P}{Q} \div \frac{R}{S} = \frac{P}{Q} \times \frac{S}{R} \quad (Q, S, R \neq 0)$$

An algebraic expression containing the root of a variable is called a **radical expression**. A root is denoted by the radical sign, $\sqrt{}$, or by a fractional exponent. In the radical expression $\sqrt[3]{x}$, 3 is the **index** and x is the **radicand**. Using a fractional exponent, this expression can be written $x^{\frac{1}{3}}$.

For each replacement of its variable(s), every radical expression represents a number. If the replacements are real numbers and if all radicands are positive, the radical expression represents a real number. In such cases, the rules on page 40 may be applied.

A radical expression may be **simplified** by expressing all radicands without fractions and without powers whose exponents are greater than or equal to the index and by rationalizing all denominators. A denominator is rationalized when it is expressed without radicals.

Simplification of Compound Fractions

$$\frac{\frac{2}{3} + \frac{4}{7}}{\frac{8}{5} - \frac{2}{7}} = \frac{\frac{26}{21}}{\frac{46}{35}}$$

$$= \frac{26}{21} \div \frac{46}{35} = \frac{65}{69}$$

$$\frac{3x + \frac{3x}{2}}{\frac{3}{xy} + \frac{6}{x}} = \frac{\frac{6x}{2} + \frac{3x}{2}}{\frac{3 + 6y}{xy}} = \frac{\frac{9x}{2}}{\frac{3 + 6y}{xy}}$$

$$= \frac{9x}{2} \div \frac{3 + 6y}{xy} = \frac{3x^2 y}{2(1 + 2y)}$$

Compound (or *complex*) fractions have fractions in their numerators and/or denominators.

Numerators and denominators of compound fractions must always be expressed as single fractions before the division is performed.

Radical Numbers   *Radical Expressions*

Simplification
(powers of the index)

$$\sqrt[3]{1296} = \sqrt[3]{3^3 \cdot 3 \cdot 2^3 \cdot 2}$$
$$= 3 \cdot 2 \sqrt[3]{3 \cdot 2} = 6\sqrt[3]{6}$$

$$\sqrt{2x^4 - 12x^3 + 18x^2}$$
$$= \sqrt{2x^2(x^2 - 6x + 9)}$$
$$= \sqrt{2x^2(x - 3)^2} = x(x - 3)\sqrt{2}$$

(rationalizing the denominator)

$$\frac{\sqrt{5}}{\sqrt{3}} = \frac{\sqrt{5}}{\sqrt{3}} \cdot \frac{\sqrt{3}}{\sqrt{3}}$$
$$= \frac{\sqrt{5 \cdot 3}}{3} = \frac{\sqrt{15}}{3}$$

$$\frac{\sqrt{x}}{\sqrt{x - y}} = \frac{\sqrt{x}}{\sqrt{x - y}} \cdot \frac{\sqrt{x - y}}{\sqrt{x - y}}$$
$$= \frac{\sqrt{x(x - y)}}{x - y} = \frac{\sqrt{x^2 - xy}}{x - y}$$

A monomial denominator is rationalized by multiplying both numerator and denominator by it.

$\sqrt{x - y}$ is a monomial. In contrast, $\sqrt{x} - \sqrt{y}$ is a binomial. Note: $\sqrt{x} - \sqrt{y} \neq \sqrt{x - y}$ just as $\sqrt{4} - \sqrt{9} \neq \sqrt{4 - 9}$.

$$\frac{\sqrt{5}}{\sqrt{7} - 2}$$

$$= \frac{\sqrt{5}}{\sqrt{7} - 2} \cdot \frac{\sqrt{7} + 2}{\sqrt{7} + 2}$$

$$= \frac{\sqrt{5} \cdot \sqrt{7} + \sqrt{5} \cdot 2}{7 - 4}$$

$$= \frac{\sqrt{35} + 2\sqrt{5}}{3}$$

$$\frac{\sqrt{x}}{\sqrt{5y} + \sqrt{6}}$$

$$= \frac{\sqrt{x}}{\sqrt{5y} + \sqrt{6}} \cdot \frac{\sqrt{5y} - \sqrt{6}}{\sqrt{5y} - \sqrt{6}}$$

$$= \frac{\sqrt{x} \cdot \sqrt{5y} - \sqrt{x} \cdot \sqrt{6}}{5y - 6}$$

$$= \frac{\sqrt{5xy} - \sqrt{6x}}{5y - 6}$$

A binomial denominator is rationalized by multiplying numerator and denominator by its conjugate.

To obtain the conjugate of any binomial, change the sign of the second term. Thus, $a + b$ and $a - b$ are conjugates of each other.

Addition and Subtraction

$$\begin{aligned} &3\sqrt{7} + 4\sqrt{3} \\ &- 6\sqrt{7} + 2\sqrt{3} \\ &= -3\sqrt{7} + 6\sqrt{3} \end{aligned}$$

$$3\sqrt{xy} + 8\sqrt{x} + 2\sqrt{x} - \sqrt{xy}$$
$$= 2\sqrt{xy} + 10\sqrt{x}$$

SOLUTIONS TO EXERCISES

I

Given: $\dfrac{2x^3 - x^2 - 28x}{x^3 - 9x^2 + 20x}$

i. The given expression is not defined for $\{x \mid x^3 - 9x^2 + 20x = 0\}$.

$$x^3 - 9x^2 + 20x = 0$$
$$x(x^2 - 9x + 20) = 0$$
$$x(x - 5)(x - 4) = 0$$
$$x = 0$$
$$x - 5 = 0 \to x = 5$$
$$x - 4 = 0 \to x = 4$$

If $x = 0$, 5, or 4, the expression is not defined.

ii. To simplify a fractional expression, factor the numerator and denominator and use the rule:

$$\dfrac{PR}{QR} = \dfrac{P}{Q}$$

$$\dfrac{2x^3 - x^2 - 28x}{x^3 - 9x^2 + 20x}$$

$$= \dfrac{x(2x + 7)(x - 4)}{x(x - 5)(x - 4)} = \dfrac{2x + 7}{x - 5}$$

iii. $\dfrac{2x + 7}{x - 5}$ is not defined for $x = 5$.

iv. $\dfrac{2x^3 - x^2 - 28x}{x^3 - 9x^2 + 20x} = \dfrac{2x + 7}{x - 5}$ is not defined for $x = 0$, 5, or 4.

II

i. $\dfrac{3a}{x} - \dfrac{5b}{y} + \dfrac{7c}{z}$

$$= \dfrac{3a}{x} \cdot \dfrac{yz}{yz} - \dfrac{5b}{y} \cdot \dfrac{xz}{xz} + \dfrac{7c}{z} \cdot \dfrac{xy}{xy}$$

$$= \dfrac{3ayz}{xyz} - \dfrac{5bxz}{xyz} + \dfrac{7cxy}{xyz}$$

$$= \dfrac{3ayz - 5bxz + 7cxy}{xyz}$$

ii. $\dfrac{a - b}{a + b} + \dfrac{7b}{a - b}$

$$= \dfrac{a - b}{a + b} \cdot \dfrac{a - b}{a - b} + \dfrac{7b}{a - b} \cdot \dfrac{a + b}{a + b}$$

$$= \dfrac{a^2 - 2ab + b^2}{(a + b)(a - b)} + \dfrac{7ab + 7b^2}{(a - b)(a + b)}$$

$$= \dfrac{a^2 - 2ab + b^2 + 7ab + 7b^2}{(a + b)(a - b)}$$

$$= \dfrac{a^2 + 5ab + 8b^2}{(a + b)(a - b)}$$

iii. $\dfrac{y^3 + 9y}{y^2 - 9} \cdot \dfrac{y + 3}{y}$

$$= \dfrac{\cancel{y}(y^2 + 9)}{\cancel{(y + 3)}(y - 3)} \cdot \dfrac{\cancel{y + 3}}{\cancel{y}} = \dfrac{y^2 + 9}{y - 3}$$

iv. $\dfrac{8x^6 y}{3z} \div \dfrac{4xy^2}{6z} = \dfrac{\overset{2x^5}{\cancel{8x^6 y}}}{\cancel{3z}} \cdot \dfrac{\overset{2}{\cancel{6z}}}{\underset{y}{\cancel{4xy^2}}} = \dfrac{4x^5}{y}$

v. $\dfrac{\dfrac{x^2 - 7x - 18}{x - 3}}{\dfrac{x + 2}{x^2 - 8x + 15}}$

$$= \dfrac{x^2 - 7x - 18}{x - 3} \div \dfrac{x + 2}{x^2 - 8x + 15}$$

$$= \dfrac{(x - 9)\cancel{(x + 2)}}{\cancel{x - 3}} \cdot \dfrac{\cancel{(x - 3)}(x - 5)}{\cancel{x + 2}}$$

$$= (x - 9)(x - 5) = x^2 - 14x + 45$$

III

i. $\dfrac{2\sqrt{x}}{\sqrt{x - 7}} \cdot \dfrac{\sqrt{x - 7}}{\sqrt{x - 7}} = \dfrac{2\sqrt{x(x - 7)}}{x - 7}$

ii. $3a\sqrt{a^3} + 2a^2\sqrt{a} - \sqrt{a^7}$

$$= 3a\sqrt{a^2 \cdot a} + 2a^2\sqrt{a} - \sqrt{a^2 \cdot a^2 \cdot a}$$

$$= 3a^2\sqrt{a} + 2a^2\sqrt{a} - a^3\sqrt{a}$$

$$= 5a^2\sqrt{a} - a^3\sqrt{a} = (5a^2 - a^3)\sqrt{a}$$

11 RELATIONS AND FUNCTIONS

SELF-TEST

DIRECTIONS: Write your answers in the numbered regions to the right or use the graph paper in the back of the book. To check your answers, turn the page. Study the solutions to the problems you missed, and do the exercises following any problem or group of problems in which you had errors.

1. Does the set $\{(-4, 6), (-1, 2), (4, 0), (5, 2), (8, -1)\}$ define a function?

Let x and y represent any of the real numbers, and let f be a function relating x and y in pairs (x, y). Find the domain and range of the functions given in 2 and 3.

2. $y = |x|$

3. $y = \dfrac{x}{x - 1}$

4. Find the inverse of the function defined by $y = 2x - 3$.

5. Write the following statement in the form of an equation.

    y varies inversely as x and $y = 5$ when $x = 3$.

6. Express the area of a square as a function of its side.

7. Given $f(x) = 2x - 3$, find $\dfrac{f(x) - f(a)}{x - a}$.

8. Is $f(x) = x^4 + 2x^2 + 7$ an odd or even function?

9. Draw the graph of $s = f(t) = -16t^2 + 32t$ for $0 \le t \le 2$.

10. Draw the graph of the function defined by
$$y = \begin{cases} -1 & \text{if } x < 0 \\ \dfrac{1}{2} & \text{if } x = 0 \\ x & \text{if } x > 0. \end{cases}$$

11. Find $f + g$ and fg if $f(x) = x + 2$ and $g(x) = x - 2$.

12. Find the composite function $f[g(x)]$ if $f(x) = 3x^2 + 5x$ and $g(x) = x - 1$.

1

2
$d_f = $ _____
$r_f = $ _____

3
$d_f = $ _____
$r_f = $ _____

4

5

6

7

8

9 Use graph paper, page 181.

10 Use graph paper, page 181.

11

12

SOLUTIONS

1. Since no two of the ordered pairs have the same first member, the set defines a function.

EXERCISES I

In i–ii, a function f is defined by a rule of formation. Find $f(x)$ for $x = 1, 2, -4$.

i. $f: y = |x - 1|$

ii. $f = \{(1, 3), (2, 5), (4, 7), (-4, 1)\}$

iii. Does the set $\{(10, -1), (3, 4), (10, 8), (5, 4)\}$ define a function?

---

2. Since the equation $y = |x|$ is defined for all real numbers (R), the domain is all real numbers. Since $|x|$ is positive or zero for all x, the range includes all non-negative real numbers.

3. The denominator of $\dfrac{x}{x-1}$ is zero when $x = 1$. Consequently, the domain is $\{x|\text{ reals}, x \neq 1\}$. By solving for x, $x = \dfrac{y}{y-1}$, we see that the range is $\{y|\text{ reals}, y \neq 1\}$.

4. $x = 2y - 3$

   To find the inverse of a function, interchange the variables and solve for the new dependent variable. $y = \dfrac{1}{2}(x+3)$ defines the inverse of $x = 2y - 3$.

EXERCISES II

In i and ii, let x and y represent any real number and let f be a function relating x and y. Find the domain and range of each of the functions.

i. $y = \sqrt{4 - x^2}$

ii. $f = \{(1, 2), (\pi, -3)\}$

iii. Find the inverse of $y = \sqrt{4 - x^2}$, $-2 \leq x \leq 0$.

---

5. $y = \dfrac{k}{x}$, $5 = \dfrac{k}{3}$, $15 = k$, $y = \dfrac{15}{x}$.

EXERCISES III

In i–iii, write the statements in the form of equations.

i. y varies directly as x and $y = 3$ when $x = 2$.

ii. z varies jointly as the square of x and the square root of y. When $x = 3$ and $y = 4$, then $z = 6$.

iii. w varies directly as the product of x and y^2 and inversely as z. When $x = 3$, $y = 2$, and $z = 18$ we find that $w = 4$.

---

6. Let x be a side of the square. Let A be the area of the square. The area of a square is the product of its two sides.
$$A = (x)(x) = x^2; \{(x, A) | A = x^2\}$$

7. $f(x) = 2x - 3$; $f(a) = 2a - 3$
$f(x) - f(a) = (2x - 3) - (2a - 3)$
$\qquad\qquad\quad = 2(x - a)$
$$\dfrac{f(x) - f(a)}{x - a} = \dfrac{2(x - a)}{x - a} = 2$$

8. $f(x) = x^4 + 2x^2 + 7$
$f(-x) = (-x)^4 + 2(-x)^2 + 7$
$\qquad\quad = x^4 + 2x^2 + 7$
$f(x)$ is an even function.

9. $s = -16t^2 + 32t$; Fig. 11-1.

t	0	$\frac{1}{2}$	1	$\frac{3}{2}$	2
s	0	12	16	12	0

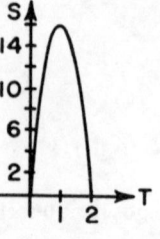

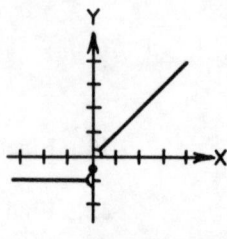

Fig. 11-1 Fig. 11-2

10. $y = -1$ if $x < 0$, $\dfrac{1}{2}$ if $x = 0$, x if $x > 0$

EXERCISES IV

i. Find the inverse of $y = x$.

ii. Show that $y = x^3 - 3x$ is an odd function.

Use the graph paper on page 181.

iii. Join the following points.

x	-1	0	1	2	3
y	2	-1	-2	-1	2

iv. Draw the graph of $f: y = x + 2$ if the domain is the integers.

11. $f(x) = x + 2$, $g(x) = x - 2$
$f + g = (x + 2) + (x - 2) = 2x$
$fg = (x + 2)(x - 2) = x^2 - 4$

12. $f(x) = 3x^2 + 5x$, $g(x) = x - 1$
$f[g(x)] = 3(x - 1)^2 + 5(x - 1)$
$= 3x^2 - 6x + 3 + 5x - 5$
$= 3x^2 - x - 2$

EXERCISES V

i. Is the graph of $x = 4$ parallel to the X- or Y-axis?

Are statements ii and iii true or false?

ii. The function $f(x) = x^2 - x$ is an odd function.

iii. The graphs of all functions are smooth curves.

iv. Find $\dfrac{f}{g}$ if $f(x) = 2x^2 + x - 3$ and $g(x) = 2x^2 + 5x + 3$

Solutions to the exercises appear on page 66.

ANSWERS

1. yes

2. $d_f = R$
   $r_f = y \geq 0$

3. $d_f = \{x | \text{reals}, x \neq 1\}$
   $r_f = \{y | \text{reals}, y \neq 1\}$

4. $y = \dfrac{1}{2}(x + 3)$

5. $y = \dfrac{15}{x}$

6. $\{(x, A) | A = x^2\}$

7. 2

8. Even

9. Fig. 11-1

10. Fig. 11-2

11. $f + g = 2x$, $fg = x^2 - 4$

12. $f[g(x)] = 3x^2 - x - 2$

1

2 $d_f =$ _____
  $r_f =$ _____

3 $d_f =$ _____
  $r_f =$ _____

4

5

6

7

8

9 Use graph paper, page 181.

10 Use graph paper, page 181.

11

12

BASIC FACTS

A symbol which, throughout a discussion, represents a fixed number is called a **constant**.

A symbol which, throughout a discussion, may assume different values is called a **variable**.

If two variables, x and y, are associated in a set of ordered pairs, we have a dyadic relation. If for every value assigned to x a unique value of y is determined, then we have a function, which is a special kind of dyadic relation.

In terms of sets we say that if with each element of a set X there is associated *exactly* one element of another set Y, then this association is a **function** from X to Y, "$f: X \longrightarrow Y$." A **function** is also defined as a set of ordered pairs (x, y) such that to each $x \in X$, there corresponds a *unique* $y \in Y$.

These definitions are all identical and only the terms of reference differ. The set of independent variables $x \in X$ is called the **domain**, and the set of dependent variables $y \in Y$ is called the **range**.

Notations for functions:

$$f: X \longrightarrow Y \text{ or } f: x \longrightarrow y$$

$$\{(x, y) | y = f(x)\}$$

The symbol $f(x)$, read "f of x," is the value of f at x.

If $f(-x) = f(x)$, the function is called an **even function**. If $f(-x) = -f(x)$, the function is called an **odd function**. Note that it is possible for a function to be neither odd or even.

The **inverse** function, f^{-1}, of a function f is the collection of ordered pairs obtained from f by interchanging the first and second elements in each ordered pair.

If two variables, y and x, are so related that the ratio $\frac{y}{x}$ is always constant, then y is said to **vary directly** as x. We write, $y = kx$.

ADDITIONAL INFORMATION

In the definition of a **function**, there must be a rule which determines the value of y for each assigned value of x. This rule may take many forms. It may be a formula, an equation, a graph, a table of data, or a statement of a principle. If the correspondence between the two variables, x and y, is such that for each x there is one or more y, we have a **relation**. A function is a special kind of relation.

The notation $y = f(x)$ is used to indicate the value of the function for a given value of x. In this relation, x is called the **independent variable** and y is called the **dependent variable**. Since for a given value of x there is determined a value of y, we have an **ordered pair** (x, y) and a function is also defined as an ordered pair in which for each x there is a unique y. The **domain** of a function is the collection of all the values of the *independent* variable for which the function is defined. The **range** of a function is the collection of all the values of the *dependent* variable defined by the function.

The **inverse** of a given function is obtained by interchanging the roles of the two variables. If the inverse of a function satisfies the definition of *function*, it is an **inverse function**; if not, it is an **inverse relation**. If a function, f, has an inverse function, f^{-1}, the range of f is the domain of f^{-1} and the domain of f is the range of f^{-1}.

The language of **variation** is a direct application of the functional relationship of two or more quantities. The main problem for the student is to interpret the verbal statement of a variation, and express this statement in terms of functional notation. Once the equation has been written or the formula has been found, given data can be substituted in it. For example, the well-known formula for the area of a circle, $A = \pi r^2$, is a direct variation which can be expressed, "the area of a circle varies directly as the square of its radius." We can use this formula to find the area of a circle if the radius is given, or to find the radius of a circle if the area is given. The accuracy of the answer depends upon the value used for the irrational number π, the constant of variation in $A = \pi r^2$.

$A = f(r) = \pi r^2$ is a function which generates pairs (r, A) for a given value of r. If we solve this equation for r, we have $r = \pm\sqrt{A/\pi}$, which is not a function since for each value of A there are two values of r. However, the formula is related to a circle and a negative value of r would have no physical meaning. Therefore, we limit the domain of the original function to non-negative values of r: $A = f(r)$, $r > 0$. This specifies the range of the inverse and we can write $f^{-1}: r = +\sqrt{A/\pi}$, which is a function.

The **graph** of the equation $y = f(x)$ is the totality of all points whose coordinates (x, y) satisfy the relation. In reference to this definition, it should be very clear that:

1. Every point whose coordinates satisfy the equation lies on the graph.
2. The coordinates of every point on the graph satisfy the equation.

If two variables, x and y, are so related that y varies directly as the reciprocal of x, then y is said to **vary inversely** as x. We write, $y = \frac{k}{x}$.

If $z = kxy$, then z is said to **vary jointly** as x and y.

Variations can be combined in a number of ways; for example, $z = k\frac{x}{y}$ is the formula if z varies directly as x, and inversely as y.

In all types of variation, k is called the **constant of variation**.

The joint variation $z = kxy$ is an example of a function of two variables; that is, x and y are independent variables and for each x and y there is a unique dependent variable z.

Since the set of ordered pairs (x, y) can be plotted on a Cartesian coordinate system, functions may be studied by considering their graphs. The graph is obtained by calculating a set of pairs giving arbitrary values to the independent variable and calculating the corresponding values of the dependent variable from the given definition of the function. These pairs are then plotted. Although it may not be possible to plot all of the points defined by the function $\{(x, y) | y = f(x)\}$, enough can usually be plotted to provide a good picture of the function. Be certain to use only those values of the independent variable for which the function is defined.

Using the two functions given by $\{(x, y) | y = f(x)\}$ and $\{(x, y) | y = g(x)\}$, we then define the operations

$$f \pm g = f(x) \pm g(x)$$

$$fg = [f(x)][g(x)]$$

$$\frac{f}{g} = \frac{f(x)}{g(x)}$$

The **composite function** of f by g, defined by $f[g(x)]$, is obtained by substituting $g(x)$ for the independent variable in f.

The graph of a given function will exist only for those pairs (x, y) whose abscissas are included in the domain of the function. The graph of the function defined by $y = x$, $(x = 2, 4, 6, \ldots)$, for example, is the collection of isolated points $(2, 2)$, $(4, 4)$, $(6, 6)$, $\ldots$. Points designated by ordered pairs in which x is some real number other than an even integer are not included in the graph.

To further illustrate this point consider the graph of the function defined by the equation $y = \frac{2x}{x - 1}$. Since this function is defined for all values of x except $x = 1$, the points $P(x, y)$ for $x < 1$ may be joined by a smooth curve and all the points $P(x, y)$ for $x > 1$ may be joined by another smooth curve. Thus the graph has two branches.

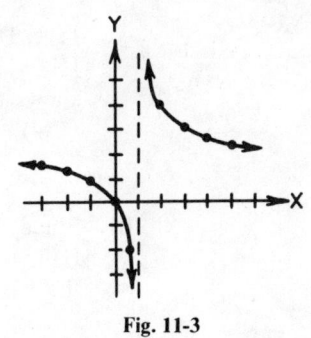

Fig. 11-3

The abscissas of the points at which a curve crosses the X-axis are called the **x-intercepts**. The ordinates of the points at which a curve crosses the Y-axis are called the **y-intercepts**.

In finding the **inverse** of a relation or function, multiplication by the variables may introduce extraneous material. To test for these extraneous values use domains and ranges. Consider the function defined by $y = \sqrt{9 - 4x^2}$ with a domain of $-\frac{3}{2} \leq x \leq \frac{3}{2}$ and a range of $0 \leq y \leq 3$, since only the positive square root is indicated. The inverse of this function is $y = \pm \frac{1}{2} \sqrt{9 - x^2}$ with a domain of $0 \leq x \leq 3$ and a range of $-\frac{3}{2} \leq y \leq \frac{3}{2}$. We notice that the inverse is a relation since for each x there are two values of y. If we had chosen the domain of f to be $-\frac{3}{2} \leq x \leq 0$, then the range of the inverse would be the same interval, i.e., $y < 0$. The plus sign of the inverse must be discarded as extraneous yielding $y = -\frac{1}{2} \sqrt{9 - x^2}$ as the inverse, which is a function.

The graph of the inverse may be obtained by direct plotting of the ordered pairs (x, y) from the equation. It may also be obtained by reflecting the graph of f in the straight line $y = x$. See Figure 11-4. For this example, what is the inverse function if the domain of $f: y = x^2$ is $0 \leq x \leq 2$?

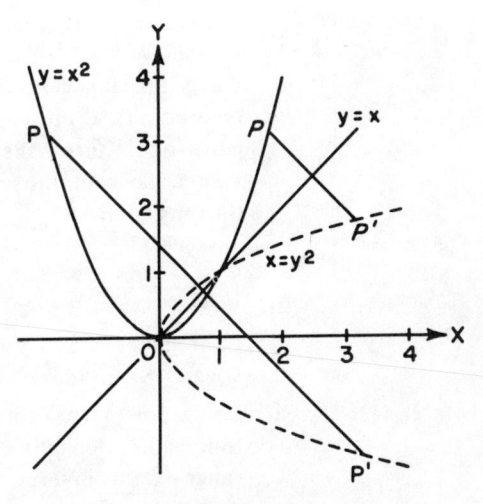

Fig. 11-4

SOLUTIONS TO EXERCISES

I

i. $f: |x - 1|$, $f(1) = |1 - 1| = 0$
$f(2) = |2 - 1| = 1$, $f(-4) = |-4 - 1| = 5$

ii. $f: (1, 3), (2, 5), (4, 7), (-4, 1), (-3, 7)$
$f(1) = 3$ since the first ordered pair shows that when $x = 1$, $y = 3$. $f(2) = 5$, $f(-4) = 1$.

iii. Since the ordered pairs $(10, -1)$ and $(10, 8)$ have the same first element, the set does not define a function.

II

i. $y = \sqrt{4 - x^2}$
If x^2 were greater than 4, the quantity $\sqrt{4 - x^2}$ would be imaginary. Since y must be a real number, we must restrict the domain to values for which $x^2 \leq 4$; that is, to values in $-2 \leq x \leq 2$. The given equation specifies that y is either zero or positive. It is zero when $x = 2$ and its largest value is at $x = 0$ when $y = \sqrt{4} = 2$. Therefore the range is $0 \leq y \leq 2$.

ii. $f = \{(1, 2), (\pi, -3)\}$
Since the function is defined by two ordered pairs, the values of x in those pairs constitutes the domain; and the values of y in those pairs constitute the range.

iii. $y = \sqrt{4 - x^2}$, $-2 \leq x \leq 0$
Interchange x and y in $y = \sqrt{4 - x^2}$ and square both sides to obtain $x^2 = 4 - y^2$. Solve for y to obtain $y = \pm\sqrt{4 - x^2}$. Since the domain of f becomes the range of the inverse of f, we have f^{-1}: $y = -\sqrt{4 - x^2}$, $-2 \leq y \leq 0$.

III

i. $y = kx$
$3 = 2k$
$\frac{3}{2} = k$
$y = \frac{3}{2}x$

ii. $z = kx^2 \sqrt{y}$
$6 = k(9)(2)$
$\frac{6}{18} = \frac{1}{3} = k$
$z = \frac{1}{3} x^2 \sqrt{y}$

iii. $w = k \dfrac{xy^2}{z}$
$4 = k\dfrac{3(4)}{18}$, $\dfrac{18}{3} = 6 = k$, $w = 6 \dfrac{xy^2}{z}$

IV

i. $y = x$. Interchange x and y. $x = y$. Solve for y: $y = x$. The function is its own inverse.

ii. $f(x) = x^3 - 3x$
$f(-x) = (-x)^3 - 3(-x)$
$= -x^3 + 3x = -f(x)$

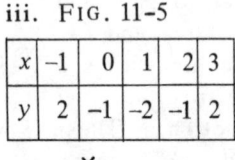

iii. FIG. 11-5

x	-1	0	1	2	3
y	2	-1	-2	-1	2

iv. FIG. 11-6: $y = x + 2$, $x =$ integers

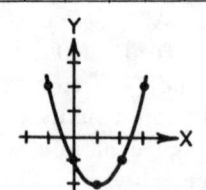

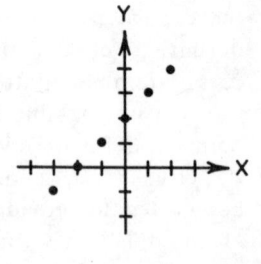

V

i. The graph of $x = 4$ is parallel to the Y-axis.

ii. $f(x) = x^2 - x$
$f(-x) = (-x)^2 - (-x)$
$= x^2 + x \neq -f(x)$
The function is neither odd nor even.

iii. No.

iv. $f(x) = 2x^2 + x - 3 = (2x + 3)(x - 1)$
$g(x) = 2x^2 + 5x + 3 = (2x + 3)(x + 1)$
$\dfrac{f}{g} = \dfrac{(2x + 3)(x - 1)}{(2x + 3)(x + 1)} = \dfrac{x - 1}{x + 1}$

12 EQUATIONS

SELF-TEST

DIRECTIONS: Write your answers in the numbered regions to the right. To check your answers, turn the page. Study the solutions to the problems you missed, and do the exercises following any problem or group of problems in which you had errors.

In **1** and **2**, solve for x.

1. $2x - 7 = 6x + 2$

2. $2x - \dfrac{3 - x}{5} = \dfrac{x + 4}{3} - 7$

In **3** and **4**, solve for y.

3. $m^2 y - m^2 + 3 = my + 6y - 2m$

4. $.3y - 1.4 = .2(y + 3)$

In **5** and **6**, find the positive acute angles which satisfy the equations.

5. $\sin x = \dfrac{1}{2}$

6. $\cot (2x + 21°) = 1$

7. Find two consecutive even integers such that twice the smaller exceeds the larger by 18.

8. If an equation has no solution, it is called a _____ equation.

9. The derived equation $x^2 - 25 = 0$ is obtained by multiplying $x - 5 = 0$ by $x + 5$. This derived equation is _____ with respect to the equation $x - 5 = 0$.

10. The equation $\sin x = \sin(x + 360°)$ is an _____.

1 _____
2 _____
3 _____
4 _____
5 _____
6 _____
7 _____
8 _____
9 _____
10 _____

SOLUTIONS

1. Given: $2x - 7 = 6x + 2$
   Add 7: $2x - 7 + 7 = 6x + 2 + 7$
   Collect terms: $2x = 6x + 9$
   Subtract $6x$: $2x - 6x = 6x - 6x + 9$
   Collect terms: $-4x = 9$
   Divide by -4: $x = -\frac{9}{4}$

2. Given: $2x - \frac{3-x}{5} = \frac{x+4}{3} - 7$
   Multiply by 15: $30x - 9 + 3x = 5x + 20 - 105$
   Collect terms: $(33 - 5)x = 29 - 105$
   Simplify: $28x = -76$
   Divide by 28: $x = -\frac{19}{7}$

EXERCISES I

In i–iii, solve for x.

i. $2\left(x - \frac{1}{3}\right) = 3\left(\frac{2}{7} - 2x\right)$

ii. $\frac{2}{x} + \frac{1}{x-1} = \frac{-5}{4(x^2 - x)}$

iii. $\frac{x+1}{x+2} - \frac{x-2}{x-1} = \frac{x-3}{x-1} - \frac{x+1}{x+3}$

---

3. $m^2 y - m^2 + 3 = my + 6y - 2m$
   $(m^2 - m - 6)y = m^2 - 2m - 3$
   $y = \frac{(m-3)(m+1)}{(m-3)(m+2)}$
   $= \frac{m+1}{m+2}$

4. $.3y - 1.4 = .2(y + 3)$
   $3y - 14 = 2y + 6$
   $y = 20$

EXERCISES II

In i–iii, solve for y.

i. $4y - b = by + 4$

ii. $3y + b = c - 2y$

iii. $\frac{cy}{a^2 - b^2} - \frac{c}{a+b} = y + b - a$

Although trigonometric equations have infinitely many solutions, the answers to 5 and 6 are just the positive-acute-angle solutions.

5. $\sin x = \frac{1}{2}$, $x = 30°$

6. $\cot (2x + 21°) = 1$
   $2x + 21° = 45°$
   $2x = 24°$, $x = 12°$

EXERCISES III

In i–iii, find the positive acute angles which satisfy the equations.

i. $\tan 3x = 1$

ii. $\cos 3x = -1$

iii. $\sin x = 2$

---

7. Let x and $x + 2$ be two consecutive even integers, then $2x = (x + 2) + 18$ states that twice the smaller integer exceeds the larger by 18. Solving this equation, we find $x = 20$ and $x + 2 = 22$.

EXERCISES IV

i. Cheryl is 6 years older than her brother Bob. Four years ago she was twice as old as Bob was. Find their present ages.

ii. Find a number such that when 6 is subtracted from it, 2 plus $\frac{1}{5}$ of the remainder is equal to $\frac{1}{3}$ of the original number.

iii. If $D^2 = \frac{(f+s)}{(f-r)}$ find the value of f when $D = 4$, $r = 3$, and $s = -18$.

iv. Two angles are said to be complementary if their sum is $90°$. Find the complementary angles whose difference is $22°$.

8. If an equation has no solution, it is called a null equation.

9. The derived equation $x^2 - 25 = 0$ obtained by multiplying $x - 5 = 0$ by $x + 5$ is redundant with respect to the equation $x - 5 = 0$.

10. The equation $\sin x = \sin(x + 360°)$ is an identity.

EXERCISES V

i. The equation $4 - x^2 = (2 - x)(2 + x)$ is an _____.

ii. Since the degree of the equation $2x^2 - 3x = 5$ is 2, it is called a _____ equation.

iii. If a root of the derived equation does not satisfy the given equation, the root is said to be _____.

iv. The equation $x - 3 = 0$ is _____ with respect to $x^2 - 9 = 0$.

v. Is $x = 3$ a root of the following equation?

$$\frac{3x}{x-3} - \frac{9}{x-3} = 0$$

vi. If x is the variable, then the equation $ax + bx^3 = x$ is a _____ degree equation.

vii. In general, equations involving trigonometric functions of the variable have an _____ number of roots.

Solutions to the exercises appear on page 72.

ANSWERS

1. $-\dfrac{9}{4}$

2. $-\dfrac{19}{7}$

3. $\dfrac{m+1}{m+2}$

4. 20

5. 30°

6. 12°

7. 20, 22

8. null

9. redundant

10. identity

Equations

BASIC FACTS

An **equation** is a statement that two expressions are equal. The expressions are called **members** or **sides** of the equality.

An equality which contains a variable is called a **conditional equality** or simply an **equation**. If the equality is true for all values of the variable, it is called an **identity**.

The variables in an equation are frequently called **unknowns**. A value of the variable which, when substituted into the equation, makes the equality true is called a **solution** or a **root** of the equation. The root is said to satisfy the equation.

The form of an equation may be changed by algebraic operations. The resulting form is called a **derived equation**. A derived equation is equivalent to the original equation if it contains all the roots of that equation and no more.

There are two important **operations** which lead to equivalent equations.

1. We may add the same number or expression to, or subtract it from, both members of the equation.
2. We may multiply or divide both members of an equation by the same number or expression providing it is not zero or does not contain the unknown.

The first of these operations leads to the concept of **transposing** terms from one side of an equality to the other. When a term is transposed, it is removed from one side of an equality, placed on the other side, and its sign is changed.

Multiplying both members of an equation by an expression which contains the unknown may result in a **redundant derived equation**. The roots of the redundant derived equation which are not roots of the original equation are called **extraneous roots**.

ADDITIONAL INFORMATION

Although the multiplication or division of both members of an equation by an expression containing the variable does not lead to equivalent equations, the operations are sometimes performed. A derived equation is said to be **redundant** with respect to the original equation if it has roots that are not roots of the original equation. The additional roots are called **extraneous roots**. A derived equation is said to be **defective** with respect to the original equation if it does not have all the roots of that equation.

An example of a redundant derived equation is that obtained by multiplying $x + 2 = 3$ (which has the root $x = 1$) by $x + 1$. The result is $x^2 + 3x + 2 = 3x + 3$, or $x^2 = 1$, which has roots $x = \pm 1$. The root $x = -1$ is extraneous.

An example of a defective derived equation is that obtained by dividing $x^2 - 4 = 0$ by $x + 2$. We obtain $x - 2 = 0$ which has the root $x = 2$. However, $x = -2$ also satisfies the original equation. Therefore, the derived equation is defective.

The procedure of multiplying by the variable is used frequently in the solution of equations which are in fractional form. In such cases, we multiply both members by the lowest common denominator (LCD) to clear the equation of fractions, and sometimes the LCD contains the variable. For example, if we multiply both members of the equation, $\frac{x}{x-3} - \frac{3x}{x+3} = 2 - \frac{4x^2}{x^2-9}$, by the lowest common denominator, $(x-3)(x+3)$, we obtain

$$x(x+3) - 3x(x-3) = 2(x+3)(x-3) - 4x^2.$$

The solution obtained from this equation, $x = \frac{3}{2}$, must be checked in the original equation to insure that it was not introduced in the multiplication operation. We must also be careful to limit our solution to the numbers for which the equation is defined. In the above example, the equation is *not defined* for $x = \pm 3$ since for these numbers we would have a division by zero in two of the fractions.

Consider the equation,

$$\frac{2x}{x-2} + \frac{1}{x+2} = 2 + \frac{16}{x^2-4}.$$

If we multiply by the LCD and solve for x, we obtain $x = 2$. However, when we check this solution we find that the equation is not defined for $x = 2$. Since there are no solutions which satisfy this equation, it is a **null equation**.

In solving **word problems**, it is necessary to convert the words in the problem into the symbols and equations of algebra. A thorough reading and understanding of the problem is the first step. Also, some facts must be recalled from previous studies. In transferring word statements into algebraic statements, be very specific in the definitions of the variables. Consider the following example. Let A and B start from the same place and travel in the same direction with B starting 20 minutes after A. If A travels at an average rate of

Dividing both members of an equation by an expression which contains the unknown may result in a **defective derived equation**.

To solve an equation is to find its roots. This is accomplished by applying the appropriate algebraic operations until the value of the variable is clearly indicated.

One of the simpler types of equations is the polynomial equation which has one variable and in which the variable occurs only to the first power. An equation of this type is a **linear equation**. An example is $ax + b = 0$, where x is the variable and a and b are constants. A linear equation has one and only one solution.

The **general polynomial equation in one variable** is given by

$$a_0 x^n + a_1 x^{n-1} + \ldots + a_{n-1} x + a_n = 0$$

where n is a positive integer and the coefficients $a_0, a_1, a_2, \ldots, a_n$ are constants. The degree of the equation is the highest degree of the variable.

Another simple type of equation is the trigonometric equation—an equation which involves a trigonometric function. In this case, there are infinitely many solutions since the values of the trigonometric functions of α and $\alpha + 2n\pi$ are the same. The fundamental solutions are those which lie in the interval from $0°$ to $360°$. There are also trigonometric equations for which we can limit the solutions to acute angles; but it should be remembered that, in general, trigonometric equations have an infinite number of solutions. Once the fundamental solutions have been found, the rest can be indicated by adding $2n\pi$, where n is an integer, to the fundamental solutions.

Equations which are not true for any value of the variable are called **null equations**.

60 miles per hour and B, 75 miles per hour, how far have they traveled when B overtakes A?

First define some variables. Let t be the number of hours that B has traveled. Then, since 20 minutes is $\frac{1}{3}$ of an hour, $t + \frac{1}{3}$ is the number of hours A has traveled. Let d be the distance traveled.

Now apply the physical law. In this case, the law is that the distance traveled equals the average speed times the time. Write two equations.

$$A \text{ travels: } d = 60 \left(t + \tfrac{1}{3}\right)$$

$$B \text{ travels: } d = 75t$$

B overtakes A when they have both traveled the same distance. Therefore,

$$75t = 60 \left(t + \tfrac{1}{3}\right) \text{ or } t = 1\tfrac{1}{3} \text{ hours.}$$

We can compute the distance from either equation.

$$A: \ d = 60 \left(1\tfrac{1}{3} + \tfrac{1}{3}\right) = 100 \text{ miles}$$

$$B: \ d = 75 \left(\tfrac{4}{3}\right) = 100 \text{ miles}$$

Trigonometric equations are solved by finding the fundamental solutions in the interval from $0°$ to $360°$ and then adding integral multiples of $360°$ if we wish to express the infinite number of solutions. The procedure is best explained by considering an example.*

Find the solution of $2 \cos \theta - 1 = 0$.

First rewrite the equation in the form $\cos \theta = \frac{1}{2}$. Now recall that $\cos 60° = \frac{1}{2}$ and that the cosine is positive in the first and fourth quadrants; therefore $\theta = 60°$ and $300°$ both satisfy the equation. The general solutions are expressed by $\theta = 60° + n360°$ and $\theta = 300° + n360°$ where n is an integer.

If the variable of a trigonometric equation is expressed in multiple form, solve for the multiple angle and then perform the necessary algebraic operations to find the values of the variable.

$3x^4 + 5x - 4 = 0$ is an example of a **polynomial equation in one variable**. Its degree is 4. The degree of the equation $2^5 x^3 + 3x - 4^2 = 0$ is 3.

Special names are given to some of the **lower-degree equations**.

degree	name	equation
first	linear	$ax + b = 0$
second	quadratic	$ax^2 + bx + c = 0$
third	cubic	$ax^3 + bx^2 + cx + d = 0$
fourth	quartic	$ax^4 + bx^3 + cx^2 + dx + e = 0$

*See Appendix A, page 159, for a brief review of trigonometric functions.

SOLUTIONS TO EXERCISES

I

i. Given: $2\left(x - \frac{1}{3}\right) = 3\left(\frac{2}{7} - 2x\right)$
Simplify: $2x - \frac{2}{3} = \frac{6}{7} - 6x$
Transpose terms: $8x = \frac{6}{7} + \frac{2}{3} = \frac{32}{21}$
Divide by 8: $x = \frac{4}{21}$

ii. Given: $\frac{2}{x} + \frac{1}{x-1} = \frac{-5}{4(x^2 - x)}$
Multiply by the LCD $= 4x(x - 1)$:
$$8x - 8 + 4x = -5$$
Collect terms: $12x = 3$, $x = \frac{1}{4}$
Check: $8 + \frac{4}{-3} = \frac{20}{4-3}$, $20 = 20$

iii. $\frac{x+1}{x+2} - \frac{x-2}{x-1} = \frac{x-3}{x-1} - \frac{x+1}{x+3}$
Place each side over its LCD:
$$\frac{(x+1)(x-1) - (x-2)(x+2)}{(x+2)(x-1)}$$
$$= \frac{(x-3)(x+3) - (x+1)(x-1)}{(x-1)(x+3)}$$
Simplify: $\frac{x^2 - 1 - (x^2 - 4)}{(x+2)(x-1)}$
$$= \frac{x^2 - 9 - (x^2 - 1)}{(x-1)(x+3)}$$
Multiply by $(x-1)(x+2)(x+3)$:
$$(x+3)(x^2 - 1 - x^2 + 4)$$
$$= (x+2)(x^2 - 9 - x^2 + 1)$$
$$(x+3)(3) = (x+2)(-8)$$
$$3x + 9 + 8x + 16 = 0$$
$$11x = -25, \quad x = -\frac{25}{11}$$

II

i. $4y - b = by + 4$
$4y - by = b + 4$
$y = \frac{b+4}{4-b}$

ii. $3y + b = c - 2y$
$5y = c - b$
$y = \frac{c-b}{5}$

iii. $\frac{cy}{a^2 - b^2} - \frac{c}{a+b} = y + b - a$
$cy - c(a - b)$
$= (a^2 - b^2)y + (a^2 - b^2)(b - a)$
$(c - a^2 + b^2)y = (a - b)(c - a^2 + b^2)$
$y = a - b$

III

i. $\tan 3x = 1$, $3x = 45°$, $x = 15°$
$3x = 225°$, $x = 75°$

ii. $\cos 3x = -1$, $3x = 180°$, $x = 60°$

iii. Since $\sin x \not> 1$, $\sin x = 2$ has no solution.

IV

i. Let x be Bob's age. Then Cheryl's age is $x + 6$.
$$(x + 6) - 4 = 2(x - 4)$$
$$x + 2 = 2x - 8$$
$$10 = x = \text{Bob's present age}$$
Cheryl's present age is $x + 6 = 16$.

ii. $2 + \frac{1}{5}(x - 6) = \frac{1}{3}x$
$\frac{1}{5}(10 + x - 6) = \frac{1}{3}x$
$3(x + 4) = 5x$, $12 = 2x$, $x = 6$

iii. $D^2(f - r) = f + s$, $D = 4$, $r = 3$, $s = 8$
$16(f - 3) = f - 18$
$15f = 48 - 18 = 30$, $f = 2$

iv. Let x be one angle. Then its complement is $90° - x$.
$$x - (90° - x) = 22°, \quad 2x = 112°$$
$$x = 56° \text{ and } 90° - x = 34°$$

V

i. The equation $4 - x^2 = (2 - x)(2 + x)$ is an <u>identity</u>.

ii. Since the degree of the equation $2x^2 - 3x = 5$ is 2, it is called a <u>quadratic</u> equation.

iii. If a root of the derived equation does not satisfy the given equation, the root is said to be <u>extraneous</u>.

iv. The equation $x - 3 = 0$ is <u>defective</u> with respect to $x^2 - 9 = 0$.

v. $x = 3$ is *not* a root of $\frac{3x}{x-3} - \frac{9}{x-3} = 0$.

vi. If x is the variable, then the equation $ax + bx^3 = x$ is a <u>third</u> degree equation.

vii. In general, equations involving trigonometric functions of the variable have an <u>infinite</u> number of roots.

13 THE LINEAR EQUATION

SELF-TEST

DIRECTIONS: Write your answers in the numbered regions to the right. To check your answers, turn the page. Study the solutions to the problems you missed, and do the exercises following any problem or group of problems in which you had errors.

The linear equation $4x + 3y - 12 = 0$ is to be used in **1-5**.

1. Find the *x*-intercept.

2. Find the *y*-intercept.

3. Find the slope of the line defined by the equation.

4. Find the equation of a line through the point $P(0, 2)$ and parallel to the given line.

5. Find the equation of a line through the point $P(0, 2)$ and perpendicular to the given line.

6. Find the distance from the point $P(2, 5)$ to the line $3x + 4y - 6 = 0$.

7. Show that the points $(1, 2)$, $(4, -1)$, and $(-2, 5)$ lie on a line. Find the equation of the line.

8. The general form of an equation of a straight line is given by $Ax + By + C = 0$. Change this equation to the intercept form and write the coefficients of *x* and *y* in the answer column.

9. If *y* varies directly as *x*, and $y = 3$ when $x = 9$, write a linear equation relating *x* and *y*.

1 _____

2 _____

3 _____

4 _____

5 _____

6 _____

7 _____

8 _____

9 _____

SOLUTIONS

The equation for **1-5** is $4x + 3y - 12 = 0$.

1. The x-intercept is $4x = 12$ or $x = 3$.

2. The y-intercept is $3y = 12$ or $y = 4$.

3. The given equation can be changed to $y = -\frac{4}{3}x + 4$ and the slope is $-\frac{4}{3}$.

4. The line through $P(0, 2)$ and parallel to $4x + 3y - 12 = 0$ is $y - 2 = -\frac{4}{3}(x - 0)$ or $3y + 4x - 6 = 0$.

5. The line through $P(0, 2)$ and perpendicular to $4x + 3y - 12 = 0$ is $y - 2 = \frac{3}{4}(x - 0)$ or $4y - 3x - 8 = 0$.

EXERCISES I

Are the statements of i-iii true or false?

i. The x-intercept of the line $3x + 5y = 2$ is $\frac{2}{5}$.

ii. The straight line $2x + y^2 = 4$ crosses the X-axis at $x = 2$.

iii. The two lines $y = x - 4$ and $2y + 8 = 3x$ have the same y-intercept.

iv. Find the rate of change of the linear function defined by $3y = 12x - 6$.

Are the statements of v-vii true or false?

v. The slope of the line $y = x - 2$ is 1.

vi. The line $y = 3$ is perpendicular to the line $x = 2$.

vii. The lines $2x + 3y = 2$ and $3y + 2x = 7$ are parallel.

---

6. $d = \dfrac{Ax_1 + By_1 + C}{\pm\sqrt{A^2 + B^2}}$

$\sqrt{A^2 + B^2} = \sqrt{9 + 16} = 5$

$Ax_1 + By_1 + C = 3(2) + 4(5) - 6 = 20$

$d = \dfrac{20}{5} = 4$

7. The equation of the line through $(1, 2)$ and $(4, -1)$ is $y - 2 = \dfrac{-1-2}{4-1}(x - 1)$ or $y + x = 3$. Substitute the coordinates of the point $(-2, 5)$ into this equation to obtain $5 - 2 = 3 = 3$. Since the equation is satisfied by $(1, 2)$, $(4, -1)$, and $(-2, 5)$, these three points lie on the line $y + x = 3$.

EXERCISES II

i. Is it true or false that the distance from $P(2, 0)$ to the line $y = 3$ is 3?

The two points, $P_1(2, 3)$ and $P_2(4, -5)$, are to be used in ii-vi.

ii. Find the midpoint of the line segment $\overline{P_1 P_2}$.

iii. Find the slope of the line through P_1 and P_2.

iv. Find the equation of the line through P_1 and P_2.

v. Find the equation of a line through P_1 and parallel to the X-axis.

vi. Find the equation of a line through P_2 and parallel to the Y-axis.

---

8. $Ax + By + C = 0$

The intercepts are $a = -\dfrac{C}{A}$ and $b = -\dfrac{C}{B}$. Using the intercept form, the desired equation is $\dfrac{x}{-C/A} + \dfrac{y}{-C/B} = 1$. The coefficient of x is $(-C/A)^{-1}$, and that of y is $(-C/B)^{-1}$.

9. Direct variation is given by $y = kx$. If $y = 3$ when $x = 9$, $k = \dfrac{1}{3}$. The equation relating x and y is $3y = x$.

EXERCISES III

i. Is the formula for the area of a circle linear?

Are statements ii and iii true or false?

ii. The equation $y = \dfrac{k}{x}$ is linear.

iii. The function defined by $d = vt$ is linear if v is an average constant velocity.

iv. If $3y = x$, find y when $x = 12$.

v. Assuming that an article depreciates 5% of its original value each year and that it originally cost $5000, what is the slope of the linear equation that expresses the value of the article at the end of t years?

vi. A car travels 40 miles the first hour and 35 miles each hour thereafter. Write an equation for the distance travelled in t hours.

vii. A collection of coins totals $11.00. If there are 10 times as many nickels as dimes and twice as many quarters as dimes, find the number of dimes.

Solutions to the exercises appear on page 78.

ANSWERS

1. 3
2. 4
3. $-\dfrac{4}{3}$
4. $3y + 4x = 6$
5. $4y - 3x = 8$
6. 4
7. $y + x = 3$
8. $(-C/A)^{-1}$
   $(-C/B)^{-1}$
9. $3y = x$

76 The Linear Equation

BASIC FACTS

An equation of the form $Ax + By + C = 0$ is called a **linear equation**. The equation is of the first degree in the variables.

The geometric representation of a linear equation is a **straight line**.

The tangent of the angle of inclination of the line with the positive X-axis is called the slope of the line. It can be found by the formula,

$$m = \frac{y_2 - y_1}{x_2 - x_1}.$$

The slope of a line parallel to the X-axis is zero. The slope of a line parallel to the Y-axis is undefined.

There are many forms of the linear equation. In the following forms, k, A, B, and C are constants; m is the slope; a is the x-intercept; b is the y-intercept; and the variables with subscripts are constants representing the coordinates of specific points.

The Line Parallel to the Y-axis

$$x = k$$

The Line Parallel to the X-axis

$$y = k$$

The Point-slope form

$$y - y_1 = m(x - x_1)$$

The Two-point form

$$y - y_1 = \frac{y_2 - y_1}{x_2 - x_1}(x - x_1)$$

The Slope-intercept form

$$y = mx + b$$

The General form

$$Ax + By + C = 0$$

The Intercept form

$$\frac{x}{a} + \frac{y}{b} = 1$$

ADDITIONAL INFORMATION

The Coordinates of a Point on a Line Segment

In Figure 13-1, the point P divides the line segment $P_1 P_2$ in a given ratio, k, so that we have $P_1 P = k(P_1 P_2)$. From the similar triangles $P_1 P_2 R_2$ and $P_1 PR$,

$$\frac{N_1 N}{N_1 N_2} = \frac{P_1 P}{P_1 P_2} = k$$

so that $N_1 N = k N_1 N_2$ and since $N_1 N = x - x_1$ and $N_1 N_2 = x_2 - x_1$ we have $x - x_1 = k(x_2 - x_1)$ which, when solved for x, gives the abscissa of P. The y-coordinate can be found in a similar manner by constructing the perpendiculars to the Y-axis.

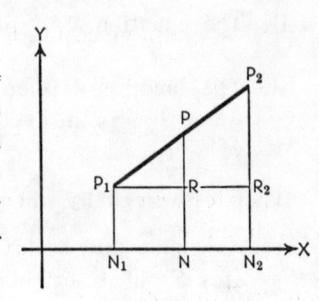

Fig. 13-1

The Slope of a Line

If we choose two points $P_1(x_1, y_1)$ and $P_2(x_2, y_2)$ on a line, as shown in Figure 13-2, we see that $P_1 R P_2$ is a right triangle and that

$$\tan \alpha = \frac{RP_2}{P_1 R} = \frac{y_2 - y_1}{x_2 - x_1} = m$$

since $\tan \alpha = m$ by definition.

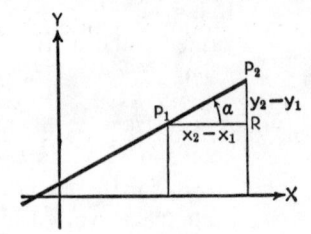

Fig. 13-2

Forms of the Linear Equation

The **point-slope** form of the equation of a line is obtained from the definition of the slope by letting P_2 be a general point $P(x, y)$. Then,

$$\tan \alpha = m = \frac{y - y_1}{x - x_1} \quad \text{and} \quad y - y_1 = m(x - x_1).$$

The **two-point** form of the equation of a line is obtained by substituting into the point-slope form the formula for the slope in terms of the coordinates of two points, P_1 and P_2.

$$y - y_1 = m(x - x_1) \quad \text{and} \quad y - y_1 = \frac{y_2 - y_1}{x_2 - x_1}(x - x_1)$$

If $x_1 = x_2$, then $x_2 - x_1 = 0$ and the above form is undefined. However, the two-point form may be written,

$$(y - y_1)(x_2 - x_1) = (y_2 - y_1)(x - x_1)$$

Then if $x_1 = x_2$,
Dividing by $(y_2 - y_1)$,

$$0 = (y_2 - y_1)(x - x_1)$$
$$0 = x - x_1; \quad x = x_1$$

which is a line parallel to the Y-axis.

The **slope-intercept** form of the equation of a line is obtained from the point-slope form. The point where the line crosses the Y-axis is $P_1(0, b)$. Thus we have $y - b = m(x - 0)$ or $y = mx + b$.

If the linear equation is written in the **general form** $Ax + By + C = 0$, the following properties hold:

If $A = 0$, the line is parallel to the X-axis.

If $B = 0$, the line is parallel to the Y-axis.

The slope is given by $m = -\frac{A}{B}$.

The intercepts are
$$a = -\frac{C}{A}, \quad b = -\frac{C}{B}.$$

The following properties hold for any linear equation(s):

If two lines with slopes m_1 and m_2 are **parallel**, $m_1 = m_2$.

If two lines with slopes m_1 and m_2 are **perpendicular**, the slopes are negative reciprocals of each other.
$$m_1 m_2 = -1$$

The **angle between two lines** can be found from the slopes of the lines.
$$\tan \theta = \frac{m_2 - m_1}{1 + m_1 m_2}.$$

The **distance** from a point $P_1(x_1, y_1)$ to a line $Ax + By + C = 0$ is given by
$$d = \left| \frac{Ax_1 + By_1 + C}{\sqrt{A^2 + B^2}} \right|.$$

A **line segment** is defined as that part of a line which lies between two specified points, $P_1(x_1, y_1)$ and $P_2(x_2, y_2)$. The coordinates of a point $P(x, y)$ which lies on the line so that the portion $P_1 P$ is a given fraction of the segment $P_1 P_2$ are given by $x = x_1 + k(x_2 - x_1)$, $y = y_1 + k(y_2 - y_1)$, where $k = \overline{P_1 P}/\overline{P_1 P_2}$.

The **midpoint** of a line segment has coordinates $x = \frac{1}{2}(x_1 + x_2)$ and $y = \frac{1}{2}(y_1 + y_2)$.

Any function defined by an equation of the form $Ax + By + C = 0$ is called a **linear function**. The **rate of change** of a linear function is constant and equals the slope of its straight line.

The **intercept form** is obtained by finding the equation of a line through the points $P_1(a, 0)$ and $P_2(0, b)$. We have $y - 0 = \frac{b}{-a}(x - a)$ or $ay + bx = ab$, which reduces to the intercept form if we divide by ab.

Distance from a Point to a Line

Let the line l_1 be defined by the equation $Ax + By + C = 0$, and let $P_1(x_1, y_1)$ be a point not on the line. Figure 13-3 shows l_1 with an angle of inclination α, $P_1 R \perp l_1$, and $PN \perp X$-axis. Since the sides of angle $NP_1 R$ are respectively perpendicular to the sides of the angle α, we have $\angle NP_1 R = \alpha$. The (perpendicular) distance from P_1 to l_1, then, can be expressed in terms of $\cos \alpha$.

$$\cos \alpha = \frac{P_1 R}{P_1 M}$$
$$P_1 R = d = P_1 M \cos \alpha$$
$$= (P_1 N - MN) \cos \alpha$$
$$= (y_1 - MN) \cos \alpha$$

Values must now be found for $(y_1 - MN)$ and $\cos \alpha$ such that when these values are substituted in the above equation, the formula for the distance from a point to a line will have been obtained.

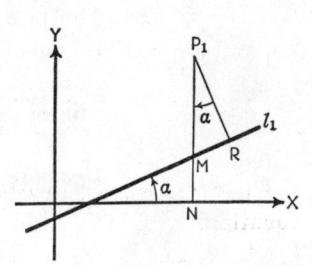

Fig. 13-3

Let us begin by evaluating $(y_1 - MN)$. The coordinates of M are (x_1, y) and since M is on the line l we have $Ax_1 + By + C = 0$. Solving for y, $y = \frac{Ax_1 + C}{B}$, if $B \neq 0$. From the figure it can be seen that $MN = y$. Therefore,
$$y_1 - MN = y_1 + \frac{Ax_1 + C}{B} = \frac{1}{B}(By_1 + Ax_1 + C).$$

Now evaluate $\cos \alpha$. Since α is the angle of inclination, we have, by the definition of slope, $\tan \alpha = m = -\frac{A}{B}$. Then by the theorem of Pythagoras, $\cos \alpha = \pm \frac{B}{\sqrt{A^2 + B^2}}$. Substituting these values into the formula for d we have

$$d = \frac{1}{B}(By_1 + Ax_1 + C)\left(\pm \frac{B}{\sqrt{A^2 + B^2}}\right)$$
$$= \pm \frac{Ax_1 + By_1 + C}{\sqrt{A^2 + B^2}}.$$

One needs only two points **to draw a straight line**. The easiest points to find are usually the intercepts, the coordinates of which are found by setting each variable, in turn, equal to zero and finding the corresponding value of the other variable. This procedure fails if the line passes through the origin; but once it has been learned that a given line passes through $(0,0)$, it is necessary to find just one more point. This additional point can be found by setting one of the variables equal to a small integer, and finding the corresponding value of the other variable.

SOLUTIONS TO EXERCISES

I

i. The x-intercept of $3x + 5y = 2$ is $x = \frac{2}{3}$.

ii. The graph of $2x + y^2 = 4$ is not a straight line.

iii. Given: $y = x - 4$ and $2y + 8 = 3x$. y-intercepts: $y = -4$ and $y = -\frac{8}{2} = -4$.

iv. The rate of change of a linear function is equal to the slope of its straight line. Solve $3y = 12x - 6$ for y to obtain $m = 4$.

v. Compare $y = x - 2$ with $y = mx + b$ to find $m = 1$.

vi. $y = 3$ is parallel to the X-axis, and $x = 2$ is parallel to the Y-axis. Thus, the lines are $\perp$.

vii. For $2x + 3y = 2$, $m_1 = -\frac{2}{3}$.

For $3y + 2x = 7$, $m_2 = -\frac{2}{3}$.

Since $m_1 = m_2$, the lines are parallel.

II

i. The line defined by $y = 3$ is parallel to the X-axis at 3 units above it. The point $P(2, 0)$ is on the X-axis; therefore $d = 3$.

The points for ii-vi are $P_1(2, 3)$ and $P_2(4, -5)$.

ii. The midpoint of $\overline{P_1 P_2}$ has coordinates $x = \frac{1}{2}(2 + 4) = 3$ and $y = \frac{1}{2}(3 - 5) = -1$.

iii. The slope of a line through P_1 and P_2 is $m = \frac{y_2 - y_1}{x_2 - x_1} = \frac{-5 - 3}{4 - 2} = -4$.

iv. The line through P_1 and P_2 is $y - 3 = -4(x - 2)$ or $y + 4x = 11$.

v. The line through P_1 and parallel to the X-axis is $y = 3$.

vi. The line through P_2 and parallel to the Y-axis is $x = 4$.

III

i. The area of a circle is given by the formula $A = \pi r^2$. Since the variable, r, is squared the equation is not linear.

ii. Since $y = \frac{k}{x}$ simplifies to $yx = k$, and since the degree of yx is 2, $y = \frac{k}{x}$ is not linear.

iii. If v is constant, then $d = vt$ is of the first degree in d and t.

iv. $y = \frac{1}{3}x = \frac{1}{3}(12) = 4$

v. Value = Original cost − Depreciation
$V = 5000 - (.05)(5000)t = 5000 - 250t$
The slope is −250.

vi. Distance = (Speed)(Time)
$d = 40(1) + 35(t - 1) = 5 + 35t$

vii. Let $x =$ number of dimes, then $10x =$ number of nickels, and $2x =$ number of quarters. The sum of x dimes, $10x$ nickels, and $2x$ quarters is $11.00.
$.10(x) + .05(10x) + .25(2x) = 11.00$
$1.1x = 11.00$
$x = 10$

14 MATRICES AND DETERMINANTS

SELF-TEST

DIRECTIONS: Write your answers in the numbered regions to the right. To check your answers, turn the page. Study the solutions to the problems you missed, and do the exercises following any problem or group of problems in which you had errors.

1. Find the values of w, x, y, and z if
$$\begin{pmatrix} 7 & w & x+4 \\ 3y & 2 & 5z \end{pmatrix} = \begin{pmatrix} 7 & 2w-6 & 3x \\ 9 & 2 & z+12 \end{pmatrix}$$

Refer to matrices A–F for 2-12.

$$A = \begin{pmatrix} 2 & -2 \\ 4 & 1 \end{pmatrix} \qquad B = \begin{pmatrix} 6 & 15 \\ -2 & -5 \end{pmatrix}$$

$$C = \begin{pmatrix} 3 & 6 & 2 \\ -4 & 8 & 1 \end{pmatrix} \qquad D = \begin{pmatrix} 4 & 6 & 2 \\ -3 & 1 & 7 \end{pmatrix}$$

$$E = \begin{pmatrix} 3 & 2 & -5 \\ 6 & 0 & 1 \\ 4 & 2 & 7 \end{pmatrix} \qquad F = \begin{pmatrix} 3 & 0 & 0 & 2 \\ 0 & 1 & -3 & 2 \\ 4 & -5 & 1 & 8 \\ 2 & 3 & -6 & 4 \end{pmatrix}$$

Is it possible to express each of the following as a single matrix?

2. $2C$
3. $2C + D$
4. $B + C$
5. AB
6. $(AB)C$
7. ED
8. $A - B$

$\delta(A)$ denotes the determinant of A. Evaluate each of the following determinants.

9. $\delta(A)$
10. $\delta(C)$
11. $\delta(E)$
12. $\delta(F)$

In 13 and 14, answer true or false.

13. Multiplication of matrices is commutative.

14. $\begin{pmatrix} 1 & 0 \\ 0 & 1 \end{pmatrix}$ is the additive identity for the set of 2 by 2 matrices.

15. Can matrix A be reduced to matrix B by row operations? Show why the reduction can or cannot be performed.

$$A = \begin{pmatrix} 2 & 1 & 0 \\ 0 & 0 & 7 \\ 0 & 1 & -1 \end{pmatrix} \qquad B = \begin{pmatrix} 1 & 0 & 0 \\ 0 & 1 & 0 \\ 0 & 0 & 1 \end{pmatrix}$$

1. $w = $
   $x = $
   $y = $
   $z = $
2.
3.
4.
5.
6.
7.
8.
9.
10.
11.
12.
13.
14.
15.

SOLUTIONS

1. $\begin{pmatrix} 7 & w & x+4 \\ 3y & 2 & 5z \end{pmatrix} = \begin{pmatrix} 7 & 2w-6 & 3x \\ 9 & 2 & z+12 \end{pmatrix}$

Equate corresponding elements.

$7 = 7$
$w = 2w - 6 \rightarrow w = 6$
$x + 4 = 3x \rightarrow x = 2$
$3y = 9 \rightarrow y = 3$
$2 = 2$
$5z = z + 12 \rightarrow z = 3$

EXERCISES I

Find the values of a, b, c, and d.

i. $\begin{pmatrix} a+3 \\ 2b-2 \\ c+7 \end{pmatrix} = \begin{pmatrix} -3 \\ 7 \\ 9 \end{pmatrix}$

ii. $\begin{pmatrix} 3a & 4 \\ 2b+3 & c \\ 3d-2 & b \end{pmatrix} = \begin{pmatrix} 12+a & 4 \\ 11 & 3c-6 \\ 5d+7 & 4 \end{pmatrix}$.

---

2. $2C = 2\begin{pmatrix} 3 & 6 & 2 \\ -4 & 8 & 1 \end{pmatrix} = \begin{pmatrix} 2(3) & 2(6) & 2(2) \\ 2(-4) & 2(8) & 2(1) \end{pmatrix}$

$= \begin{pmatrix} 6 & 12 & 4 \\ 8 & 16 & 2 \end{pmatrix}$

3. $2C + D = \begin{pmatrix} 6 & 12 & 4 \\ 8 & 16 & 2 \end{pmatrix} + \begin{pmatrix} 4 & 6 & 2 \\ -3 & 1 & 7 \end{pmatrix}$

$= \begin{pmatrix} 10 & 18 & 6 \\ -11 & 17 & 9 \end{pmatrix}$

4. Addition of matrices is defined only for matrices of the same dimension. Since B and C are not of the same dimension, they cannot be added.

5. $AB = \begin{pmatrix} 2 & -2 \\ 4 & 1 \end{pmatrix}\begin{pmatrix} 6 & 15 \\ -2 & -5 \end{pmatrix}$

$= \begin{pmatrix} 2(6) + (-2)(-2) & 2(15) + (-2)(-5) \\ 4(6) + (1)(-2) & 4(15) + (1)(-5) \end{pmatrix}$

$= \begin{pmatrix} 16 & 40 \\ 22 & 55 \end{pmatrix}$

6. $(AB)C = \begin{pmatrix} 16 & 40 \\ 22 & 55 \end{pmatrix}\begin{pmatrix} 3 & 6 & 2 \\ -4 & 8 & 1 \end{pmatrix}$

$= \begin{pmatrix} -112 & 416 & 72 \\ -154 & 572 & 99 \end{pmatrix}$

7. Multiplication of matrices is defined only when the number of columns in the left factor equals the number of rows in the right factor.

8. $A - B = \begin{pmatrix} 2 & -2 \\ 4 & 1 \end{pmatrix} + \begin{pmatrix} -6 & -15 \\ 2 & 5 \end{pmatrix}$

$= \begin{pmatrix} -4 & -17 \\ 6 & 6 \end{pmatrix}$

EXERCISES II

$A = \begin{pmatrix} 2 & -5 \\ 6 & 7 \end{pmatrix}$ $B = \begin{pmatrix} -4 & -1 \\ 3 & 5 \end{pmatrix}$ $C = \begin{pmatrix} 7 & 3 \\ 8 & -4 \\ 1 & 2 \end{pmatrix}$

i. Find $A + B$.   **iv.** Find AB.
ii. Find $-C$.   **v.** Find CB.
iii. Find $2B + A$.   **vi.** Find BC.

---

9. $\delta(A) = \begin{pmatrix} 2 & -2 \\ 4 & 1 \end{pmatrix} = 2(1) - 4(-2)$

$= 2 + 8 = 10$

10. The domain of δ is the set of square matrices.

11. $\delta(E) = \begin{vmatrix} 3 & 2 & -5 & 3 & 2 \\ 6 & 0 & 1 & 6 & 0 \\ 4 & 2 & 7 & 4 & 2 \end{vmatrix}$

$= 0 + 8 - 60 + 0 - 6 - 84$
$= -142$

12. Expanding by cofactors of the first row.

$3\begin{vmatrix} 1 & -3 & 2 \\ -5 & 1 & 8 \\ 3 & -6 & 4 \end{vmatrix} - 0 + 0 - 2\begin{vmatrix} 0 & +1 & -3 \\ 4 & -5 & 1 \\ 2 & 3 & -6 \end{vmatrix}$

$= 3(-26) - 0 + 0 - 2(-40)$
$= -78 + 80 = 2$

Matrices and Determinants   81

EXERCISES III

Evaluate each of the following determinants.

i. $\begin{vmatrix} 5 & -6 \\ 9 & -3 \end{vmatrix}$

ii. $\begin{vmatrix} 8 & 7 & 2 \\ -3 & 1 & 6 \\ 4 & 5 & 3 \end{vmatrix}$

iii. $\begin{vmatrix} 5 & 9 & 0 & 1 \\ 0 & 5 & 1 & -2 \\ 6 & 3 & -4 & 10 \\ 0 & 5 & 3 & 1 \end{vmatrix}$

13. Multiplication of matrices is not commutative.

14. The additive identity for the set of 2 by 2 matrices is $\begin{pmatrix} 0 & 0 \\ 0 & 0 \end{pmatrix}$, since adding this to any 2 by 2 matrix leaves that matrix unchanged.

EXERCISES IV

Answer true or false.

i. The product AB is 2 by 3 if A is 2 by 5 and B is 5 by 3.

ii. $\begin{pmatrix} 2 & 3 \\ 4 & 5 \end{pmatrix} = \begin{vmatrix} 2 & 3 \\ 4 & 5 \end{vmatrix}$

iii. $\begin{vmatrix} 5 & 7 & -2 & 1 \\ 4 & 6 & 0 & 3 \\ \pi & 4 & \frac{1}{2} & 7 \\ 6 & 3 & 1 & 10 \end{vmatrix} = - \begin{vmatrix} 5 & -2 & 7 & 1 \\ 4 & 0 & 6 & 3 \\ \pi & \frac{1}{2} & 4 & 7 \\ 6 & 1 & 3 & 10 \end{vmatrix}$

15. $\begin{pmatrix} 2 & 1 & 0 \\ 0 & 0 & 7 \\ 0 & 1 & -1 \end{pmatrix}$

$\equiv \begin{pmatrix} 2 & 0 & 1 \\ 0 & 0 & 1 \\ 0 & 7 & 0 \end{pmatrix}$ (row 1) − (row 3)
$\frac{1}{7}$ (row 2)
7 (row 3) + (row 2)

ANSWERS

1. $w = 6$
   $x = 2$
   $y = 3$
   $z = 3$

2. yes

3. yes

4. no

5. yes

6. yes

7. no

8. yes

9. $\delta(A) = 10$

10. not defined

11. $\delta(E) = -142$

12. $\delta(F) = 2$

13. false

14. false

15. yes

$\equiv \begin{pmatrix} 2 & 0 & 0 \\ 0 & 0 & 1 \\ 0 & 1 & 0 \end{pmatrix}$ (row 1) − (row 2)
$\frac{1}{7}$ (row 3)

$\equiv \begin{pmatrix} 1 & 0 & 0 \\ 0 & 1 & 0 \\ 0 & 0 & 1 \end{pmatrix}$ $\frac{1}{2}$ (row 1)
(row 2) & (row 3) interchanged

Note: solutions vary.

BASIC FACTS

A **matrix** is a set of numbers arranged in a rectangular array. It is usually denoted by a capital letter. The actual matrix is enclosed in parentheses. Sometimes square brackets or double vertical bars are used to denote matrices.

The numbers of the array are the **elements** or **entries** of the matrix. Each horizontal line of elements is called a **row**; each vertical line of elements is called a **column**. In general, a matrix which has m rows and n columns is called an m by n matrix. In this terminology, the number of rows is always stated first. A matrix which has n rows and n columns is called a **square matrix** of order n. Its principal diagonal is the diagonal starting in the upper left corner.

With every square matrix there is associated a real number called the **determinant** of that matrix. This association is a function, denoted δ, from the set of square matrices to the set of real numbers. The functional value, $\delta(A)$, the value of the determinant, is found by combining the elements of the matrix in a certain manner. The determinant of a matrix is symbolized by enclosing the elements of the matrix in single vertical bars.

$$A = \begin{pmatrix} a_1 & b_1 \\ a_2 & b_2 \end{pmatrix} \qquad A = \begin{vmatrix} a_1 & b_1 \\ a_2 & b_2 \end{vmatrix}$$

The **value of a second-order determinant** is defined as

$$\begin{vmatrix} a_1 & b_1 \\ a_2 & b_2 \end{vmatrix} = a_1 b_2 - a_2 b_1.$$

This value is easily remembered by the following scheme.

$$\begin{matrix} & + & \\ \begin{bmatrix} a_1 & b_1 \\ a_2 & b_2 \end{bmatrix} & & \\ & - & \end{matrix}$$

ADDITIONAL INFORMATION

Algebraic Properties of Matrices

Two matrices are **equal** if and only if they have the same dimensions and their corresponding elements are equal.

Addition of matrices is defined only on sets of matrices which have the same dimensions. The **sum** of two m by n matrices is an m by n matrix whose elements are the sums of the corresponding elements of the given matrices.

If $A = \begin{pmatrix} a & b \\ c & d \\ e & f \end{pmatrix}$ and $B = \begin{pmatrix} u & v \\ w & x \\ y & z \end{pmatrix}$, then $A + B = \begin{pmatrix} a+u & b+v \\ c+w & d+x \\ e+y & f+z \end{pmatrix}$.

The m by n matrix each of whose elements is zero is called the **zero matrix** for that dimension and denoted by 0. Since $0 + A = A + 0 = A$, for any given matrix A, the zero matrix is the **identity for addition**.

Any matrix whose elements are the negatives of the corresponding elements of a given matrix A is called the negative of A and denoted by $-A$. Since $-A + A = A + (-A) = 0$, $-A$ is the **inverse for addition**.

Scalar multiplication is a binary operation which is performed on elements of unlike sets (real numbers and matrices) and yields an element in one of them (matrix). The **scalar product** of a real number k and a matrix A is a matrix whose elements are k times the corresponding elements of A.

If $A = \begin{pmatrix} a & b & c \\ d & e & f \end{pmatrix}$, then $kA = k\begin{pmatrix} a & b & c \\ d & e & f \end{pmatrix} = \begin{pmatrix} ka & kb & kc \\ kd & ke & kf \end{pmatrix}$.

Multiplication of two matrices is defined only when the number of columns of the first matrix equals the number of rows of the second matrix. Let A be an m by k matrix and B be a k by n matrix. The **product** of the two matrices, denoted AB, is an m by n matrix whose elements are such that the entry in any row i and any column j is the sum of the products formed by multiplying each element in row i of A by the corresponding element in column j of B.

$A = \begin{pmatrix} a & b \\ c & d \\ e & f \end{pmatrix}$ and $B = \begin{pmatrix} g & h \\ i & j \end{pmatrix}$, then $AB = \begin{pmatrix} ag+bi & ah+bj \\ cg+di & ch+dj \\ eg+ei & eh+fj \end{pmatrix}$.

The n by n matrix whose elements in the principal diagonal are all ones and whose other elements are all zeros is called the **identity matrix** for that dimension and denoted by I. Since $IA = AI = A$, for any square matrix A, I is the **identity for multiplication**. This identity is defined only for sets of square matrices of the same dimension.

The **multiplicative inverse** of an n by n matrix A is an n by n matrix, denoted by A^{-1}, which satisfies the equations $A^{-1}A = I$ and $AA^{-1} = I$. Inverses are defined only for square matrices and

From the product of the elements of the principal diagonal subtract the product of the elements of the other diagonal.

The **value of a third-order determinant** is defined as

$$\begin{vmatrix} a_1 & b_1 & c_1 \\ a_2 & b_2 & c_2 \\ a_3 & b_3 & c_3 \end{vmatrix} = a_1b_2c_3 + a_2b_3c_1 + a_3b_1c_2 \\ - a_3b_2c_1 - a_2b_1c_3 - a_1b_3c_2$$

Schematically:

$$\begin{array}{ccc|cc} \overset{+}{a_1} & \overset{+}{b_1} & \overset{+}{c_1} & a_1 & b_1 \\ a_2 & b_2 & c_2 & a_2 & b_2 \\ a_3 & b_3 & c_3 & a_3 & b_3 \\ \underset{-}{} & \underset{-}{} & \underset{-}{} & & \end{array}$$

Rewrite the first two columns as shown. Combine the products of the diagonals, prefixing the indicated signs to each product.

THEOREM 14-1: If all corresponding rows and columns of a matrix are interchanged, its determinant is unchanged.

$$\begin{vmatrix} a_1 & b_1 \\ a_2 & b_2 \end{vmatrix} = \begin{vmatrix} a_1 & a_2 \\ b_1 & b_2 \end{vmatrix}$$

THEOREM 14-2: If any two columns (or any two rows) of a matrix are interchanged, the sign of the determinant is changed.

$$\begin{vmatrix} a_1 & b_1 & c_1 \\ a_2 & b_2 & c_2 \\ a_3 & b_3 & c_3 \end{vmatrix} = -\begin{vmatrix} b_1 & a_1 & c_1 \\ b_2 & a_2 & c_2 \\ b_3 & a_3 & c_3 \end{vmatrix}$$

THEOREM 14-3: If two columns (or two rows) of a matrix are identical, its determinant is zero.

$$\begin{vmatrix} a_1 & a_1 & b_1 \\ a_2 & a_2 & b_2 \\ a_3 & a_3 & b_3 \end{vmatrix} = 0$$

THEOREM 14-4: If each element of a column (or a row) of a matrix is multiplied by the same number, the determinant is multiplied by that number.

$$\begin{vmatrix} a_1 & kb_1 \\ a_2 & kb_2 \end{vmatrix} = k\begin{vmatrix} a_1 & b_1 \\ a_2 & b_2 \end{vmatrix}$$

exist if and only if the determinant of the matrix is not equal to zero.

It can be shown that addition of matrices is both commutative and associative and that multiplication is distributive over addition. However, multiplication is not commutative and does not satisfy the cancellation law.

Two m by n matrices are **row-equivalent** if each can be obtained from the other by row operations. A **row operation** is an operation performed on each entry of a row of a matrix. There are three basic row operations:

1. Multiplying each entry in a row by a nonzero constant.
2. Adding to each entry in one row the same multiple of corresponding entries in another row.
3. Interchanging any two rows.

The following example shows the use of row operations in transforming a matrix into the identity matrix.

$$\begin{pmatrix} 0 & 1 & 2 \\ 1 & -3 & 0 \\ -1 & 1 & 0 \end{pmatrix} \equiv \begin{pmatrix} 0 & 1 & 2 \\ 1 & 0 & 6 \\ 0 & -2 & 0 \end{pmatrix} \quad \begin{array}{l} 3 \text{ (row 1)} + \text{row 2} \\ \text{row 2} + \text{row 3} \end{array}$$

$$\equiv \begin{pmatrix} 0 & 1 & 2 \\ 1 & 0 & 6 \\ 0 & 0 & 4 \end{pmatrix} \quad 2\text{ (row 1)} + \text{row 3} \quad \equiv \begin{pmatrix} 0 & 1 & 2 \\ 1 & 0 & 6 \\ 0 & 0 & 1 \end{pmatrix} \quad \tfrac{1}{4}\text{(row 3)}$$

$$\equiv \begin{pmatrix} 0 & 1 & 0 \\ 1 & 0 & 0 \\ 0 & 0 & 1 \end{pmatrix} \quad \begin{array}{l} -2 \text{ (row 3)} + \text{row 1} \\ -6 \text{ (row 3)} + \text{row 2} \end{array} \equiv \begin{pmatrix} 1 & 0 & 0 \\ 0 & 1 & 0 \\ 0 & 0 & 1 \end{pmatrix}$$

The last matrix is obtained by interchanging rows 1 and 2 of the preceding matrix.

Evaluating Determinants of Higher Order

The **minor** of any element of an nth-order determinant is the $(n-1)$st-order determinant obtained by deleting the row and column of that element. The **cofactor** of an element is the product of the minor of that element and $(-1)^{i+j}$, where i is the number of the row and j is the number of the column of the element.

The **value of any determinant** is the sum of the products obtained by multiplying the elements of any row (or any column) by their cofactors.

$$\begin{vmatrix} a_1 & b_1 & c_1 & d_1 \\ a_2 & b_2 & c_2 & d_2 \\ a_3 & b_3 & c_3 & d_3 \\ a_4 & b_4 & c_4 & d_4 \end{vmatrix} = a_1\begin{vmatrix} b_2 & c_2 & d_2 \\ b_3 & c_3 & d_3 \\ b_4 & c_4 & d_4 \end{vmatrix} - a_2\begin{vmatrix} b_1 & c_1 & d_1 \\ b_3 & c_3 & d_3 \\ b_4 & c_4 & d_4 \end{vmatrix}$$
$$+ a_3\begin{vmatrix} b_1 & c_1 & d_1 \\ b_2 & c_2 & d_2 \\ b_4 & c_4 & d_4 \end{vmatrix} - a_4\begin{vmatrix} b_1 & c_1 & d_1 \\ b_2 & c_2 & d_2 \\ b_3 & c_3 & d_3 \end{vmatrix}$$

SOLUTIONS TO EXERCISES

I

i. $\begin{pmatrix} a+3 \\ 2b-2 \\ c+7 \end{pmatrix} = \begin{pmatrix} -3 \\ 7 \\ 9 \end{pmatrix}$ $a+3 = -3 \longrightarrow a = -6$
$2b-2 = 7 \longrightarrow b = \frac{9}{2}$
$c+7 = 9 \longrightarrow c = 2$

ii. $\begin{pmatrix} 3a & 4 \\ 2b+3 & c \\ 3d-2 & b \end{pmatrix} = \begin{pmatrix} 12+a & 4 \\ 11 & 3c-6 \\ 5d+7 & 4 \end{pmatrix}$

$3a = 12 + a \longrightarrow a = 6$
$4 = 4$
$2b + 3 = 11 \longrightarrow b = 4$
$c = 3c - 6 \longrightarrow c = 3$
$3d - 2 = 5d + 7 \longrightarrow d = -\frac{9}{2}$
$b = 4$

II

i. $A + B = \begin{pmatrix} 2 & -5 \\ 6 & 7 \end{pmatrix} + \begin{pmatrix} -4 & -1 \\ 3 & 5 \end{pmatrix} = \begin{pmatrix} -2 & -6 \\ 9 & 12 \end{pmatrix}$

ii. $-C = \begin{pmatrix} -7 & -3 \\ -8 & 4 \\ -1 & -2 \end{pmatrix}$

iii. $2B + A = \begin{pmatrix} -8 & -2 \\ 6 & 10 \end{pmatrix} + \begin{pmatrix} 2 & -5 \\ 6 & 7 \end{pmatrix} = \begin{pmatrix} -6 & -7 \\ 12 & 17 \end{pmatrix}$

iv. $AB = \begin{pmatrix} 2 & -5 \\ 6 & 7 \end{pmatrix} \begin{pmatrix} -4 & -1 \\ 3 & 5 \end{pmatrix}$

$= \begin{pmatrix} 2(-4) - 5(3) & 2(-1) - 5(5) \\ 6(-4) + 7(3) & 6(-1) + 7(5) \end{pmatrix}$

$= \begin{pmatrix} -23 & -27 \\ -3 & 29 \end{pmatrix}$

v. $CB = \begin{pmatrix} 7 & 3 \\ 8 & -4 \\ 1 & 2 \end{pmatrix} \begin{pmatrix} -4 & -1 \\ 3 & 5 \end{pmatrix}$

$= \begin{pmatrix} 7(-4) + 3(3) & 7(-1) + 3(5) \\ 8(-4) - 4(3) & 8(-1) - 4(5) \\ 1(-4) + 2(3) & 1(-1) + 2(5) \end{pmatrix}$

$= \begin{pmatrix} 19 & 8 \\ -44 & -28 \\ 2 & 9 \end{pmatrix}$

vi. BC is not defined.

III

i. $\begin{vmatrix} 5 & -6 \\ 9 & -3 \end{vmatrix} = 5(-3) - 9(-6) = 39$

ii.

$\begin{vmatrix} 8 & 7 & 2 \\ -3 & 1 & 6 \\ 4 & 5 & 3 \end{vmatrix}$

$= 8(1)(3) + 7(6)(4) + 2(-3)(5)$
$\quad - 4(1)(2) - 5(6)(8) - 3(-3)(7)$
$= -23$

iii. Expand by the first column.

$\begin{vmatrix} 5 & 9 & 0 & 1 \\ 0 & 5 & 1 & -2 \\ 6 & 3 & -4 & 10 \\ 0 & 5 & 3 & 1 \end{vmatrix} = 5 \begin{vmatrix} 5 & 1 & -2 \\ 3 & -4 & 10 \\ 5 & 3 & 1 \end{vmatrix} + 6 \begin{vmatrix} 9 & 0 & 1 \\ 5 & 1 & -2 \\ 5 & 3 & 1 \end{vmatrix}$

$= 5(-181) + 6(73) = -467$

IV

i. If A is 2 by 5 and B is 5 by 3 then AB is 2 by 3.

ii. $\begin{pmatrix} 2 & 3 \\ 4 & 5 \end{pmatrix} \neq \begin{vmatrix} 2 & 3 \\ 4 & 5 \end{vmatrix}$ since the left expression is a matrix, which is an array of numbers, and the right expression is its determinant, which is a real number.

iii. If two columns of a matrix are interchanged, its determinant changes in sign only.

15 SYSTEMS OF LINEAR EQUATIONS

SELF-TEST

DIRECTIONS: Write your answers in the numbered regions to the right. To check your answers, turn the page. Study the solutions to the problems you missed, and do the exercises following any problem or group of problems in which you had errors.

1. Solve the following system of linear equations by addition or subtraction.

$$2x - 3y = 6$$
$$x + y = 3$$

2. Solve the following system of equations by substitution.

$$\frac{1}{x} - \frac{3}{y} = 2$$
$$\frac{3}{x} + \frac{4}{y} = -7$$

3. Use elimination to find x and substitute to find the other variables.

$$x + 2y - z = 3$$
$$2x - y + 3z = -4$$
$$3x - 2y + 3z = -1$$

In 4 and 5, solve the systems of linear equations by using determinants.

4. $3x + 2y = 5$
   $2x + 3y = 7$

5. $x + y = 1$
   $x - z = 0$
   $y - 2w = 2$
   $z + w = 1$

6. Use matrices to solve $AX = C$ where

$$A = \begin{pmatrix} 3 & 5 \\ 5 & 6 \end{pmatrix}, X = \begin{pmatrix} x \\ y \end{pmatrix}, C = \begin{pmatrix} 2 \\ 8 \end{pmatrix}$$

In 7-10, determine whether the systems of linear equations are consistent, inconsistent, or dependent.

7. $2x + 3y = 7$
   $x + y = 5$

8. $2x + 3y = 5$
   $2x + 3y = 7$

9. $ax + ay = 2a$
   $2ax + 2ay = 4a$

10. $3x - 2y + 4z = 0$
    $x - y + z = 0$
    $2x - 3y + z = 0$

1. $x =$
   $y =$

2. $x =$
   $y =$

3. $x =$
   $y =$
   $z =$

4. $x =$
   $y =$

5. $x =$
   $y =$
   $z =$
   $w =$

6.

7.

8.

9.

10.

85

SOLUTIONS

1.
$$2x - 3y = 6 \quad (1)$$
$$x + y = 3 \quad (2)$$

Rewrite (1): $\quad 2x - 3y = 6$
Multiply (2) by 3: $\quad \underline{3x + 3y = 9}$
Add: $\quad 5x = 15, \; x = 3$
Multiply (2) by 2: $\quad 2x + 2y = 6$
Subtract from (1): $\quad -5y = 0, \; y = 0$

EXERCISES I

Solve the following systems of linear equations by addition or subtraction.

i. $\quad 2x - 3y = -6$
$\quad\;\; 3x + 2y = 4$

ii. $\quad \dfrac{2}{x} + \dfrac{3}{y} = 29$
$\quad\;\; \dfrac{1}{x} - \dfrac{2}{y} = 4$

---

2. Let $v = \dfrac{1}{x}$ and $w = \dfrac{1}{y}$, then we have

$$v = 2 + 3w \quad (1)$$
$$3v + 4w = -7 \quad (2)$$

Substitute
(1) into (2): $\quad 3(2 + 3w) + 4w = -7$
Solve for w: $\quad 13w = -13, \; w = -1$
Solve for v: $\quad v = 2 + 3w = 2 - 3$
$\quad\quad\quad\quad\quad = -1$
Then: $\quad x = \dfrac{1}{v} = -1$
$\quad y = \dfrac{1}{w} = -1$

3.
$$x + 2y - z = 3 \quad (1)$$
$$2x - y + 3z = -4 \quad (2)$$
$$3x - 2y + 3z = -1 \quad (3)$$
$$3(1): \; 3x + 6y - 3z = 9 \quad (4)$$
$$(2) + (4): \; 5x + 5y = 5 \quad (5)$$
$$(3) + (4): \; 6x + 4y = 8 \quad (6)$$

$(6) - \dfrac{4}{5} (5): \; 2x = 4, \; x = 2$
Sub. in (5): $5y = 5 - 10, \; y = -1$
Sub. in (1): $2 - 2 - 3 = z, \; z = -3$

EXERCISES II

Solve the following systems of equations by substitution.

i. $\quad 3x + y = 0$
$\quad\;\; 2x + 3y - z = 8$
$\quad\;\; 4x - y + 5z = 16$

ii. $\quad ax + y = 4$
$\quad\;\; 2ax - 3y = 3$

iii. $\quad y + 2x = 7$
$\quad\;\; 3x - 2y = 7$

---

4. $3x + 2y = 5 \quad \Delta = \begin{vmatrix} 3 & 2 \\ 2 & 3 \end{vmatrix} = 5$
$2x + 3y = 7$

$N_x = \begin{vmatrix} 5 & 2 \\ 7 & 3 \end{vmatrix} = 1; \; N_y = \begin{vmatrix} 3 & 5 \\ 2 & 7 \end{vmatrix} = 11$

$x = \dfrac{N_x}{\Delta} = \dfrac{1}{5}; \; y = \dfrac{N_y}{\Delta} = \dfrac{11}{5}$

5. $\Delta = \begin{vmatrix} 1 & 1 & 0 & 0 \\ 1 & 0 & -1 & 0 \\ 0 & 1 & 0 & -2 \\ 0 & 0 & 1 & 1 \end{vmatrix} = -1$

$N_x = -3, \; N_y = 2, \; N_z = -3, \; N_w = 2$

$x = \dfrac{N_x}{\Delta} = \dfrac{-3}{-1} = 3; \; y = \dfrac{N_y}{\Delta} = \dfrac{2}{-1} = -2$

$z = \dfrac{N_z}{\Delta} = \dfrac{-3}{-1} = 3; \; w = \dfrac{N_w}{\Delta} = \dfrac{2}{-1} = -2$

EXERCISES III

Solve the following systems of linear equations by using determinants.

i. $\quad \dfrac{1}{x} - \dfrac{3}{y} = 2$

$\quad\;\; \dfrac{3}{x} + \dfrac{2}{y} = 5$

ii. $\quad 2x - y + 3z = 0$
$\quad\;\; x + 2y - 2z = 5$
$\quad\;\; 3x - 3y + z = 1$

6. $A = \begin{pmatrix} 3 & 5 \\ 5 & 6 \end{pmatrix}$, $X = \begin{pmatrix} x \\ y \end{pmatrix}$, $C = \begin{pmatrix} 2 \\ 8 \end{pmatrix}$

$\delta(A) = \begin{vmatrix} 3 & 5 \\ 5 & 6 \end{vmatrix} = 18 - 25 = -7$

$A^{-1} = -\dfrac{1}{7} \begin{pmatrix} 6 & -5 \\ -5 & 3 \end{pmatrix}$

$X = A^{-1} C = -\dfrac{1}{7} \begin{pmatrix} 6 & -5 \\ -5 & 3 \end{pmatrix} \begin{pmatrix} 2 \\ 8 \end{pmatrix}$

$= -\dfrac{1}{7} \begin{pmatrix} 12 - 40 \\ -10 + 24 \end{pmatrix} = -\dfrac{1}{7} \begin{pmatrix} -28 \\ 14 \end{pmatrix} = \begin{pmatrix} 4 \\ -2 \end{pmatrix}$

EXERCISE IV

i. Write the inverse of
$$A = \begin{pmatrix} 1 & 3 \\ 2 & 4 \end{pmatrix}$$

ii. Write $AX - C = 0$ as a system of linear equations if
$$A = \begin{pmatrix} -1 & 7 \\ 3 & 2 \end{pmatrix}, X = \begin{pmatrix} x \\ y \end{pmatrix}, C = \begin{pmatrix} a \\ b \end{pmatrix}.$$

iii. Solve $AX = C$ if
$$A = \begin{pmatrix} 1 & a \\ 1 & b \end{pmatrix}, X = \begin{pmatrix} x \\ y \end{pmatrix}, C = \begin{pmatrix} 1 \\ 1 \end{pmatrix}$$

---

7. $\Delta = \begin{vmatrix} 2 & 3 \\ 1 & 1 \end{vmatrix} = -1$, consistent

8. $\Delta = 0$ and $N_x = -6$, inconsistent

9. $\Delta = 0$, $N_x = 0$, $N_y = 0$; dependent

10. $\Delta = 0$, $N_x = 0$, $N_y = 0$, $N_z = 0$, dependent

EXERCISE V

Determine whether the following system of linear equations is consistent, inconsistent, or dependent.

$x + z = 1$
$z + y = 2$
$x + y = 1$

ANSWERS

1. $x = 3$
   $y = 0$

2. $x = -1$
   $y = -1$

3. $x = 2$
   $y = -1$
   $z = -3$

4. $x = \frac{1}{5}$
   $y = \frac{11}{5}$

5. $x = 3$
   $y = -2$
   $z = 3$
   $w = -2$

6. $X = \begin{pmatrix} 4 \\ -2 \end{pmatrix}$

7. consistent

8. inconsistent

9. dependent

10. dependent

---

1 $x =$
  $y =$

2 $x =$
  $y =$

3 $x =$
  $y =$
  $z =$

4 $x =$
  $y =$

5 $x =$
  $y =$
  $z =$
  $w =$

6

7

8

9

10

Systems of Linear Equations

BASIC FACTS

Two (or more) linear equations in the same two (or more) unknowns are said to form a system of **simultaneous linear equations**. A solution of such a system is a set of values of the variables which satisfy all equations of the system. There are four methods for obtaining such solutions.

Elimination Method

Eliminate each variable in turn by making the absolute value of its coefficient the same in both equations, and then adding or subtracting the equations. The values of the coefficients are made identical by multiplying each equation by an appropriate numerical factor.

Substitution Method

Solve one of the equations for one variable in terms of the others and substitute this value into the other equations. Continue this process until one linear equation in one unknown obtained. Solve this equation for the value of its variable, and substitute that value into a derived equation containing this variable and one other variable. Then solve this equation for the value of the other unknown. Continue the process until the values of all the variables have been obtained.

Graphing Method

If the system is limited to two equations in two unknowns, we can draw the graphs of both equations on the same rectangular coordinate system. The coordinates of the point of intersection of the two lines is the solution.

Determinant Method

Write the system of equations so that only the constant terms are in the right member of each equation. Let Δ be the determinant obtained by placing the coefficients of each variable in a column. If a variable does

ADDITIONAL INFORMATION

Any number of linear equations, each written in terms of the same number of the same variables, is said to form a **system of linear equations**. In general, a **solution** of a system of n linear equations in n variables is a set of n values (one for each variable) which satisfies every equation in the system.

Consider the system of linear equations in two variables shown at the right. To solve this system by the **elimination**

$$a_1 x + b_1 y = c_1 \quad (1)$$
$$a_2 x + b_2 y = c_2 \quad (2)$$

method, multiply (1) by b_2, (2) by $-b_1$, and add the results. This eliminates y, and the resulting equation can be solved for x.

$$(a_1 b_2 - a_2 b_1) x = b_2 c_1 - b_1 c_2 \text{ or } x = \frac{b_2 c_1 - b_1 c_2}{a_1 b_2 - a_2 b_1}, \ a_1 b_2 - a_2 b_1 \neq 0.$$

Similarly, x can be eliminated, and the equations solved for y.

$$(a_1 b_2 - a_2 b_1) y = a_1 c_2 - a_2 c_1 \text{ or } y = \frac{a_1 c_2 - a_2 c_1}{a_1 b_2 - a_2 b_1}, \ a_1 b_2 - a_2 b_1 \neq 0$$

The **substitution method** may be used in combination with any of the other methods. For example, after eliminating y and solving for x, we could substitute the value of x into either equation to find the value of y. This procedure should be used carefully, for if a mistake is made in solving for x, the value found for y will be incorrect. Solutions must be checked in all equations of the system.

The **graphing method** is illustrated in Figure 15-1 in which the equations $y + 2x = 7$ and $2y + 2x = 10$ are graphed. $(2, 3)$, the point of intersection of the two lines, is the solution of the system.

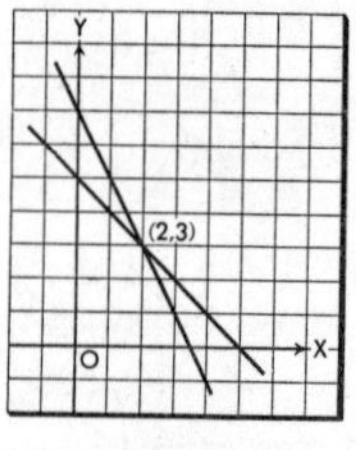

Fig. 15-1

The **determinant method** of solving a system of two linear equations can be justified in terms of the expressions for x and y obtained by the elimination method. The denominators are the same in both equations, and are equivalent to the determinant,

$$\Delta = \begin{vmatrix} a_1 & b_1 \\ a_2 & b_2 \end{vmatrix} = a_1 b_2 - a_2 b_1.$$

This determinant, Δ, is called the **determinant of the system**.

The numerators in the solutions for x and y can also be expressed as determinants.

$$b_2 c_1 - b_1 c_2 = \begin{vmatrix} c_1 & b_1 \\ c_2 & b_2 \end{vmatrix} = N_x \text{ and } a_1 c_2 - a_2 c_1 = \begin{vmatrix} a_1 & c_1 \\ a_2 & c_2 \end{vmatrix} = N_y$$

The determinant method can be extended, and used to solve a system of n linear equations in n unknowns. Let us consider the system and the determinant of the system, Δ.

$$\begin{aligned} a_{11} x_1 + a_{12} x_2 + \ldots + a_{1n} x_n &= c_1 \\ a_{21} x_1 + a_{22} x_2 + \ldots + a_{2n} x_n &= c_2 \\ &\vdots \\ a_{n1} x_1 + a_{n2} x_2 + \ldots + a_{nn} x_n &= c_n \end{aligned} \qquad \Delta = \begin{vmatrix} a_{11} & a_{12} & \ldots & a_{1n} \\ a_{21} & a_{22} & \ldots & a_{2n} \\ \vdots & \vdots & \ldots & \vdots \\ a_{n1} & a_{n2} & \ldots & a_{nn} \end{vmatrix}$$

not appear in an equation, its coefficient is zero.

Suppose the variables of a given system are x, y, z, etc. Let N_x be the determinant obtained by replacing the column of the coefficients of x in Δ by the column of constant terms. Proceed similarly with the remaining variables. Then the solution of the system is $x = \dfrac{N_x}{\Delta}$, $y = \dfrac{N_y}{\Delta}$, $z = \dfrac{N_z}{\Delta}$, etc.

This is known as **Cramer's Rule**.

Equations in Matrix Form

The equation $AX = C$ represents a system of n linear equations in n unknowns, where $A = (a_{ij})$, an $n \times n$ matrix of the coefficients of the variables; $X = (x_i)$, a column matrix of variables; $C = (c_i)$, a column matrix of constants; and $(i, j = 1, \ldots, n)$.

The solution may be expressed by $X = A^{-1}C$ where A^{-1} is the inverse of the matrix A.

A system which has a unique solution is said to be **consistent**; a system which has no solution is said to be **inconsistent**; and a system which has an unlimited number of solutions is said to be **dependent**.

Homogeneous Equations

A linear equation is said to be homogeneous if the constant term is equal to zero. A system of homogeneous linear equations is satisfied if each variable is zero. The solution set whose elements are all zero is called a trivial solution. However, if $\Delta = 0$ in a homogeneous system, then the system has an infinite number of solutions. If one variable is chosen as a parameter, then the other variables may be expressed in terms of it.

More Variables Than Equations

A system of m linear equations in n variables where $n > m$ usually has infinitely many solutions.

Let N_i be the determinant of the system with the column of the coefficients of x_i replaced by the column of constant terms c_i ($i = 1, \ldots, n$). The solutions are then given by

$$x_i = \frac{N_i}{\Delta} \text{ for } i = 1, \ldots, n.$$

In evaluating the determinants recall their properties and obtain zero elements wherever possible.

Matrix Method

Consider the equation $AX = C$ where A, X, and C are matrices. Multiplying the equation by A^{-1} to obtain $A^{-1}AX = IX = X = A^{-1}C$ yields an expression for the solution of the system, $\{x_i = k_i\}$, where k_i is the ith element of the column matrix $A^{-1}C$.

To solve equations by the matrix method it is necessary to determine the **inverse of a matrix**. For a 2×2 matrix, the inverse is defined as follows:

$$A = \begin{pmatrix} a_{11} & a_{12} \\ a_{21} & a_{22} \end{pmatrix} \qquad A^{-1} = \frac{1}{\delta A} \begin{pmatrix} a_{22} & -a_{12} \\ -a_{21} & a_{11} \end{pmatrix}$$

The computation of inverses of higher order matrices is more involved and beyond the scope of this guide.

The matrix method of solution is frequently used when the computation is done with electronic computers.

Two equations in two unknowns are **consistent** if the slopes of the two lines are not equal. In a system of two equations in two unknowns, consistency can also be determined by considering the graphs. If the two lines intersect, there can be only one point of intersection, and therefore only one point $P_1(x_1, y_1)$ which is common to both lines and which satisfies both equations. If the two lines are parallel, they have no point in common and the equations are **inconsistent**. If the two lines coincide, in which case all points satisfy both equations, the system has an infinite number of solutions. In such a case, the equations are **dependent**.

For systems of linear equations with more than two variables, we use the determinant of the system to check the consistency of the equations in the system. If $\Delta \neq 0$, the system is consistent. If $\Delta = 0$ and one $N_i \neq 0$, where $i = x, y, z, \ldots$, the equations are inconsistent. If $\Delta = 0$, and all $N_i = 0$, the equations are dependent.

SOLUTIONS TO EXERCISES

I

$2x - 3y = -6$ (1)
$3x + 2y = 4$ (2)

i. Multiply (1) by 2: $4x - 6y = -12$
Multiply (2) by 3: $9x + 6y = 12$
Add and solve: $13x = 0,\ x = 0$
Multiply (1) by 3: $6x - 9y = -18$
Multiply (2) by 2: $6x + 4y = 8$
Subtract and solve: $-13y = -26$
           $y = 2$

ii. Let $v = \frac{1}{x}$ and $w = \frac{1}{y}$:

$2v + 3w = 29$
$v - 2w = 4$
Eliminate v: $7w = 21,\ w = 3$
Eliminate w: $7v = 70,\ v = 10$

Then $x = \frac{1}{v} = \frac{1}{10}$ and $y = \frac{1}{w} = \frac{1}{3}$.

II

i.
$3x + y = 0$ (1)
$2x + 3y - z = 8$ (2)
$4x - y + 5z = 16$ (3)

Solve (1) for y: $y = -3x$
Substitute into
(2) and (3): $2x - 9x - z = 8$
             $4x + 3x + 5z = 16$
Solve (2) for z: $z = -7x - 8$
Substitute into (3): $7x - 35x - 40 = 16$
Solve for x: $x = -2$
Substitute: $z = -7(-2) - 8 = 6$
Substitute: $y = -3(-2) = 6$

ii.
$ax + y = 4$ (1)
$2ax - 3y = 3$ (2)

Solve (1) for y: $y = 4 - ax$
Substitute in (2): $2ax - 12 + 3ax = 3$

$x = \frac{3}{a}$

Solve for y: $y = 4 - a\left(\frac{3}{a}\right) = 1$

iii.
$y + 2x = 7$
$3x - 2y = 7$
$y = 7 - 2x$
$3x - 14 + 4x = 7,\ x = 3$
$y = 7 - 2(3) = 7 - 6 = 1$

III

i. $\Delta = \begin{vmatrix} 1 & -3 \\ 3 & 2 \end{vmatrix} = 11$

$N_x = \begin{vmatrix} 2 & -3 \\ 5 & 2 \end{vmatrix} = 19$

$\frac{1}{x} = \frac{N_x}{\Delta} = \frac{19}{11},\ x = \frac{11}{19}$

$N_y = \begin{vmatrix} 1 & 2 \\ 3 & 5 \end{vmatrix} = -1,\ \frac{1}{y} = \frac{-1}{11},\ y = -11$

ii. $\Delta = \begin{vmatrix} 2 & -1 & 3 \\ 1 & 2 & -2 \\ 3 & -3 & 1 \end{vmatrix} = -28$

$N_x = -44,\ N_y = -28,\ N_z = 20$

$x = \frac{44}{28} = \frac{11}{7},\ y = \frac{28}{28} = 1,\ z = \frac{20}{-28} = -\frac{5}{7}$

IV

i. $A = \begin{pmatrix} 1 & 3 \\ 2 & 4 \end{pmatrix};\ \delta A = -2$

Hence

$A^{-1} = -\frac{1}{2}\begin{pmatrix} 4 & -3 \\ -2 & 1 \end{pmatrix} = \begin{pmatrix} -2 & \frac{3}{2} \\ 1 & -\frac{1}{2} \end{pmatrix}$

ii. $AX - C = 0$

$AX = \begin{pmatrix} -1 & 7 \\ 3 & 2 \end{pmatrix}\begin{pmatrix} x \\ y \end{pmatrix} = \begin{pmatrix} -x + 7y \\ 3x + 2y \end{pmatrix}$

$-x + 7y - a = 0$
$3x + 2y - b = 0$

iii. $AX = C$

$A = \begin{pmatrix} 1 & a \\ 1 & b \end{pmatrix},\ X = \begin{pmatrix} x \\ y \end{pmatrix},\ C = \begin{pmatrix} 1 \\ 1 \end{pmatrix}$

$\delta(A) = \begin{vmatrix} 1 & a \\ 1 & b \end{vmatrix} = b - a$

$X = A^{-1}C = \frac{1}{b-a}\begin{pmatrix} b & -a \\ -1 & 1 \end{pmatrix}\begin{pmatrix} 1 \\ 1 \end{pmatrix}$

$= \frac{1}{b-a}\begin{pmatrix} b & -a \\ -1 & +1 \end{pmatrix} = \begin{pmatrix} 1 \\ 0 \end{pmatrix}$

V

$\Delta = 2,\ \therefore \text{consistent}$

16 THE QUADRATIC EQUATION

SELF-TEST

DIRECTIONS: Write your answers in the numbered regions to the right. To check your answers, turn the page. Study the solutions to the problems you missed, and do the exercises following any problem or group of problems in which you had errors.

1. Solve $x^2 - x - 6 = 0$ for x by factoring.

2. Solve $a = 2bx - x^2$ for x by completing the square.

In 3 and 4, solve for x by using the quadratic formula.

3. $x^2 - 4x - 8 = 0$

4. $\dfrac{2x}{x-2} + \dfrac{3}{2+x} = 1$

5. Find k if the two roots of $x^2 - kx + 4 = 0$ are equal.

6. Find k if the discriminant of $x^2 - 4x + k = 0$ is the smallest possible perfect square such that k is an integer.

7. Find the x-intercepts and the vertex of the parabola defined by $y = -x^2 + 2x + 3$.

In 8–11, complete the given statements.

8. If the discriminant of a quadratic equation is zero, the roots are _____ and _____.

9. The graph of the quadratic function defined by $y = 3x^2 + bx + c$ is a _____.

10. If $c = 0$, the roots of $ax^2 + bx + c = 0$ are _____ and _____.

11. The equation $\sqrt{2x+1} + 1 = 0$ has _____ roots.

SOLUTIONS

1. Recall the typical products from Chapter 9.
$x^2 - x - 6 = 0$
$(x - 3)(x + 2) = 0$
$x = 3, x = -2$

EXERCISES I

In i–iv, solve for x by factoring.

i. $6x^2 = 18x$

ii. $4x^2 - 25 = 0$

iii. $6x^2 - x - 12 = 0$

iv. $15x^2 + 16x - 15 = 0$

---

2. $a = 2bx - x^2$
$x^2 - 2bx + b^2 = b^2 - a$
$(x - b)^2 = b^2 - a$
$x - b = \pm \sqrt{b^2 - a}$
$x = b \pm \sqrt{b^2 - a}$

EXERCISES II

In i–iv, solve for x by completing the square.

i. $x^2 - 2x - 8 = 0$

ii. $2x^2 - 4x - 7 = 0$

iii. $x^2 + ax - 4 = 0$

iv. $3x^2 + x + 3 = 0$

The quadratic formula for the equation $ax^2 + bx + c = 0$ is $\dfrac{-b \pm \sqrt{b^2 - 4ac}}{2}$.

3. $x^2 - 4x - 8 = 0$
$x = \dfrac{4 \pm \sqrt{16 + 32}}{2} = \dfrac{4 \pm 4\sqrt{3}}{2}$
$= 2 \pm 2\sqrt{3}$

4. $\dfrac{2x}{x - 2} + \dfrac{3}{2 + x} = 1$. Clear fractions:
$2x(x + 2) + 3(x - 2) = x^2 - 4$
$4x^2 + 4x + 3x - 6 = x^2 - 4$
$3x^2 + 7x - 2 = 0$
$x = \dfrac{-7 \pm \sqrt{49 + 24}}{6} = \dfrac{-7 \pm \sqrt{73}}{6}$

EXERCISES III

In i–vi, solve for x by using the quadratic formula.

i. $7x^2 + 3x = 0$

ii. $6x^2 - x - 2 = 0$

iii. $2x^2 + 3x = 2$

iv. $ax^2 + 2ax = 3$

v. $3x^2 - 6x + 2 = 0$

vi. $x^2 + x + 1 = 0$

---

5. If $b^2 - 4ac = 0$, the roots are equal.
$k^2 - 16 = 0$ or $k = \pm 4$

6. $b^2 - 4ac = 16 - 4k$, k is an integer.

Values of k	0	1	2	3	4
$b^2 - 4ac$	16	12	8	4	0

Smallest perfect square > 0: 4 at $k = 3$

7. $y = -x^2 + 2x + 3$
$V(1, 4)$; x-intercepts at $x = 3, -1$

EXERCISES IV

i. Find k if the sum of the roots of $3x^2 + k^2x - 5 = 0$ equals -3.

ii. Find k if the product of the roots of $kx^2 - 3x + k^2 - 28 = 0$ equals 3.

iii. Find k and m if the sum of the roots of $2x^2 + kx + mx + 3m - k = 0$ is 5 and the product of the roots is 7.

iv. Find the quadratic equation whose roots are 5 and -3.

v. Find the vertex of the parabola defined by the equation $y = x^2 - x - 6$.

vi. Find the values of x which satisfy the equation $3\sqrt{2x + 1} - \sqrt{x - 3} = 3\sqrt{x + 4}$.

8. If the discriminant of a quadratic equation is zero, the roots are <u>real</u> and <u>equal</u>.

9. The graph of the quadratic function defined by $y = 3x^2 + bx + c$ is a <u>parabola</u>.

10. If $c = 0$, the roots of $ax^2 + bx + c = 0$ are $\underline{0}$ and $\underline{-\dfrac{b}{a}}$.

11. The equation $\sqrt{2x+1} + 1 = 0$ has <u>no</u> roots.

EXERCISES V

In i–v, complete the given statements.

i. A quadratic equation has _____ roots.

ii. If the discriminant of a quadratic equation is a perfect square, the roots are _____.

iii. The product of the roots of the equation $x^2 + bx + c = 0$ equals _____.

iv. The roots of a quadratic equation will be complex numbers if the discriminant is _____.

v. If $a = 0$, the equation $ax^2 + bx + c = 0$ is _____.

Solutions to the exercises appear on page 96.

ANSWERS

1. $3, -2$

2. $b \pm \sqrt{b^2 - a}$

3. $2 \pm 2\sqrt{3}$

4. $\dfrac{-7 \pm \sqrt{73}}{6}$

5. ± 4

6. 3

7. $V(1, 4)$

   $3, -1$

8. real, equal

9. parabola

10. $0, -\dfrac{b}{a}$

11. no

The Quadratic Equation

BASIC FACTS

The equation,
$$ax^2 + bx + c = 0$$
where $a \neq 0$ and a, b, and c are constants, is called the **general quadratic equation** in x. The highest degree of the variable is two.

If $b = 0$, the equation becomes
$$ax^2 + c = 0$$
and is called a **pure quadratic equation**.

If $a = 0$, the equation reduces to a linear equation.

The values of x which satisfy the quadratic equation can be found by substituting into the **quadratic formula**.

$$x = \frac{-b \pm \sqrt{b^2 - 4ac}}{2a}$$

The expression, $b^2 - 4ac$, is called the **discriminant** of the quadratic equation. If the coefficients a, b, and c are real numbers, some characteristics of the roots of the equation can be determined from the value of the discriminant.

if	then the roots are
$b^2 - 4ac > 0$	real and unequal
$b^2 - 4ac = 0$	real and equal
$b^2 - 4ac < 0$	not real

If the roots are real and a, b, and c are rational numbers, further characteristics of the roots of the equation can be determined from the discriminant.

if $b^2 - 4ac$	the roots are
is a perfect square	rational
is not a perfect square	irrational

The sum of the roots of a quadratic equation

$$r_1 + r_2 = -\frac{b}{a}$$

ADDITIONAL INFORMATION

Any equation in which the highest degree of the variable is two can be put into the form of the **general quadratic equation** by collecting the quadratic terms, the linear terms, and the constants. There are three algebraic methods of finding the **solutions** of quadratic equations.

The Factoring Method

Set the given quadratic equation equal to zero, and transform it into the product of two linear factors. Since a product is equal to zero if either or both of its factors is zero, we may set each factor equal to zero, and find the roots of the resulting two linear equations.

Completing the Square

The process of completing the square facilitates the solution of a quadratic equation by making a perfect square of the member of the equation which contains the variable. Since an expression of the form $x^2 + ax + \left(\frac{1}{2}a\right)^2$ is a perfect square, the completion of the square can be accomplished by writing an equation in such a way that the coefficient of the x^2-term is 1 and the constant term appears alone in the right member of the equation; then one need only add the square of half the coefficient of the x-term to both members of the equation in order to make the left member a perfect square. If one then takes the square root of both members of the equation, the left member will be x plus some constant, and one need only transpose to solve for x.

The Quadratic Formula

By applying the process of completing the square to the general quadratic equation, it is possible to derive a formula which can be used to solve all quadratic equations.

The General Quadratic Equation:	$ax^2 + bx + c = 0$, $a \neq 0$
Transpose the constant term to the right side:	$ax^2 + bx = -c$
Divide by the coefficient of x^2:	$x^2 + \frac{b}{a}x = -\frac{c}{a}$
Take $\frac{1}{2}$ the coefficient of x, square it, and add to both sides:	$x^2 + \frac{b}{a}x + \left(\frac{b}{2a}\right)^2 = \frac{b^2}{4a^2} - \frac{c}{a}$
Factor the left side and simplify the right side:	$\left(x + \frac{b}{2a}\right)^2 = \frac{b^2 - 4ac}{4a^2}$
Extract the square root of both members:	$x + \frac{b}{2a} = \frac{\pm\sqrt{b^2 - 4ac}}{2a}$
Solve for x:	$x = \frac{-b \pm \sqrt{b^2 - 4ac}}{2a}$

If the two **roots** given by the quadratic formula are **added**, we find that $r_1 + r_2 = -\frac{b}{a}$. If the roots are **multiplied**, we find that $r_1 r_2 = \frac{c}{a}$.

The graph of the quadratic function defined by $y = f(x) = ax^2 + bx + c$ is a **parabola**. The graph may be obtained by calculating a table of ordered pairs (x, y), plotting these points on the rectangular

The product of the roots of a quadratic equation

$$r_1 r_2 = \frac{c}{a}$$

The graph of the function, $y = f(x) = ax^2 + bx + c$, $a \neq 0$ is a **parabola**. The **vertex** of the parabola is the point,

$$\left(-\frac{b}{2a}, \frac{4ac - b^2}{4a}\right)$$

and the graph is symmetric with respect to the line,

$$x = -\frac{b}{2a}.$$

If a is positive, the parabola opens upward, and the vertex is the lowest (**minimum**) point on the graph. If a is negative, the parabola opens downward, and the vertex is the highest (**maximum**) point on the graph.

The parabola must either cross the X-axis in two distinct points, be tangent to the X-axis, or not touch the X-axis at all.

The x-intercepts of the parabola are the values of x which satisfy the quadratic equation $ax^2 + bx + c = 0$. The x-intercepts are the roots of the equation and can be approximated graphically.

Some equations of degree higher than two may be written in quadratic form in a new variable. For example, the equation $px^4 + qx^2 + r = 0$, which can be written in the form $pv^2 + qv + r = 0$, is a quadratic equation in $v = x^2$.

Fractional and irrational equations may be changed to quadratic equations by performing the usual operations of multiplying by the lowest common denominator or squaring both sides of the equation.

When an equation is multiplied by one of its variables, be sure to check the solutions in the original equation.

coordinate system, and joining the points with a smooth curve. Two examples are shown in Figures 16-1 and 16-2.

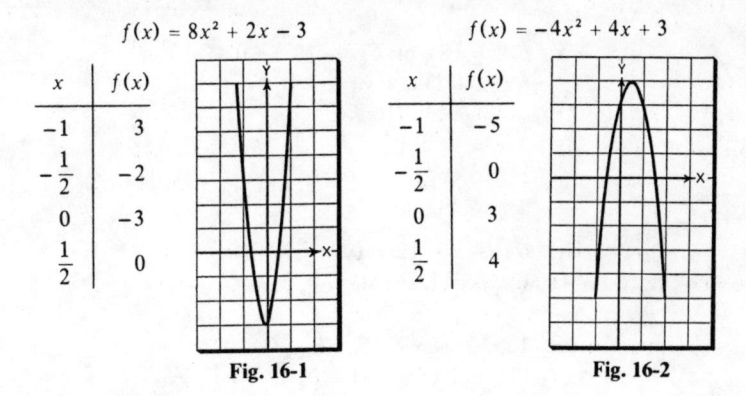

$f(x) = 8x^2 + 2x - 3$

x	$f(x)$
-1	3
$-\frac{1}{2}$	-2
0	-3
$\frac{1}{2}$	0

$f(x) = -4x^2 + 4x + 3$

x	$f(x)$
-1	-5
$-\frac{1}{2}$	0
0	3
$\frac{1}{2}$	4

Fig. 16-1   Fig. 16-2

Some characteristics of the roots of $y = f(x)$ can be determined graphically.

if the graph of $y = f(x)$	the roots of $f(x) = 0$ are
crosses the X-axis in two points	real and unequal
touches the X-axis in only one point	real and equal
does not touch the X-axis at all	not real

The techniques for solving quadratic equations can be applied to equations which can be put into quadratic form. For example, consider the equation $(2x + 1)^4 - 4(2x + 1)^2 - 5 = 0$. Let $v = (2x + 1)^2$, then the equation is $v^2 - 4v - 5 = 0$ with solutions $v = 5$ and $v = -1$. We have,

$$(2x + 1)^2 = 5 \qquad (2x + 1)^2 = -1$$
$$4x^2 + 4x - 4 = 0 \qquad 4x^2 + 4x + 2 = 0$$
$$x = -\frac{1}{2} \pm \frac{\sqrt{5}}{2} \qquad x = -\frac{1}{2} \pm \frac{1}{2}i.$$

Fractional Equations

Consider the equation $\frac{x-4}{x-2} + \frac{x}{x+2} = 1$. If we multiply by the LCD, $x^2 - 4$, we have $x^2 - 2x - 8 + x^2 - 2x = x^2 - 4$; or $x^2 - 4x - 4 = 0$ with solutions $x = 2 \pm 2\sqrt{2}$. Since we multiplied by the variable, we may have introduced extraneous roots; and the solutions must be checked in the original equation.

Irrational Equations

Equations in which the variable appears in one or more radicals are called irrational equations. The solutions of these equations are obtained by raising both members of the equation to a power a sufficient number of times to remove all the radicals. Each such operation is a multiplication with the variable and thus the solutions must be checked to eliminate any extraneous roots.

SOLUTIONS TO EXERCISES

I

i. $6x^2 = 18x$ or $6x^2 - 18x = 0$
$6x(x - 3) = 0$ or $x = 0, 3$

ii. $4x^2 - 25 = 0$
$(2x - 5)(2x + 5) = 0$, $x = \frac{5}{2}, -\frac{5}{2}$

iii. $6x^2 - x - 12 = 0$
$(2x - 3)(3x + 4) = 0$, $x = \frac{3}{2}, -\frac{4}{3}$

iv. $15x^2 + 16x - 15 = 0$
$(3x + 5)(5x - 3) = 0$, $x = -\frac{5}{3}, \frac{3}{5}$

II

i. $x^2 - 2x - 8 = 0$
Add $\left(\frac{1}{2} \text{ coefficient of } x\right)^2$:
$x^2 - 2x + 1 = 8 + 1 = 9$
$(x - 1)^2 = 9$, $x = 4, -2$

ii. $2x^2 - 4x - 7 = 0$
$2(x^2 - 2x + 1) = 7 + 2 = 9$
$2(x - 1)^2 = 9$, $x = 1 \pm \frac{9\sqrt{2}}{2}$

iii. $x^2 + ax - 4 = 0$
$x^2 + ax + \frac{a^2}{4} = 4 + \frac{a^2}{4}$
$\left(x + \frac{a}{2}\right)^2 = \frac{1}{4}(16 + a^2)$
$x + \frac{a}{2} = \pm \frac{1}{2}\sqrt{16 + a^2}$
$x = -\frac{a \pm \sqrt{16 + a^2}}{2}$

iv. $3x^2 + x + 3 = 0$
$3\left(x^2 + \frac{1}{3}x + \frac{1}{36}\right) = -3 + \frac{3}{36}$
$x = \frac{-1 \pm \sqrt{35}\, i}{6}$

III

i. $7x^2 + 3x = 0$
$x = \frac{-3 \pm \sqrt{9 - 0}}{14} = \frac{-3 \pm 3}{14} = 0, -\frac{3}{7}$

ii. $6x^2 - x - 2 = 0$
$x = \frac{1 \pm \sqrt{1 + 48}}{12} = \frac{1 \pm 7}{12} = \frac{2}{3}, -\frac{1}{2}$

iii. $2x^2 + 3x - 2 = 0$
$x = \frac{-3 \pm \sqrt{9 + 16}}{4} = \frac{-3 \pm 5}{4} = -2, \frac{1}{2}$

iv. $ax^2 + 2ax - 3 = 0$
$x = \frac{-2a \pm \sqrt{4a^2 + 12a}}{2a}$
$= -1 \pm \frac{\sqrt{a^2 + 3a}}{a}$

v. $3x^2 - 6x + 2 = 0$
$x = \frac{6 \pm \sqrt{36 - 24}}{6} = 1 \pm \frac{\sqrt{3}}{3}$

vi. $x^2 + x + 1 = 0$
$x = \frac{-1 \pm \sqrt{1 - 4}}{2} = -\frac{1}{2} \pm \frac{\sqrt{3}}{2} i$

IV

i. $r_1 + r_2 = -\frac{b}{a}$, $-\frac{k^2}{3} = -3$, $k = \pm 3$

ii. $r_1 r_2 = \frac{c}{a}$, $\frac{k^2 - 28}{k} = 3$
$(k - 7)(k + 4) = 0$, $k = 7, -4$

iii. $-\frac{k + m}{2} = 5$ and $\frac{3m - k}{2} = 7$, $m = 1$, $k = -11$

iv. $(x - 5)(x + 3) = 0$, $x^2 - 2x - 15 = 0$

v. Vertex at $\left(-\frac{b}{2a}, \frac{4ac - b^2}{4a}\right)$
$-\frac{b}{2a} = \frac{1}{2}$, $\frac{4ac - b^2}{4a} = -\frac{25}{4}$

vi. $3\sqrt{2x + 1} - \sqrt{x - 3} = 3\sqrt{x + 4}$
Square both sides, isolate the radical, and square both sides again. Then $x = 3$ and 12 and both check.

V

i. A quadratic equation has <u>two</u> roots.
ii. If the discriminant of a quadratic equation is a perfect square, the roots are <u>rational</u>.
iii. <u>The</u> product of the roots of the equation $x^2 + bx + c = 0$ equals <u>c</u>.
iv. The roots of a quadratic equation will be complex numbers if the discriminant is <u>negative</u>.
v. If <u>$a = 0$</u>, the equation $ax^2 + bx + c = 0$ is <u>linear</u>.

17 SYSTEMS OF QUADRATIC EQUATIONS

SELF-TEST

DIRECTIONS: Write your answers in the numbered regions to the right. To check your answers, turn the page. Study the solutions to the problems you missed, and do the exercises following any problem or group of problems in which you had errors.

Using the following classification, identify the case to which the systems of **1–4** belong. It is not necessary to solve the systems.

CASE I: One equation is linear.

CASE II: $ax^2 + cy^2 + f = 0$
$bxy + g = 0$

CASE III: Both equations are of the form $ax^2 + cy^2 + f = 0$.

CASE IV: $ax^2 + cy^2 + f = 0$
$Ax^2 + Ey + F = 0$ or $Cy^2 + Dx + F = 0$

CASE V: Both equations are of the form $ax^2 + bxy + cy^2 + f = 0$.

CASE VI: Both equations are of the form
$A(x^2 + y^2) + Bxy + D(x + y) = F$.

1. $x^2 + y^2 = 3x + 3y - 7$
   $5x + 5y = 2x^2 + 2y^2 - 3$

2. $x - 3y^2 = 5$
   $y + x = 7$

3. $16xy = 2$
   $3x^2 + 144y^2 = 15$

4. $x^2 - y^2 = 17$
   $3x^2 + 2y^2 = 15$

In **5–10**, find the pairs (x, y) which satisfy the given system of equations.

5. $y^2 + 3x - 5y = 30$
   $3x - 2y = 12$

6. $x^2 + 4y^2 = 20$
   $xy = 4$

7. $x^2 + y^2 = 25$
   $x^2 - y^2 = 7$

8. $x^2 + 2y^2 - 2 = 0$
   $x^2 - y - 1 = 0$

9. $2x^2 + 3xy + 2y^2 = -1$
   $x^2 + 2xy + 3y^2 = -2$

10. $x^2 - 2xy + y^2 + x + y = 8$
    $2x^2 - 4xy + 2y^2 - x - y = 4$

97

SOLUTIONS

1. $x^2 + y^2 = 3x + 3y - 7$
   $5x + 5y = 2x^2 + 2y^2 - 3$, Case VI

2. $x - 3y^2 = 5$
   $y + x = 7$, Case I

3. $16xy = 2$
   $3x^2 + 144y^2 = 15$, Case II

4. $x^2 - y^2 = 17$
   $3x^2 + 2y^2 = 15$, Case III

EXERCISES I

Using the classification given on page 97, identify the case to which each system belongs. It is not necessary to solve the system.

i. $17x^2 = 14y^2 - 3$
   $3y + 2 = 3x^2$

ii. $y^2 = 2x^2 - 3xy + 12$
    $2y^2 + x^2 = 3xy$

iii. $2xy - 45 = 0$
     $2x = 5y$

iv. $16xy + y^2 = x^2 - 3$
    $x^2 - 3xy = y^2 + 2$

v. $3x^2 + 2xy - 2x = 2y - 3y^2 + 3$
   $2y^2 + 3xy + 3y = 3xy - 2x^2 + 7$

vi. $2x^2 + 2y^2 + 3xy + 5x + 5y = 2$
    $x + 3y = 2$

vii. $3x^2 = 2y^2 - 5$
     $2y^2 = x + 7$

---

5. The system belongs to Case I.

   $3x - 2y = 12$ $y^2 + 3x - 5y = 30$
   $x = \frac{2}{3}y + 4$ $y^2 + 3\left(\frac{2}{3}y + 4\right) - 5y = 30$
                  $y^2 - 3y - 18 = 0$
                  $(y - 6)(y + 3) = 0$
   $y = 6$ $y = -3$
   $x = \frac{2}{3}(6) + 4 = 8$ $x = \frac{2}{3}(-3) + 4 = 2$
   (8, 6)   (2, -3)

6. The system belongs to Case II.

   $xy = 4$ $x^2 + 4y^2 = 20$
   $y = \frac{4}{x}$ $x^2 + 4\left(\frac{16}{x^2}\right) = 20$
              $x^4 + 64 = 20x^2$
              $(x^2 - 16)(x^2 - 4) = 0$
   $x^2 = 16$ $x^2 = 4$
   $x = \pm 4$ $x = \pm 2$
   $y = \frac{4}{+4} = +1$ $y = \frac{4}{+2} = +2$
   $y = \frac{4}{-4} = -1$ $y = \frac{4}{-2} = -2$
   (4, 1), (-4, -1)   (2, 2), (-2, -2)

7. The system belongs to Case III.

   $x^2 + y^2 = 25$
   $x^2 - y^2 = 7$
   $\overline{2x^2 = 32}$ $y^2 = 25 - x^2$
       $x^2 = 16$ $y^2 = 25 - 16 = 9$
       $x = \pm 4$ $y = \pm 3$

   Solutions: (4, 3), (4, -3), (-4, 3), (-4, -3)

8. The system belongs to Case IV.

   $x^2 + 2y^2 - 2 = 0$ (1)
   $x^2 - y - 1 = 0$ (2)

   Subtract (2) from (1) to eliminate x^2. Then $2y^2 + y - 1 = 0$ and $y = \frac{1}{2}$ and -1. Substitute these values in (2); then $y = \frac{1}{2}$ yields $x = \pm\frac{\sqrt{6}}{2}$, and $y = -1$ yields $x = 0$.

   Solutions: $\left(\pm\frac{\sqrt{6}}{2}, \frac{1}{2}\right)$, $(0, -1)$

9. The system belongs to Case V.

   $2x^2 + 3xy + 2y^2 = -1$ (1)
   $x^2 + 2xy + 3y^2 = -2$ (2)

   Multiply (1) by 2 and (2) by -1 and add to eliminate the constant terms. Then

   $3x^2 + 4xy + y^2 = 0$ $(3x + y)(x + y) = 0$
   $y = -3x$ $y = -x$
   $x^2 - 6x^2 + 27x^2 = -2$ $x^2 - 2x^2 + 3x^2 = -2$
   $x = \pm\frac{\sqrt{11}}{11} i$ $x = \pm i$
   $y = -3\left(\pm\frac{\sqrt{11}}{11} i\right)$ $y = -(\pm i)$
   $\left(\pm\frac{\sqrt{11}}{11} i, \mp\frac{3\sqrt{11}}{11} i\right)$ $(\pm i, \mp i)$

10. The system belongs to Case VI.
$$x^2 - 2xy + y^2 + x + y = 8$$
$$2x^2 - 4xy + 2y^2 - x - y = 4$$
Substitute: $x = u + v$ and $y = u - v$
$$0u^2 + 4v^2 + 2u = 8$$
$$0u^2 + 8v^2 - 2u = 4$$
Eliminate v^2: $6u = 12$, $u = 2$
When $u = 2$: $4v^2 = 8 - 4 = 4$, $v = \pm 1$
$$x = u + v = 2 \pm 1 = 3, 1$$
$$y = u - v = 2 \mp 1 = 1, 3$$

Solutions: (3,1), (1,3)

EXERCISES II

In i–xii, find the pairs (x, y) which satisfy the given system of equations.

i. $xy = 6$
$\quad 2y - x = 1$

ii. $2x^2 + xy + y^2 = 2$
$\quad 2y = x - 3$

iii. $4x^2 + y^2 = 4$
$\quad xy = 1$

iv. $5x^2 + \frac{1}{4} y^2 = 6$
$\quad xy = 2$

v. $4x^2 + 5y^2 = 16$
$\quad 13y^2 - 5x^2 = 57$

vi. $\frac{1}{5} x^2 - \frac{1}{7} y^2 = 2$
$\quad \frac{1}{3} x^2 + \frac{3}{7} y^2 = 8$

vii. $2x^2 + 3y^2 = 3$
$\quad y^2 - 3x = 4$

viii. $x^2 + y^2 + 4 = 0$
$\quad x^2 - 8y + 13 = 0$

ix. $2x^2 + 2xy + 3y^2 = 3$
$\quad x^2 - xy + 2y^2 = 4$

x. $x^2 - 5xy + 3y^2 = 3$
$\quad x^2 + 3xy - 2y^2 = -2$

xi. $2x^2 + 3xy + 2y^2 + 2x + 2y = -1$
$\quad 3(x^2 + y^2) + 2xy + 3(x + y) = -4$

Solutions to the exercises appear on page 102.

ANSWERS

1. VI

2. I

3. II

4. III

5. (2, −3), (8, 6)

6.
$(4, 1), (−4, −1)$

$(2, 2), (−2, −2)$

7.
$(4, 3), (4, −3)$

$(−4, 3), (−4, −3)$

8.
$\left(\pm \dfrac{\sqrt{6}}{2}, \dfrac{1}{2} \right)$

$(0, −1)$

9.
$\left(\pm \dfrac{\sqrt{11}}{11} i, \mp \dfrac{3\sqrt{11}}{11} i \right)$

$(\pm i, \mp i)$

10. (3,1), (1,3)

Systems of Quadratic Equations

BASIC FACTS

The general quadratic equation in two variables, x and y, is given by
$$ax^2 + bxy + cy^2 + dx + ey + f = 0$$
where a, b, c, d, e, and f are constants, and where a, b, and c are not all zero.

The solutions of a system of two quadratic equations in two unknowns are those pairs of corresponding values of the unknowns which satisfy both equations.

Only the value of y which has been determined by substituting a certain value of x into an equation can be associated with that value of x.

There may be four, three, two, one, or no solutions to a system of quadratic equations.

The general procedure for solving all systems of quadratic equations is to solve one of the equations for one of the variables in terms of the other and then to substitute the expression for the solved-for variable into the other equation. This procedure, however, often yields a fourth-degree equation which, though in one variable, may be difficult to solve. We shall consider six types whose solutions can be determined with relative ease.

Case I

If one of the given equations is linear, solve the linear equation for one unknown in terms of the other, and substitute into the quadratic equation. Then solve the resulting quadratic equation, and substitute the solutions into the linear equation to find the corresponding values of the second unknown.

Case II

$$ax^2 + cy^2 + f = 0$$
$$bxy + g = 0$$

ADDITIONAL INFORMATION

Whether or not the solutions to a system of quadratic equations are real numbers can be determined by graphing the two relations that are defined by the equations to see if their graphs intersect. Consider, for example, the system $y - x^2 + 4 = 0$ and $4x + 4y + 1 = 0$. From Chapter 16, we know that the graph of $y = x^2 - 4$ is a parabola; and from Chapter 13, that the graph of $4x + 4y + 1 = 0$ is a straight line. Both curves are plotted on the same coordinate system and shown as solid lines in Figure 17-1. From the figure, we see that the two graphs intersect in two points and that the solutions of the system, $\left(-\frac{5}{2}, \frac{9}{4}\right)$ and $\left(\frac{3}{2}, -\frac{7}{4}\right)$, are real numbers. On the other hand, the graphs of the system, $y - x^2 + 4 = 0$ and $x + y + 5 = 0$, which are the (solid) parabola and the dotted line of Figure 17-1, do not intersect; and the solutions $\left(-\frac{1}{2} \pm \frac{\sqrt{3}}{2}i, -\frac{11}{2} \mp \frac{\sqrt{3}}{2}i\right)$ are not real numbers.

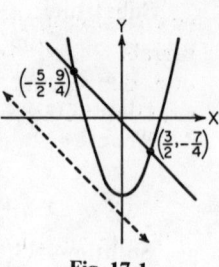

Fig. 17-1

In finding the solutions to a system of quadratic equations, *it is important that the values be properly paired.* Consider, for example, the solutions of the system
$$x^2 - 4x + y^2 - 6y + 4 = 0$$
$$x + y = 8 \qquad y = 8 - x.$$

Substitute $8 - x$ for y in the quadratic equation to obtain $x^2 - 7x + 10 = 0$, which has solutions $x = 5$ and $x = 2$. When these values are substituted into the linear equation, we obtain two solutions $x = 5$, $y = 3$ and $x = 2$, $y = 6$; but the pair $x = 5$ and $y = 6$ is *not* a solution.

The solutions to systems of quadratic equations are frequently written in the form of **ordered pairs** (x, y). Solutions are also written in the form $(\pm x, \mp y)$, a notation which indicates the two solutions $(x, -y)$ and $(-x, y)$. Note that *neither* $(-x, -y)$ *nor* (x, y) is indicated by this notation.

Case IV: Example

Solve:
$$x^2 + 4y^2 - 25 = 0 \tag{1}$$
$$x^2 - 2y - 5 = 0 \tag{2}$$

Subtract (2) from (1): $\quad 4y^2 + 2y - 20 = 0 \tag{3}$

Solve (3) for y: $\quad (2y + 5)(y - 2) = 0$

Substitute each value of y in (1):

	$y = -\frac{5}{2}$	$y = 2$
	$x^2 = 25 - 4\left(\frac{25}{4}\right)$	$x^2 = 25 - (4)(4)$
	$= 0$	$= 9$
Solve for x:	$x = 0$	$x = \pm 3$
Solutions:	$\left(0, -\frac{5}{2}\right)$	$(3, 2), (-3, 2)$

Solve the second equation for one variable in terms of the other and substitute into the first. Then solve the new first equation for the value of its variable, and substitute that value into the second equation to find the value of the other variable.

Case III

If both equations are of the form,
$$ax^2 + cy^2 + f = 0,$$
the system can be solved as a system of linear equations in the unknowns x^2 and y^2.

Case IV

$$ax^2 + cy^2 + f = 0$$
and $$Ax^2 + Ey + F = 0$$
or $$Cy^2 + Dx + F = 0$$

Eliminate the variable which is quadratic in both equations, and solve the resulting quadratic equation in one unknown. Substitute this value into either equation to obtain the value of the second unknown.

Case V

If both equations are of the form,
$$ax^2 + bxy + cy^2 + f = 0,$$
solve by eliminating the constant term, f. Factor the resulting equation into two linear factors, and substitute each factor into one of the given equations.

Case VI
The symmetrical case

If both equations are of the form,
$$A(x^2 + y^2) + Bxy + D(x + y) = F,$$
let $x = u + v$ and let $y = u - v$. Then the given equations take the form,
$$(2A + B)u^2 + (2A - B)v^2 + 2Du = F.$$
Eliminate v^2 and solve for u. Substitute to obtain v; and from the values of u and v, obtain the values of x and y.

Case V: Example

In addition to eliminating the constant (see problem 9), Case V systems can also be solved by substituting mx for y, and eliminating x by solving both equations for x^2.

Solve the system: $\quad x^2 + 3xy - 5 = 0 \quad$ and $\quad x^2 - y^2 = 3$

Let $y = mx$: $\quad x^2 + 3mx^2 = 5 \quad$ and $\quad x^2 - m^2x^2 = 3$

Solve for x^2: $\quad x^2 = \dfrac{5}{1 + 3m} \quad$ and $\quad x^2 = \dfrac{3}{1 - m^2}$

Equate the values of x^2:
$$\dfrac{5}{1 + 3m} = \dfrac{3}{1 - m^2}$$
$$5m^2 + 9m - 2 = 0$$
$$\left(m - \tfrac{1}{5}\right)(m + 2) = 0$$

Solve for m: $\quad m = \tfrac{1}{5} \qquad\qquad m = -2$

Solve for x: $\quad x^2 = \tfrac{25}{8} \qquad\qquad x^2 = -1$
(Use either equation.) $\quad x = (5^2 \cdot 2^{-4} \cdot 2)^{\frac{1}{2}} \qquad x = \pm i$
$\quad x = \pm \tfrac{5}{4}\sqrt{2}$

Solve for y: $\quad y = mx \qquad\qquad\qquad\qquad y = mx$
$\quad = \tfrac{1}{5}\left(\pm\tfrac{5}{4}\sqrt{2}\right) = \pm\tfrac{1}{4}\sqrt{2} \qquad = -2(\pm i) = \mp 2i$

Solutions: $\quad \left(\tfrac{5}{4}\sqrt{2}, \tfrac{1}{4}\sqrt{2}\right), \left(-\tfrac{5}{4}\sqrt{2}, -\tfrac{1}{4}\sqrt{2}\right) \ (i, -2i), (-i, 2i)$

Case VI: Example

Solve: $\quad x^2 + 3xy + y^2 + 2x + 2y - 5 = 0 \quad (1)$
$\quad\quad\quad 3x^2 - 2xy + 3y^2 - 5x - 5y - 26 = 0 \quad (2)$

Let $x = u + v$
$\quad y = u - v$
$$(2 + 3)u^2 + (2 - 3)v^2 + 2(2)u - 5 = 0$$
$$(6 - 2)u^2 + (6 + 2)v^2 + 2(-5)u - 26 = 0$$
$$5u^2 - v^2 + 4u - 5 = 0 \quad (3)$$
$$4u^2 + 8v^2 - 10u - 26 = 0 \quad (4)$$

Add 8 times (3) to (4):
$$44u^2 + 22u - 66 = 0$$
$$2u^2 + u - 3 = 0$$
$$(2u + 3)(u - 1) = 0$$

Solve for u: $\quad u = -\dfrac{3}{2} \qquad\qquad u = 1$

Solve for v in (3):
$\quad 5\left(\tfrac{3}{4}\right) - v^2 + 4\left(-\tfrac{3}{2}\right) - 5 = 0 \qquad 5(1) - v^2 + 4(1) - 5 = 0$
$\quad v^2 = \tfrac{1}{4} \ \text{and} \ v = \pm\tfrac{1}{2} \qquad\qquad v^2 = 4 \ \text{and} \ v = \pm 2$

Thus, with $x = u + v$ and $y = u - v$, the solutions are:
$$(-1, -2), (-2, -1), (3, -1), (-1, 3)$$

SOLUTIONS TO EXERCISES

I

i. $17x^2 = 14y^2 - 3$
$3y + 2 = 3x^2$, Case IV

ii. $y^2 = 2x^2 - 3xy + 12$
$2y^2 + x^2 = 3xy$, Case V

iii. $2xy - 45 = 0$
$2x = 5y$, Case I

iv. $16xy + y^2 = x^2 - 3$
$x^2 - 3xy = y^2 + 2$, Case V

v. $3x^2 + 2xy - 2x = 2y - 3y^2 + 3$
$2y^2 + 3xy + 3y = 3xy - 2x^2 + 7$
Case VI

vi. $2x^2 + 2y^2 + 3xy + 5x + 5y = 2$
$x + 3y = 2$, Case I

vii. $3x^2 = 2y^2 - 5$
$2y^2 = x + 7$, Case IV

II

i. Case I
$2y - x = 1$
$x = 2y - 1$
$xy = 6$
$(2y - 1)y = 6$
$2y^2 - y - 6 = 0$
$(2y + 3)(y - 2) = 0$
$y = -\frac{3}{2}$
$y = 2$
$x = -4$
$x = 3$
Solutions: $\left(-4, -\frac{3}{2}\right)$, $(3, 2)$

ii. $2x^2 + xy + y^2 = 2$
$2y = x - 3$, Case I
Solutions: $(1, -1)$, $\left(\frac{1}{11}, \frac{16}{11}\right)$

iii. Case II
$xy = 1$ $\quad$ $4x^2 + y^2 = 4$
$y = 1/x$ $\quad$ $4x^2 + 1/x^2 = 4$
$\quad$ $4x^4 + 1 - 4x^2 = 0$
$\quad$ $(2x^2 - 1)(2x^2 - 1) = 0$
$x = \pm\frac{\sqrt{2}}{2}$, $y = \pm\sqrt{2}$

iv. $5x^2 + \frac{1}{4}y^2 = 6$
$xy = 2$, Case II
Solutions: $\left(\pm\frac{\sqrt{5}}{5}, \pm 2\sqrt{5}\right)$ and $(\pm 1, \pm 2)$

v. $4x^2 + 5y^2 = 16$ $\quad$ (1)
$-5x^2 + 13y^2 = 57$ $\quad$ (2), Case III
Multiply (1) by 5 and (2) by 4 and add.
$77y^2 = 308$, $y = \pm 2$
$4x^2 + 5(4) = 16$ $\quad$ (1)
$x = \pm i$
Solutions: $(i, \pm 2)$, $(-i, \pm 2)$

vi. $\frac{1}{5}x^2 - \frac{1}{7}y^2 = 2$
$\frac{1}{3}x^2 + \frac{3}{7}y^2 = 8$, Case III
Solutions: $(\sqrt{15}, \pm\sqrt{7})(-\sqrt{15}, \pm\sqrt{7})$

vii. $2x^2 + 3y^2 = 3$ $\quad$ (1)
$y^2 - 3x = 4$ $\quad$ (2), Case IV
Multiply (2) by -3 and add. Then $x = -\frac{3}{2}$ and -3. When $x = -\frac{3}{2}$, $y = \pm\frac{\sqrt{2}}{2}i$. When $x = -3$, $y = \pm\sqrt{5}i$.
Solutions: $\left(-\frac{3}{2}, \pm\frac{\sqrt{2}}{2}i\right)$, $(-3, \pm\sqrt{5}i)$

viii. $x^2 + y^2 + 4 = 0$
$x^2 - 8y + 13 = 0$, Case IV
Solutions: $(\pm\sqrt{85}i, -9)$, $(\pm\sqrt{5}i, 1)$

ix. $2x^2 + 2xy + 3y^2 = 3$ $\quad$ (1)
$x^2 - xy + 2y^2 = 4$ $\quad$ (2), Case V
Multiply (1) by 4 and (2) by -3 and add. Then $x = -\frac{6}{5}y$ and $x = -y$.
Substitute in (2): $y = \pm\frac{5\sqrt{29}}{29}$, ± 1
$x = \mp\frac{6\sqrt{29}}{29}$, ∓ 1
Solutions: $\left(\pm\frac{6\sqrt{29}}{29}, \pm\frac{5\sqrt{29}}{29}\right)$, $(\mp 1, \pm 1)$

x. $x^2 - 5xy + 3y^2 = 3$
$x^2 + 3xy - 2y^2 = -2$, Case V
Solutions: $(\pm 1, \mp 1)$

xi. Case VI
$2x^2 + 3xy + 2y^2 + 2x + 2y = -1$
$3(x^2 + y^2) + 2xy + 3(x + y) = -4$
Substitute $x = u + v$ and $y = u - v$, and eliminate v^2. Then $u = 0$ and $-\frac{1}{2}$.
When $u = 0$, $v = \pm i$, $x = \pm i$, and $y = \mp i$.
When $u = -\frac{1}{2}$, $v = \pm\frac{\sqrt{3}}{2}i$, $x = -\frac{1}{2} \pm \frac{\sqrt{3}}{2}i$.
Solutions: $(\pm i, \mp i)$, $\left(-\frac{1}{2} \pm \frac{\sqrt{3}}{2}i, -\frac{1}{2} \mp \frac{\sqrt{3}}{2}i\right)$

18 THE CONICS

SELF-TEST

DIRECTIONS: Write your answers in the numbered regions to the right. To check your answers, turn the page. Study the solutions to the problems you missed, and do the exercises following any problem or group of problems in which you had errors.

1. Find (a) the vertex, (b) the focus, (c) the equation of the directrix, and (d) the length of the latus rectum of the parabola defined by the equation $y^2 - 4y = 12x + 32$.

2. Find the equation of the parabola with vertex $(2, -1)$ and focus $(2, -3)$.

3. Given $4x^2 + 9y^2 - 16x + 54y + 61 = 0$, find (a) center, (b) semiaxes, (c) vertices, (d) foci, (e) eccentricity, (f) directrices, and (g) the length of the latus rectum.

4. Given $9x^2 - 16y^2 - 18x + 64y = 199$, find (a) center, (b) semiaxes, (c) vertices, (d) foci, (e) eccentricity, (f) directrices, (g) length of latus rectum, and (h) asymptotes.

5. Find the equation of the hyperbola with center at $(0,0)$, the transverse axis along X-axis, the eccentricity equal to $\sqrt{3}$, and the length of the latus rectum equal to 10.

6. Find the equation of the locus of a point which moves so that the sum of its distance from $(3, 4)$ and $(3, -2)$ is 8.

7. Show that $5x^2 + 4xy + 2y^2 - 2x + 4y = 19$ is a proper conic and determine its name.

1a
b
c
d

2

3a
b
c
d
e
f
g

4a
b
c
d
e
f
g
h

5

6

7

103

SOLUTIONS

1. Complete the square on y:
$y^2 - 4y + 4 = 12x + 32 + 4$
$(y - 2)^2 = 12(x + 3)$
Compare: $(y - k)^2 = 4a(x - h)$
Vertex: $V(h, k) = (-3, 2)$
Since $4a = 12$ we have $a = 3$
Focus: $F(h + a, k) = (0, 2)$
Directrix: $x = h - a = -3 - 3 = -6$
Latus rectum: L.R. $= 4a = 12$.

2. Given $V(2, -1)$ and $F(2, -3)$. Since the x coordinates ($x = 2$) are equal, the axis $\parallel$ Y-axis and the form of the parabola is $(x - h)^2 = 4a(y - k)$ with $h = 2$ and $k = -1$. Since $F(h, k + a)$ we have $-1 + a = -3$ and $a = -2$. The equation of the parabola is
$(x - 2)^2 = -8(y + 1)$.

EXERCISES I

i. Find the vertex of
$2x^2 + 6x - y + 3 = 0$

ii. Find the equation of a parabola with axis $\parallel$ Y-axis and passing through $(-2, 11), (0, 5), (2, 3)$.

iii. Find the equation of a parabola with vertex $(4, 2)$ and focus $(2, 2)$.

---

3. Change to standard form by completing the square on both x and y:
$4(x^2 - 4x + 4) + 9(y^2 + 6y + 9) = 36$
$\dfrac{(x - 2)^2}{9} + \dfrac{(y + 3)^2}{4} = 1$
Center: $C(h, k) = (2, -3)$
Semiaxes: $a = 3$ and $b = 2$
$V_i(h \pm a, k)$: $V_1(5, -3), V_2(-1, -3)$
$ae = \sqrt{a^2 - b^2} = \sqrt{9 - 4} = \sqrt{5}$
$F_i(h \pm ae, k)$: $F_i(2 \pm \sqrt{5}, -3)$
Eccentricity: $e = \dfrac{ae}{a} = \dfrac{\sqrt{5}}{3}$
Directricies: $y = k \pm \dfrac{a}{e} = 2 \pm \dfrac{9\sqrt{5}}{5}$
L.R. $= \dfrac{2b^2}{a} = \dfrac{2(4)}{3} = \dfrac{8}{3}$

4. Change to standard form:
$9(x^2 - 2x + 1) - 16(y^2 - 4y + 4) = 144$
$\dfrac{(x - 1)^2}{16} - \dfrac{(y - 2)^2}{9} = 1$
Center: $C(h, k) = C(1, 2)$
Semiaxes: $a = 4$ and $b = 3$
$ae = \sqrt{a^2 + b^2} = \sqrt{16 + 9} = 5$
$V_i(h \pm a, k)$: $V_1(5, 2), V_2(-3, 2)$
$F_i(h \pm ae, k)$: $F_1(6, 2), F_2(-4, 2)$
Eccentricity: $e = \dfrac{ae}{a} = \dfrac{5}{4}$
Directrices: $x = h \pm \dfrac{a}{e} = 1 \pm \dfrac{16}{5}$
L.R. $= \dfrac{2b^2}{a} = \dfrac{2(9)}{4} = \dfrac{9}{2}$
Asymptotes: $y - k = \pm \dfrac{b}{a}(x - h)$
$y - 2 = \pm \dfrac{3}{4}(x - 1)$
$4y - 8 = \pm 3(x - 1)$

EXERCISES II

i. Find the eccentricity of
$12x^2 + 9y^2 = 108$

ii. Find the foci of
$7x^2 + 2y^2 + 4y + 16 = 28x$

iii. Find the eccentricity of
$9x^2 - 16y^2 = 36x - 96y - 36$

iv. Find the equations of the asymptotes of the hyperbola
$16x^2 - 81y^2 - 16x + 270y = 257$

---

5. Given $e = \sqrt{3}$ and L.R. $= 10$
$ae = \sqrt{a^2 + b^2} = \sqrt{3}a$ or $b^2 = 2a^2$
L.R. $= \dfrac{2b^2}{a} = 10$ or $b^2 = 5a$
Then $5a = 2a^2$ and $a = \dfrac{5}{2}$
(The value $a = 0$ is extraneous.)
Then $b^2 = 5a = \dfrac{25}{2}$ and $a^2 = \dfrac{25}{4}$
Since the center is at $(0, 0)$ and the transverse axis $\parallel$ X-axis,
$\dfrac{x^2}{a^2} - \dfrac{y^2}{b^2} = 1$ or $4x^2 - 2y^2 = 25$

6. Let the moving point be $P(x, y)$. Use the distance formula to write
$\sqrt{(x - 3)^2 + (y - 4)^2}$
$\qquad + \sqrt{(x - 3)^2 + (y + 2)^2} = 8$

Transpose one radical and square:
$(x - 3)^2 + (y - 4)^2$
$\quad = 64 - 16\sqrt{(x - 3)^2 + (y + 2)^2}$
$\quad\quad + (x - 3)^2 + (y + 2)^2$

Isolate the radical:
$16\sqrt{(x - 3)^2 + (y + 2)^2}$
$\quad = 64 + (y + 2)^2 - (y - 4)^2$
$\quad = 12y + 52$

Divide by 4 and square:
$16[(x - 3)^2 + (y + 2)^2]$
$\quad\quad = 9y^2 + 78y + 169$
$16x^2 + 7y^2 - 96x - 14y + 39 = 0$

7. $5x^2 + 4xy + 2y^2 - 2x + 4y = 19$

$$\Delta = \begin{vmatrix} 2A & B & D \\ B & 2C & E \\ D & E & 2F \end{vmatrix} = \begin{vmatrix} 10 & 4 & -2 \\ 4 & 4 & 4 \\ -2 & 4 & -38 \end{vmatrix}$$

$$= (2)(4)(2) \begin{vmatrix} 3 & 0 & -3 \\ 1 & 1 & 1 \\ -3 & 0 & -21 \end{vmatrix}$$

$= 16(-63 - 9) = 16(-54) \neq 0$

Thus we have a proper conic.

$B^2 - 4AC = 16 - 40 < 0$, ellipse

EXERCISES III

Determine the names of the following curves.

i. $3x^2 - 6xy + 2y^2 - 3x - 4y = 7$
ii. $x^2 + 4xy + 4y^2 + 2x = 13$
iii. $3x^2 + 2xy + y^2 = 3x + 2$
iv. $x^2 + y^2 = 0$

Solutions to the exercises appear on page 108.

ANSWERS

1. a. $V(-3, 2)$
   b. $F(0, 2)$
   c. $x = -6$
   d. L.R. = 12

2. $(x - 2)^2 = -8(y + 1)$

3. $C(2, -3)$
   $a = 3, b = 2$
   $V_1(5, -3), V_2(-1, -3)$
   $F_i(2 \pm \sqrt{5}, -3)$
   $e = \dfrac{\sqrt{5}}{3}$
   $y = 2 \pm \dfrac{9\sqrt{5}}{5}$
   L.R. = $\dfrac{8}{3}$

4. $C(1, 2)$
   $a = 4$ and $b = 3$
   $V_1(5, 2)$ and $V_2(-3, 2)$
   $F_1(6, 2)$ and $F_2(-4, 2)$
   $e = \dfrac{5}{4}$
   $5x = 21$ and $5x + 11 = 0$
   L.R. = $\dfrac{9}{2}$
   $4y - 8 = \pm(3x - 3)$

5. $4x^2 - 2y^2 = 25$

6. $16x^2 + 7y^2 = 96x + 14y - 39$

7. ellipse

The Conics

BASIC FACTS

A conic section is the path of a point which moves so that the ratio of its distance from a fixed point to its distance from a fixed line is constant. The constant ratio is called the **eccentricity** (e); the fixed line, the **directrix**; and the fixed point, the **focus**.

The conic sections are divided into three classes.

if	the conic is
$e < 1$	an ellipse
$e = 1$	a parabola
$e > 1$	a hyperbola

The line through the focus perpendicular to the directrix is the axis of the parabola, major axis for the ellipse, transverse axis for the hyperbola. The points where the curve intersects the axis are called the **vertices**. The line through the focus parallel to the directrix intersects the curve in two points, R_1 and R_2; the line segment, R_1R_2, is called the latus rectum.

The ellipse and the hyperbola each have a second definition.

A **hyperbola** is the locus of a point which moves so that the difference of its distance from two fixed points is a constant.

An **ellipse** is the locus of a point which moves so that the sum of its distances from two fixed points is a constant.

The **circle** is a special case of the ellipse. Its general equation is

$$ax^2 + ay^2 + dx + ey + f = 0$$

which can be transformed to the form $x^2 + y^2 + Dx + Ey + F = 0$ by dividing by a and letting $D = \frac{d}{a}$, $E = \frac{e}{a}$, and $F = \frac{f}{a}$. This equation defines a relation, but not a function, between x and y.

If the center of a circle is at (h, k) and the fixed distance is r, the equation can be written,

ADDITIONAL INFORMATION

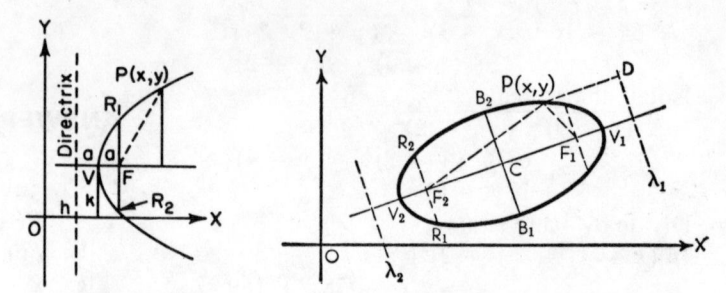

Fig. 18-1: Parabola

Fig. 18-2: Ellipse

There are eight standard forms of parabolas which have their axes parallel to the coordinate axes. These are summarized in the following table. (See Figure 18-1.)

No.	Equation	Focus	Directrix	Length of Latus Rectum	Curve Opening
\multicolumn{6}{c}{Vertex at (0,0); Axis Parallel to X-axis}				
1	$y^2 = 4ax$	$(a,0)$	$x = -a$	$4a$	to the right
2	$y^2 = -4ax$	$(-a,0)$	$x = a$	$4a$	to the left
\multicolumn{6}{c}{Vertex at (0,0); Axis Parallel to Y-axis}				
3	$x^2 = 4ay$	$(0,a)$	$y = -a$	$4a$	upward
4	$x^2 = -4ay$	$(0,-a)$	$y = a$	$4a$	downward
\multicolumn{6}{c}{Vertex at (h,k); Axis Parallel to X-axis}				
5	$(y-k)^2 = 4a(x-h)$	$(h+a,k)$	$x = h-a$	$4a$	to the right
6	$(y-k)^2 = -4a(x-h)$	$(h-a,k)$	$x = h+a$	$4a$	to the left
\multicolumn{6}{c}{Vertex at (h,k); Axis Parallel to Y-axis}				
7	$(x-h)^2 = 4a(y-k)$	$(h,k+a)$	$y = k-a$	$4a$	upward
8	$(x-h)^2 = -4a(y-k)$	$(h,k-a)$	$y = k+a$	$4a$	downward

The standard forms and the characteristic elements for the ellipse are summarized in the following table. The letters in parentheses refer to Figure 18-2. In the table the arbitrary constants a and b have been chosen so that $a^2 > b^2$.

Equation	Center (C)	Vertices (V_1 and V_2)	Foci (F_1 and F_2)	Directrices (λ_1 and λ_2)
\multicolumn{5}{c}{Major Axis ($\overline{V_1 V_2}$) Parallel to X-axis}			
$\frac{x^2}{a^2} + \frac{y^2}{b^2} = 1$	$(0,0)$	$(a,0)$ $(-a,0)$	$(ae,0)$ $(-ae,0)$	$x = \pm\frac{a}{e}$
$\frac{(x-h)^2}{a^2} + \frac{(y-k)^2}{b^2} = 1$	(h,k)	$(h+a,k)$ $(h-a,k)$	$(h+ae,k)$ $(h-ae,k)$	$x = h \pm \frac{a}{e}$

$$(x - h)^2 + (y - k)^2 = r^2.$$

If the center of the circle is at the origin, $(h, k) = (0, 0)$ and $x^2 + y^2 = r^2$.

Since the equation of a circle contains three arbitrary constants, D, E, and F, three **conditions** will determine the equation of a circle. The conditions may be the coordinates of points that lie on the circle or other geometric facts which permit the calculation of values for D, E, and F.

The **general second-degree equation in two variables**, $Ax^2 + Bxy + Cy^2 + Dx + Ey + F = 0$, represents a conic.

The discriminant of this equation is

$$\Delta = \begin{vmatrix} 2A & B & D \\ B & 2C & E \\ D & E & 2F \end{vmatrix}.$$

If $\Delta = 0$, the conic is **degenerate**, and the locus may be parallel lines, intersecting lines, or a point. If $\Delta \neq 0$, we have a proper conic.

The coefficients of x^2, xy, and y^2 in a second-degree equation in two variables may be used to determine the locus defined by the equation.

Condition	Locus
$B^2 - 4AC = 0$	a parabola
$B^2 - 4AC < 0$	an ellipse
$B^2 - 4AC > 0$	a hyperbola

The standard forms and formulas for the conics are given in the Additional Information.

\multicolumn{5}{c}{Major Axis ($\overline{V_1 V_2}$) Parallel to Y-axis}
$\dfrac{x^2}{b^2} + \dfrac{y^2}{a^2} = 1$	$(0,0)$	$(0, a)$ $(0, -a)$	$(0, ae)$ $(0, -ae)$	$y = \pm \dfrac{a}{e}$
$\dfrac{(x-h)^2}{b^2} + \dfrac{(y-k)^2}{a^2} = 1$	(h, k)	$(h, k+a)$ $(h, k-a)$	$(h, k+ae)$ $(h, k-ae)$	$y = k \pm \dfrac{a}{e}$

Semimajor axis ($\overline{V_2 C}$): a Eccentricity $\left(\dfrac{\overline{PF_1}}{\overline{PD}}\right)$: $e = \dfrac{\sqrt{a^2 - b^2}}{a} < 1$

Semiminor axis ($\overline{B_1 C}$): b Length of latus rectum ($\overline{R_1 R_2}$): $\dfrac{2b^2}{a}$

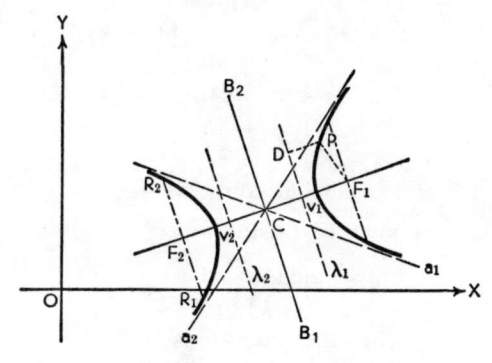

Fig. 18-3: Hyperbola

In the following table, the letters in parentheses refer to Figure 18-3.

Equation	Center (C)	Vertices (V_1 and V_2)	Foci (F_1 and F_2)	Directrices (λ_1 and λ_2)	Asymptotes (a_1 and a_2)
\multicolumn{6}{c}{Transverse Axis ($\overline{V_1 V_2}$) Parallel to the X-axis}
$\dfrac{x^2}{a^2} - \dfrac{y^2}{b^2} = 1$	$(0,0)$	$(a, 0)$ $(-a, 0)$	$(ae, 0)$ $(-ae, 0)$	$x = \pm \dfrac{a}{e}$	$y = \pm \dfrac{b}{a} x$
$\dfrac{(x-h)^2}{a^2} - \dfrac{(y-k)^2}{b^2} = 1$	(h, k)	$(h+a, k)$ $(h-a, k)$	$(h+ae, k)$ $(h-ae, k)$	$x = h \pm \dfrac{a}{e}$	$y - k = \pm \dfrac{b}{a}(x-h)$
\multicolumn{6}{c}{Transverse Axis ($\overline{V_1 V_2}$) Parallel to the Y-axis}
$-\dfrac{x^2}{b^2} + \dfrac{y^2}{a^2} = 1$	$(0,0)$	$(0, a)$ $(0, -a)$	$(0, ae)$ $(0, -ae)$	$y = \pm \dfrac{a}{e}$	$y = \pm \dfrac{a}{b} x$
$-\dfrac{(x-h)^2}{b^2} + \dfrac{(y-k)^2}{a^2} = 1$	(h, k)	$(h, k+a)$ $(h, k-a)$	$(h, k+ae)$ $(h, k-ae)$	$y = k \pm \dfrac{a}{e}$	$y - k = \pm \dfrac{a}{b}(x-h)$

Semitransverse axis ($\overline{V_2 C}$): a Eccentricity $\left(\dfrac{\overline{PF_1}}{\overline{PD}}\right)$: $e = \dfrac{\sqrt{a^2 + b^2}}{a} > 1$

Semiconjugate axis ($\overline{B_1 C}$): b Length of latus rectum ($\overline{R_1 R_2}$): $\dfrac{2b^2}{a}$

SOLUTIONS TO EXERCISES

I

i. Complete the square:
$$2\left(x^2 + 3x + \tfrac{9}{4}\right) = y + \tfrac{9}{2} - 3$$
$$\left(x + \tfrac{3}{2}\right)^2 = \tfrac{1}{2}\left(y + \tfrac{3}{2}\right)$$
$$V\left(-\tfrac{3}{2}, -\tfrac{3}{2}\right)$$

ii. Substitute the given points into the standard form
$(x - h)^2 = 4a(y - k)$:
$$4 + 4h + h^2 = 4a(11 - k) \quad (1)$$
$$h^2 = 4a(5 - k) \quad (2)$$
$$4 - 4h + h^2 = 4a(3 - k) \quad (3)$$
$(1) - (2)$: $4 + 4h = 4a$ $\quad (6)$
$(1) - (3)$: $8h = 4a$ $\quad (8)$
$1 + h = 6a$ and $h = 4a$
Then $a = \tfrac{1}{2}$ and $h = 2$

Substitute into (2):
$4 = 2(5 - k)$ or $k = 3$
Then the equation is
$(x - 2)^2 = 2(y - 3)$ or
$x^2 - 4x - 2y + 10 = 0$

iii. Given: $V(4, 2)$ and $F(2, 2)$. Since $y = 2$ for both V and F, the axis of the parabola is parallel to the X-axis and the curve opens to the left. Thus $h = 4$, $k = 2$, and $a = |2 - 4| = 2$. Then $(y - k)^2 = -4a(x - h)$ becomes
$$y^2 - 4y + 8x - 28 = 0.$$

II

i. $12x^2 + 9y^2 = 108$
Divide by 108:
$$\tfrac{x^2}{9} + \tfrac{y^2}{12} = 1$$
The graph is an ellipse with $a^2 = 12$ and $b^2 = 9$
$$ae = \sqrt{a^2 - b^2}$$
$$2\sqrt{3}\,e = \sqrt{12 - 9} = \sqrt{3}$$
$$e = \tfrac{\sqrt{3}}{2\sqrt{3}} = \tfrac{1}{2}$$

ii. $7x^2 + 2y^2 + 4y + 16 = 28x$
Put into standard form:
$$7(x^2 - 4x + 4) + 2(y^2 + 2y + 1)$$
$$= 28 + 2 - 16 = 14$$
$$\tfrac{(x - 2)^2}{2} + \tfrac{(y + 1)^2}{7} = 1$$
$h = 2, k = -1, a = \sqrt{7}, b = \sqrt{2}$
$ae = \sqrt{7 - 2} = \sqrt{5}$; $F_i(h, k \pm ae)$:
$F_1(2, -1 + \sqrt{5})$
$F_2(2, -1 - \sqrt{5})$

iii. Put into standard form
$$9(x^2 - 4x + 4) - 16(y^2 - 6y + 9)$$
$$= -36 + 36 - 144 = -144$$
$$-\tfrac{(x - 2)^2}{16} + \tfrac{(y - 3)^2}{9} = 1$$
This is a hyperbola with $a^2 = 9$ and $b^2 = 16$, so that $ae = \sqrt{9 + 16} = 5$ and $e = \tfrac{5}{3}$

iv. Put into standard form:
$$16\left(x^2 - x + \tfrac{1}{4}\right) - 81\left(y^2 - \tfrac{10}{3}y + \tfrac{25}{9}\right)$$
$$= 257 + 4 - 225 = 36$$
$$\frac{\left(x - \tfrac{1}{2}\right)^2}{\tfrac{9}{4}} - \frac{\left(y - \tfrac{5}{3}\right)^2}{\tfrac{4}{9}} = 1$$
$h = \tfrac{1}{2}, k = \tfrac{5}{3}, a = \tfrac{3}{2}, b = \tfrac{2}{3}$
Equations of Asymptotes
$$y - k = \pm \tfrac{b}{a}(x - h)$$
$$y - \tfrac{5}{3} = \pm \tfrac{4}{9}\left(x - \tfrac{1}{2}\right)$$
$$9y - 15 = \pm(4x - 2)$$

III

i. $B^2 - 4AC = 36 - 24 = 12$, hyperbola

ii. $B^2 - 4AC = 16 - 16 = 0$, parabola

iii. $B^2 - 4AC = 4 - 12 = -8$, ellipse

iv. Since the discriminant is zero the conic is degenerate, and since $x^2 + y^2 = 0$ is satisfied in the real number system only by $(0, 0)$, its curve is a point.

19 EQUATIONS OF HIGHER DEGREE

SELF-TEST

DIRECTIONS: Write your answers in the numbered regions to the right. To check your answers, turn the page. Study the solutions to the problems you missed, and do the exercises following any problem or group of problems in which you had errors.

In **1** and **2** find all the roots for the given equation.

1. $2x^4 + x^3 - 2x - 1 = 0$

2. $x^3 = 1$

Find a rational integral equation which satisfies the conditions in each of **3** and **4**. Place the term of highest degree and the constant term in the answer column.

3. One root is $2 + 3i$ and another root is $2 - \sqrt{3}$.

4. One root is 3, and the sum and product of two other roots are $\frac{2}{3}$ and $\frac{3}{2}$, respectively.

5. Locate the smallest positive root of $x^3 + 3x - 20 = 0$ to one decimal place.

Are the statements in **6–11** true or false?

6. If $f(r) = 0$, then $(x - r)$ is a factor of $f(x)$.

7. When $x^4 + 2x^3 - 3x^2 + 4x - 5$ is divided by $x + 2$, the remainder is -20.

8. If $2 + 3i$ is a root of a rational integral equation, then $2 - 3i$ is also a root.

9. The equation $6x^4 + x^3 - 26x^2 - 4x + 8 = 0$ has no more than two positive real roots.

10. The equation $x^3 + ax^2 + c = 0$, where a and c are integers, has at least one real root.

11. The roots of $x^3 - 4x^2 - 20x + 48 = 0$ are double the roots of $x^3 - 2x^2 - 5x + 6 = 0$.

SOLUTIONS

1. $2x^4 + x^3 - 2x - 1 = 0$

$$\begin{array}{rrrrr|r}
2 & 1 & 0 & -2 & -1 & \underline{1} \\
  & 2 & 3 & 3 & 1 & \\
\hline
2 & 3 & 3 & 1 & 0 & \underline{-\tfrac{1}{2}} \\
  & -1& -1& -1 &   & \\
\hline
2 & 2 & 2 & 0 &   &
\end{array}$$

$x = 1$, $x = -\tfrac{1}{2}$

Using the quadratic formula,
$x^2 + x + 1 = 0$, $x = \dfrac{-1 \pm \sqrt{1-4}}{2} = -\dfrac{1}{2} \pm \dfrac{\sqrt{3}}{2} i$.

2. $x^3 = 1$ or $x^3 - 1 = 0$
Factor: $(x-1)(x^2 + x + 1) = 0$
Solving we obtain the three cubic roots of unity, $1, -\dfrac{1}{2} \pm \dfrac{\sqrt{3}}{2}i$.

EXERCISES I
Find all the roots for each equation in i–iii.

i. $12x^3 - 4x^2 - 3x + 1 = 0$

ii. $36x^4 + 48x^3 - 23x^2 - 17x = -6$

iii. $4x^4 + 4x^3 + x^2 - 9 = 0$

---

3. If $2 + 3i$ is a root, so is $2 - 3i$. If $2 - \sqrt{3}$ is a root, so is $2 + \sqrt{3}$. We then have the two sets of factors
$[x - (2 + 3i)][x - (2 - 3i)]$
$= x^2 - 4x + 13$
$[x - (2 - \sqrt{3})][x - (2 + \sqrt{3})]$
$= x^2 - 4x + 1$
The equation is the product of these factors set equal to zero,
$x^4 - 8x^3 + 30x^2 - 56x + 13 = 0$.

4. The sum of two roots of a quadratic is equal to the negative the coefficient of the linear term, and their product is the constant term. Therefore we have
$x^2 - \dfrac{2}{3}x + \dfrac{3}{2} = \dfrac{1}{6}(6x^2 - 4x + 9)$
as one factor and $x - 3$ as the other factor. The equation is
$6x^3 - 22x^2 + 21x - 27 = 0$.

5. Given: $f(x) = x^3 + 3x - 20$
Calculate $f(0) = -20$, $f(1) = -16$, $f(2) = -6$, $f(3) = +16$. There is a root between $x = 2$ and $x = 3$. Use Horner's method to reduce the root by 2.

$$\begin{array}{rrrr|r}
1 & 0 & 3 & -20 & \underline{2} \\
  & 2 & 4 & 14 & \\
\hline
1 & 2 & 7 & -6 & \\
  & 2 & 8 &   & \\
\hline
1 & 4 & 15 &   & \\
  & 2 &   &   & \\
\hline
1 & 6 &   &   &
\end{array}$$

$f(x) = x^3 + 6x^2 + 15x - 6$

$f(.3)$:
$$\begin{array}{rrrr|r}
1 & 6 & 15 & -6.000 & \underline{.3} \\
  & .3 & 1.89 & 5.067 & \\
\hline
1 & 6.3 & 16.89 & -.933 &
\end{array}$$

$f(.4)$:
$$\begin{array}{rrrr|r}
1 & 6 & 15 & -6.000 & \underline{.4} \\
  & .4 & 2.56 & 7.024 & \\
\hline
1 & 6.4 & 17.56 & 1.024 &
\end{array}$$

$\therefore$ The root is between 2.3 and 2.4.

EXERCISES II
Find a rational integral equation which satisfies the conditions in each of i–iii.

i. The roots are $2, -1$, and $\dfrac{1}{2}$.

ii. The roots are twice the roots of $x^5 - 4x^3 + 1 = 0$.

iii. The roots are two less than the cube roots of unity.

---

6. If $f(r) = 0$, then $(x - r)$ is a factor of $f(x)$. *True* by The Factor Theorem.

7. When $x^4 + 2x^3 - 3x^2 + 4x - 5$ is divided by $x + 2$, the remainder is -20. *false*

$$\begin{array}{rrrrr|r}
1 & 2 & -3 & 4 & -5 & \underline{-2} \\
  & -2& 0 & 6 & -20 & \\
\hline
1 & 0 & -3 & 10 & -25 & = R
\end{array}$$

8. If $2 + 3i$ is a root of a rational integral equation, then $2 - 3i$ is also a root. *True* by the conjugate root property.

9. The equation $6x^4 + x^3 - 26x^2 - 4x + 8 = 0$ has no more than two positive real roots. *True* by Descartes' Law of Signs.

10. The equation $x^3 + ax^2 + c = 0$, where a and c are integers, has at least one real root. *true*
    Since complex roots occur in pairs and since the given cubic equation has three roots, one must be real.

11. The roots of $x^3 - 4x^2 - 20x + 48 = 0$ are double the roots of $x^3 - 2x^2 - 5x + 6 = 0$. *true*
    To double the roots of $x^3 - 2x^2 - 5x + 6 = 0$ we change the equation to $x^3 - 2x^2(2) - 5x(2)^2 + 6(2^3) = 0$ or $x^3 - 4x^2 - 20x + 48 = 0$.

EXERCISES III

Are the statements of **i–viii** true or false?

i. $x - 3$ is a factor of $x^4 - 10x^3 + 35x^2 - 50x + 24$.

ii. The equation $8x^4 - 18x^3 + 2x^2 + 7x - 6 = 0$ has two negative real roots.

iii. The graph of $y = x^3 - x^2 + 4x - 5$ crosses the X-axis between $x = 1$ and $x = 2$.

iv. The equation, $6x^4 + x^3 - 26x^2 - 4x + 8 = 0$, is satisfied by the value $x = -\frac{2}{3}$.

v. The equation $x^3 + 7x^2 = 4$ has four roots.

vi. If $x^5 - x^4 - x + 1 = 0$ has a rational root, that root must be $x = 1$.

vii. The roots of $x^4 - 2x^3 - 13x^2 + 14x + 24 = 0$ are the negative of the roots of $x^4 + 2x^3 - 13x^2 - 14x + 24 = 0$.

viii. $x^4 - 6x^3 + x^2 - 5x + 3 = 0$ has no negative roots.

Solutions to the exercises appear on page 114.

ANSWERS

1.
   $1, -\frac{1}{2}$

   $-\frac{1}{2} \pm \frac{\sqrt{3}}{2}i$

2. 1

   $-\frac{1}{2} \pm \frac{\sqrt{3}}{2}i$

3. x^4

   13

4. $6x^3$

   -27

5. $2.3 +$

6. true

7. false

8. true

9. true

10. true

11. true

BASIC FACTS

The equation,

$$a_0 x^n + a_1 x^{n-1} + \cdots + a_{n-1} x + a_n = 0$$

where n is a positive integer and $a_i (i = 0, 1, \ldots, n)$, $a_0 \neq 0$, are rational-number constants, is called a **rational integral equation of the nth degree in x**. The left member is a polynomial and is frequently represented by $f(x)$.

Remainder Theorem

If a polynomial, $f(x)$, is divided by $x - r$ until a remainder independent of x is obtained, this remainder is equal to $f(r)$.

Factor Theorem

If r is a root of $f(x)$; i.e., if $f(r) = 0$, then $(x - r)$ is a factor of $f(x)$.

Fundamental Theorems of Algebra

I. Every rational integral equation has at least one root.
II. Every rational integral equation of the nth degree has n roots and no more.

Properties of the roots of the rational integral equation

Non-real roots occur in **conjugate pairs**; i.e., if $a + bi$ is a root, then $a - bi$ is a root.

If a **quadratic surd**, $a + \sqrt{b}$, is a root, then $a - \sqrt{b}$ is a root.

The number of positive roots of an equation with real coefficients cannot exceed the number of **variations in sign** in the polynomial $f(x)$. The number of negative roots cannot exceed the number of variations in sign in $f(-x)$. A variation in sign occurs whenever two successive terms in $f(x)$ differ in sign.

If a **rational number**, $\frac{b}{c}$, is in its lowest terms and is a root of the rational integral equation, $f(x) = 0$, then b is a factor of a_n (the constant

ADDITIONAL INFORMATION

The **Remainder and Factor Theorems** can be proved from the definition of division and the notation for the polynomial of the nth degree.

Dividend = divisor × quotient + remainder: $\quad f(x) = (x - r) Q(x) + R$
This is an identity and holds for $x = r$: $\quad f(r) = (r - r) Q(r) + R$
Since $Q(r)$ is a number: $\quad f(r) = 0 \times Q(r) + R$
Thus we have the Remainder Theorem: $\quad f(r) = R$
Consequently we can write: $\quad f(x) = (x - r) Q(x) + f(r)$
By hypothesis for Factor Theorem, $f(r) = 0$: $\quad f(x) = (x - r) Q(x)$
Thus we have the Factor Theorem: $(x - r)$ is a factor of $f(x)$

We shall use the Factor Theorem to prove the **Fundamental Theorem of Algebra, II**, "Every rational integral equation of the nth degree has n roots and no more."

Given: $f(x) = a_0 x^n + a_1 x^{n-1} + \cdots + a_{n-1} x + a_n = 0$, $a_0 \neq 0$
By the Fundamental Theorem of Algebra, I, there is one root, say r_1:
$\qquad x = r_1$ and $f(r_1) = 0$
By the Factor Theorem: $\qquad f(x) = (x - r_1) Q_1(x)$
$Q_1(x)$ is a polynomial which has a zero (root), say r_2: $x = r_2$ and $Q_1(r_2) = 0$
By the Factor Theorem: $\qquad Q_1(x) = (x - r_2) Q_2(x)$

Continue this process n times to obtain

$$f(x) = (x - r_1)(x - r_2)(x - r_3) \cdots (x - r_n) Q_n(x).$$

The last value is $Q_n(x) = a_0 x^{n-n} = a_0$. Thus $f(x) = 0$ has exactly n roots $r_1, r_2, \ldots, r_n$. If there were another root, r, different from any of the r_i, then $f(r)$ would be equal to $a_0 (r - r_1)(r - r_2)(r - r_3) \cdots (r - r_n)$ which is not equal to zero since $r - r_i \neq 0$ and $a_0 \neq 0$. Thus, r cannot be a root of $f(x) = 0$.

The process of dividing a polynomial by a binomial is considerably shortened by the method of **synthetic division** which is accomplished through the following steps.

1. Arrange $f(x)$ in descending powers of x.
2. Write the coefficients a_i, $(i = 0, \ldots, n)$ in order on a line. Supply a zero for each missing term in the sequence of powers of x.
3. To divide by $x - r$, place r at the right on the first line.
4. Complete the following array performing the indicated arithmetic.

a_0	a_1	a_2	a_3	a_4	$\ldots$	a_{n-1}	a_n	$\underline{\quad r}$
	$b_0 r$	$b_1 r$	$b_2 r$	$b_3 r$	$\ldots$	$b_{n-2} r$	$b_{n-1} r$	
b_0	b_1	b_2	b_3	b_4		b_{n-1}	b_n	

where $b_0 = a_0$, $b_1 = a_1 + b_0 r$, $b_2 = a_2 + b_1 r$, $\ldots$, $b_n = a_n + b_{n-1} r = f(r)$. Then the quotient and remainder are $Q(x) = a_0 x^{n-1} + b_1 x^{n-2} + b_2 x^{n-3} + \cdots + b_{n-1}$, $R = b_n$.

To find the **approximate irrational roots** of $f(x) = 0$, we shall employ **Horner's method** of successive approximations. This method is one

Equations of Higher Degree

term) and c is a factor of a_0 (the coefficient of x^n).

The graph of $y = f(x)$ will cross the X-axis at all the real roots of $f(x) = 0$.

Between any two values, $x = a$ and $x = b$, for which the two corresponding values $f(a)$ and $f(b)$ are opposite in sign, there is at least one root of the equation $f(x) = 0$.

If n is an odd number, the equation, $f(x) = 0$, has at least one real root.

To obtain an equation each of whose roots is m times a corresponding root of $f(x) = 0$, multiply the successive coefficients, beginning with that of x^{n-1}, by m, m^2, m^3, ..., respectively.

To obtain an equation each of whose roots is equal in absolute value to a root of $f(x) = 0$, but opposite in sign, change the signs of the odd degree terms in $f(x) = 0$.

To obtain an equation each of whose roots is less by h than a corresponding root of a given equation, $f(x) = 0$, divide $f(x)$ by $x - h$ and denote the remainder by R_n. Divide the quotient by $x - h$ and indicate the remainder by R_{n-1}. Continue the process by n divisions. The last quotient, a_0, and the remainders, $R_1, R_2, ..., R_n$ are the coefficient of the transformed equation. The new equation is

$$a_0 z^n + a_1 z^{n-1} + \cdots + R_{n-1} z + R_n = 0.$$

Perform the divisions by repeated **synthetic division**.

If we can express the roots of an equation by an algebraic formula involving the coefficients of the equations, the equation is said to be solved **algebraically**. Rational integral equations of degree higher than the fourth cannot, in general, be solved algebraically.

NOTE: In some texts, roots are referred to as zeros.

of continuous synthetic division which diminishes the roots by the desired number. We shall demonstrate the method by an example.

Find the smallest positive root of: $x^3 - 2x^2 + 3x - 4 = 0$.
Find the values $f(0) = -4$, $f(1) = -2$, $f(2) = 2$ so that there is a root between 1 and 2. We diminish the roots by 1. The bold faced numbers are the coefficients of the transformed equation. To obtain the next approximation consider the fact that $f(1) = -2$ and $f(2) = 2$ so that the function may cross the X-axis half way between $x = 1$ and $x = 2$, say $x_1 = .6$ or $x_1 = .7$. We try both. Since $f(.6)$ is negative and $f(.7)$ is positive, the root is between 1.6 and 1.7. We continue to reduce the root by .6. The bold face numbers are the coefficients of $f_2(x_2)$.

```
    1   -2     3     -4    |1
         1    -1      2
    1   -1     2     -2
         1     0
    1    0     2
         1
    1    1
    
    1    1     2     -2    |.7
         .7   1.19  2.233
    1   1.7   3.19   .233
    
    1    1     2     -2    |.6
         .6   .96   1.776
    1   1.6  2.96   -.224
         .6  1.32
    1   2.2  4.28
         .6
    1   2.8
```

To find the next approximation ignore the higher order terms and solve $4.28 x_2 = .224$ to give $x_2 = .05+$. By synthetic division we find $f_2(.05) = -.002875$ and $f_2(.06) = .043096$, and we reduce the equation by .05. We can continue this process as far as desired. The root is $x = 1.65+$.

Consider the general cubic equation: $\quad x^3 + bx^2 + cx + d = 0 \quad (1)$

Let $x = y - \frac{1}{3} b$. Then (1) becomes: $\quad y^3 + py + q = 0 \quad (2)$

$$\text{where } p = c - \frac{b^3}{3} \text{ and } q = d - \frac{bc}{3} + \frac{2b^3}{27}$$

Let $y = u + v$ and $3uv + p = 0$, then (2) becomes: $u^6 + qu^3 - \frac{p^3}{27} = 0 \quad (3)$

Solve (3) for u^3: $\quad u^3 = -\frac{q}{2} \pm \sqrt{R}$, where $R = \frac{p^3}{27} + \frac{q^2}{4}$

Let $u^3 = -\frac{q}{2} + \sqrt{R} = N$ and solve for v^3: $\quad v^3 = -\frac{q}{2} - \sqrt{R} = M$

The solutions (u, v) are $(\sqrt[3]{N}, \sqrt[3]{M})$, $(w\sqrt[3]{N}, w^2\sqrt[3]{M})$, $(w^2\sqrt[3]{N}, w\sqrt[3]{M})$ where $w = -\frac{1}{2} + \frac{\sqrt{3}}{2} i$, $w^2 = -\frac{1}{2} - \frac{\sqrt{3}}{2} i$, and $w^3 = 1$. This gives us three solutions for $y = u + v$ and then three solutions for x.

$$x_1 = \sqrt[3]{N} + \sqrt[3]{M} - \frac{1}{3} b, \quad x_2 = w\sqrt[3]{N} + w^2\sqrt[3]{M} - \frac{1}{3} b, \quad x_3 = w^2\sqrt[3]{N} + w\sqrt[3]{M} - \frac{1}{3} b$$

SOLUTIONS TO EXERCISES

i. $12x^3 - 4x^2 - 3x + 1 = 0$
We try values, $\pm 1, \pm\frac{1}{2}, \pm\frac{1}{3}, \pm\frac{1}{4}, \pm\frac{1}{6}, \pm\frac{1}{12}$
Using synthetic division,

$$\begin{array}{rrrr|l} 12 & -4 & -3 & 1 & \underline{\frac{1}{2}} \quad x = \frac{1}{2}\\ & 6 & 1 & -1 & \underline{\frac{1}{3}} \quad x = \frac{1}{3}\\ \hline 12 & 2 & -2 & 0\\ & 4 & 2\\ \hline 12 & 6 & 0 & & 12x + 6 = 0,\ x = -\frac{1}{2}. \end{array}$$

ii. $36x^4 + 48x^3 - 23x^2 - 17x + 6 = 0$
The possible real roots are the factors of 6 divided by the factors of 36.

$$\begin{array}{rrrrr|l} 36 & 48 & -23 & -17 & 6 & \underline{\frac{1}{2}} \quad x = \frac{1}{2}\\ & 18 & 33 & 5 & -6 & \underline{\frac{1}{3}} \quad x = \frac{1}{3}\\ \hline 36 & 66 & 10 & -12 & 0\\ & 22 & 26 & 12\\ \hline 36 & 78 & 36 & 0 \end{array}$$

Now solve the quadratic equation
$6x^2 + 13x + 6 = 0$
$(3x + 2)(2x + 3) = 0, \ x = -\frac{2}{3}, -\frac{3}{2}$

iii. $4x^4 + 4x^3 + x^2 - 9 = 0$

$$\begin{array}{rrrrr|l} 4 & 4 & 1 & 0 & -9 & \underline{1} \quad x = 1\\ & 4 & 8 & 9 & 9 & \underline{-\frac{3}{2}} \quad x = -\frac{3}{2}\\ \hline 4 & 8 & 9 & 9 & 0\\ & -6 & -3 & -9\\ \hline 4 & 2 & 6 & 0 \end{array}$$

$2x^2 + x + 3 = 0,\ x = \frac{-1 \pm \sqrt{1-24}}{4} = -\frac{1}{4} \pm \frac{\sqrt{23}}{4}i$

II

i. If the roots are 2, -1, $\frac{1}{2}$, then the equation is $(x - 2)(x + 1)\left(x - \frac{1}{2}\right) = 0$
or $2x^3 - 3x^2 - 3x + 2 = 0.$

ii. To obtain an equation whose roots are twice the roots of a given equation, arrange the given equation in descending powers of the variable, then multiply the second term by 2, the third term by 2^2, the fourth by 2^3, etc.
$x^5 - 4x^3 + 1 = 0$
$x^5 + (2)(0) + (2)^2(-4x^3) + 2^3(0) + 2^4(1) = 0$
$x^5 - 16x^3 + 16 = 0$

iii. To obtain an equation each of whose roots is less by h than the corresponding root of a given equation, replace x by $x' + h$.
Given: $x^3 - 1 = 0$
Desired Equation: $(x + 2)^3 - 1 = 0$
Simplify: $x^3 + 2x^2 + 4x + 7 = 0$

III

i. $$\begin{array}{rrrrr|l} 1 & -10 & 35 & -50 & 24 & \underline{3}\\ & 3 & -21 & 42 & -24\\ \hline 1 & -7 & 14 & -8 & 0 \end{array}$$
Since $R = 0$, the statement is *true*.

ii. Given $f(x) = 8x^4 - 18x^3 + 2x^2 + 7x - 6$
$f(-x) = 8x^4 + 18x^3 + 2x^2 - 7x - 6$
There is one variation in sign of $f(-x)$. The statement is *false*.

iii. $f(x) = y = x^3 - x^2 + 4 - 5$
$f(1) = 1 - 1 + 4 - 5 = -1$
$f(2) = 8 - 4 + 4 - 5 = 3$
The statement is *true*.

iv. Use synthetic division to check the root.
$$\begin{array}{rrrrr|l} 6 & 1 & -26 & -4 & 8 & \underline{-\frac{2}{3}}\\ & -4 & 2 & 16 & -8\\ \hline 6 & -3 & -24 & 12 & 0 \end{array}$$
The statement is *true*.

v. Since a cubic equation has only 3 roots, the statement is *false*.

vi. Rational roots are of the form $\frac{a}{b}$ where a is a factor of the constant term and b is a factor of the coefficient of x^5; Since $\frac{a}{b} = \frac{1}{1} = 1$, the statement is *true*.

vii. $f(x) = x^4 + 2x^3 - 13x^2 - 14x + 24$
$f(-x) = x^4 - 2x^3 - 13x^2 + 14x + 24$
The statement is *true*.

viii. $f(x) = x^4 - 6x^3 + x^2 - 5x + 3$
$f(-x) = x^4 + 6x^3 + x^2 + 5x + 3$
Since there are no variations in signs of $f(-x)$, the statement is *true*.

20

INEQUALITIES

SELF-TEST

DIRECTIONS: Write your answers in the numbered regions to the right or use the graph paper in the back of the book. To check your answers, turn the page. Study the solutions to the problems you missed, and do the exercises following any problem or group of problems in which you had errors.

In 1–3, find the values of x which make the inequality statement true.

1. $11x + 5 > 3x + 21$

2. $\dfrac{x}{3} + \dfrac{2x}{5} - \dfrac{3}{7} < \dfrac{2x}{6} - \dfrac{x}{5} + \dfrac{3}{5}$

3. $x^2 - x - 6 < 0$

4. Draw a figure and shade the area which includes the points that satisfy the following inequalities simultaneously.

$$3x - 5y + 2 < 0$$
$$3x + 3y - 6 > 0$$
$$y - 4 < 0$$

5. Prove that if a and b are positive, $\sqrt{ab} \leq \dfrac{a+b}{2}$.

Determine whether the statements in 6–9 are true or false.

6. If $a > b$ and $c > a$, then $c > b$.

7. If $a > b$, then $a^2 > b^2$.

8. The point $(-1, -3)$ satisfies the inequality $2x - 3y > 0$.

9. If $a = b > 0$ and $c > d$, then $\dfrac{c}{a} < \dfrac{d}{b}$.

1
2
3
4 Use graph paper, page 181.
5
6
7
8
9

116 Inequalities

SOLUTIONS

1. Given: $\quad 11x + 5 > 3x + 21$
   Add $-3x$: $11x - 3x + 5 > 5x - 3x + 21$
   Simplify: $\quad 8x + 5 > 21$
   Add -5: $\quad 8x > 16$
   Divide by 8: $\quad x > 2$

2. Given: $\quad \dfrac{x}{3} + \dfrac{2x}{5} - \dfrac{3}{7} < \dfrac{2x}{6} - \dfrac{x}{5} + \dfrac{3}{5}$

   Simplify: $\quad \dfrac{11x}{15} - \dfrac{3}{7} < \dfrac{2x}{15} + \dfrac{3}{5}$

   Transpose: $\quad \dfrac{11x}{15} - \dfrac{2x}{15} < \dfrac{3}{7} + \dfrac{3}{5}$

   Simplify: $\quad \dfrac{9}{15}x < \dfrac{36}{35}$

   Multiply by $\dfrac{15}{9}$: $\quad x < \dfrac{12}{7}$

3. Given: $x^2 - x - 6 < 0$
   Solve the equation: $x^2 - x - 6 = 0$
   $(x - 3)(x + 2) = 0$
   $x = -2, 3$
   Test $x = 0$ in the inequality:
   $$0 - 0 - 6 < 0$$
   Since the inequality is satisfied and 0 is in the interval $-2, 3$, we have the inequality satisfied by $-2 < x < 3$.

EXERCISES I

In i–vii, find the values of x which make the inequality statement true.

i. $7x - \sqrt{48} > 3x - 4\sqrt{3}$

ii. $6x + 3i^2 - 5i^3 + 7i < 3i - 9i^3$, $i^2 = -1$

iii. $ax - c \geq 3ax + c$

iv. $6x^2 - 13x + 6 < 0$

v. $x^2 - 2ax + a^2 < 0$

vi. $x^2 - 2ax - a^2 < 0$

vii. $i^2 x^2 + x - 6i^2 < 0$, $i^2 = -1$

---

4. FIG. 15-1: $3x - 5y + 2 < 0$
   $3x + y - 6 > 0$
   $y - 4 < 0$

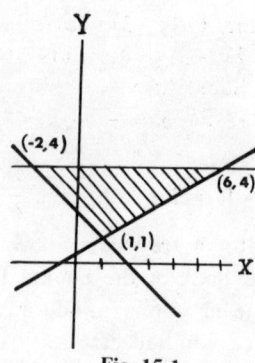

Fig. 15-1

EXERCISE II

Draw a figure and shade the area which includes the points that satisfy the following inequalities simultaneously. Use the graph paper on page 181.
$$2x + 2y - 5 < 0$$
$$2x - 2y - 1 < 0$$
$$10x - 2y - 1 > 0$$

---

5. $\sqrt{ab} \leq \dfrac{a+b}{2}$, $a > 0$, $b > 0$

Since $a > 0$ and $b > 0$, $a + b > 0$ and we can square both sides of the inequality.
$$(\sqrt{ab})^2 \leq \left(\dfrac{a+b}{2}\right)^2 \text{ or } 4ab \leq a^2 + 2ab + b^2$$
$$0 \leq a^2 - 2ab + b^2 \text{ or } 0 \leq (a-b)^2$$
which is true for all a and b. These steps are reversible and the given statement is *true*.

EXERCISES III

i. Prove that if a and b are positive and $a^2 > b^2$, then $a > b$.

ii. Show that if c is greater than a and $k < 0$, then the difference $c - a$ multiplied by k is negative.

---

6. If $a > b$ and $c > a$, then $c > b$. This statement is *true* since $c > a > b$ gives $c > b$.

7. If $a > b$, then $a^2 > b^2$. This statement is *false*. Let $a = -3$ and $b = -4$. Then $a > b$ but $a^2 \not> b^2$.

8. The point $(-1, -3)$ satisfies the inequality $2x - 3y > 0$. This statement is *true*. Substitute to obtain
$$2(-1) - 3(-3) = -2 + 9 = 7 > 0.$$

9. If $a = b > 0$ and $c > d$, then $\frac{c}{a} < \frac{d}{b}$. This statement is *false*. Let $a = b = 2$, $c = 3$, and $d = 1$. Then $\frac{c}{a} \not< \frac{d}{b}$.

EXERCISES IV
Determine whether the following statements are true or false.

i. If $c > 0$ and $x > y$, then $y - c < x - c$.

ii. If $c < 0$ and $x > y$, then $cx > cy$.

iii. If $a = b > 0$ and $c > d$, then $ac > bd$.

iv. If x is negative, then $x^3 < 0$.

v. If $a > b$, $c > d$, and $x > y$, then $2a + cx > b + dy$.

Solutions to the exercises appear on page 120.

ANSWERS

1. $x > 2$
2. $x < \frac{12}{7}$
3. $-2 < x < 3$
4. Fig. 15–1
5. true
6. true
7. false
8. true
9. false

4 Use graph paper, page 181.

Inequalities

BASIC FACTS

For each pair of real numbers x and y, one and only one of the following relations is true.

$$x < y \qquad x = y \qquad x > y$$

The symbols $<$ (is less than), $=$ (is equal to), and $>$ (is greater than) are expressions of comparison.

The definitions of the **inequalities** $<$ and $>$ are:

$a < b$ if and only if $a - b < 0$;
$a > b$ if and only if $a - b > 0$.

Two inequalities are said to have the **same sense** if their symbols for inequality point in the same direction. Otherwise, they are said to be of **opposite sense**.

The sense of an inequality is *not changed* under any of the following conditions.

both members are increased by the same number

If $a > b$, then $a + x > b + x$.

both members are diminished by the same number

If $a > b$, then $a - x > b - x$.

both members are multiplied by the same positive number

If $a > b$, then $ax > bx$ if $x > 0$.

both members are divided by the same positive number

If $a > b$, then $\dfrac{a}{x} > \dfrac{b}{x}$ if $x > 0$.

both members are positive and are raised to the same power

If $a > b$, then $a^2 > b^2$ if and only if $a > 0$ and $b > 0$.

corresponding members of inequalities in the same sense are added

If $a > b$ and $c > d$, then $a + c > b + d$.

corresponding members of inequalities in the same sense are multiplied

ADDITIONAL INFORMATION

The solution of a linear inequality is easily obtained by using the properties of inequalities stated in the Basic Facts. We illustrate by an example.

Find the values of x which satisfy the inequality $3x + 5 > 5x + 7$. First subtract the same number from both sides of the inequality, say $5x + 5$. We obtain,

$$3x + 5 - (5x + 5) > 5x + 7 - (5x + 5) \text{ or } -2x > 2.$$

Now divide both sides by -2. Since -2 is a negative number, the sense of the inequality is changed; and we obtain $x < -1$, which is true for all values of x less than -1.

To solve a quadratic inequality, we shall employ the concept of a function. Consider the inequality $x^2 - x - 12 < 0$. We are asked to find all the values of x for which the function defined by $y = x^2 - x - 12$ is negative. Let us first find the values of x for which the function is zero; i.e., the solutions of $x^2 - x - 12 = 0$. We can factor the left side into $(x - 4)(x + 3)$, and upon equating each factor to zero we have $x = 4$ and $x = -3$. Let us now consider the graph of the function $y = x^2 - x - 12$. We know that it is a parabola and can cross the X-axis in only two points; namely, $x = 4$ and $x = -3$, the values of x for which $y = 0$. For all other values of x, the graph will be either above the X-axis $(y > 0)$ or below the X-axis $(y < 0)$. Let us test the function for some value of x. Since $x = 0$ results in the simplest arithmetic, consider $f(0) = (0)^2 - (0) - 12 = -12 < 0$; i.e., the function is negative and below the X-axis. Since the graph crosses the X-axis only at $x = -3$ and $x = 4$, it will be below the X-axis for all values of x in the interval from $x = -3$ to $x = 4$, and our given inequality will be satisfied by all of these values of x. This set of values for x can be written in many forms. The two most popular are $-3 < x < 4$ and $(-3, 4)$, which are read "all x greater than -3 and less than 4" or "all x in the interval from -3 to 4."

The notation $-3 \leq x < 4$ is read "all x from -3 to 4 including -3." In notation, this is written $[-3, 4)$. See Chapter 1 for a more rigorous discussion of open and closed intervals.

Suppose that in the above example the inequality had been reversed; that is, suppose we had been given, $x^2 - x - 12 > 0$. In this case, we are looking for the values of x for which $y = x^2 - x - 12 > 0$, and the graph is above the X-axis. Let us test $x = -4$, $f(-4) = (-4)^2 - (-4) - 12 = 16 + 4 - 12 = 8 > 0$. We see that $y > 0$ for values of $x < -3$ and similarly $y > 0$ for $x > 4$. The values of x which satisfy this inequality are, therefore, $x < -3$ and $x > 4$.

Consider the **linear inequality in two variables**,

$$ax + by + c > 0.$$

This expression defines a set of pairs (x, y) which satisfies the inequality. We can consider the inequality in terms of the points in the (x, y)-plane relative to the line defined by $ax + by + c = 0$. If $a \neq 0$ (the line is not parallel to the X-axis), then the line divides the plane

If $a > b$ and $c > d$, then $ac > bd$.

The sense of an inequality is *reversed* under either of the following conditions

both members are multiplied by the same negative number

If $a > b$, then $ax < bx$ if $x < 0$.

the reciprocals of two non-zero elements of like signs are compared

If $a > b$, then $\frac{1}{a} < \frac{1}{b}$, $a, b \neq 0$.

If the sense of the inequality is the same for all values of the variables for which its members are defined, the inequality is called an **absolute or unconditional inequality**.

If the sense of the inequality holds only for certain values of the variables involved, the inequality is called a **conditional inequality**.

To solve a conditional inequality is to find those values of the variable (there are usually more than one) which satisfy the inequality, i.e., for which the inequality is true.

If an expression is either *greater than or equal to* another expression, we use the symbol $\geq$. Similarly, $\leq$ is used if an expression is either *less than or equal to* another expression.

A real number, x, is said to be **positive** if $x > 0$ and **negative** if $x < 0$.

The **absolute value** of a real number, not zero, is always positive.

If the real numbers are represented on a horizontal straight line (see Fig. 15-2), then the concept of one number being greater than another can be visualized. The greater number will always be to the right of the smaller.

Fig. 15-2

Compare, for example, the numbers 5 and 3 and also -2 and -3.

into two half planes. Consider a point $P(x_1, y_1)$ not on the line. Draw a line through P and parallel to the X-axis, and let it intersect the line at $Q(x_0, y_1)$. Then let us make the following definition.

P lies in the **right half plane** if $x_1 > x_0$;
P lies in the **left half plane** if $x_1 < x_0$.

This definition is used in the following table to state a property necessary for **graphing linear inequalities** in two variables.

	$ax + by + c$, $a \neq 0$
$ax + by + c$	if and only if
> 0	$P(x, y)$ is in the right half plane determined by the line $ax + by + c = 0$
< 0	$P(x, y)$ is in the left half plane determined by the line $ax + by + c = 0$
$= 0$	$P(x, y)$ is on the line $ax + by + c = 0$

For lines parallel to the X-axis, $y = k$, we define $y > k$ for all points above the line $y = k$ and $y < k$ for all the points below the line $y = k$.

To find the graph of $2x - 3y + 6 > 0$, for example, draw the line for $2x - 3y + 6 = 0$ by finding the intercepts $(0, 2)$ and $(-3, 0)$ and joining them with a line. All the points to the right of this line satisfy the inequality. They are shown by the shaded area of Figure 15-3.

We can now study a **system of linear inequalities in two variables**. Consider the system,

$$y - 3 < 0$$
$$x - 2y < 0$$
$$x + y - 3 > 0$$

Draw the three lines defined by setting each left member equal to zero. Then consider all the points which are simultaneously below the first line, to the left of the second line, and to the right of the third line. See Figure 15-4.

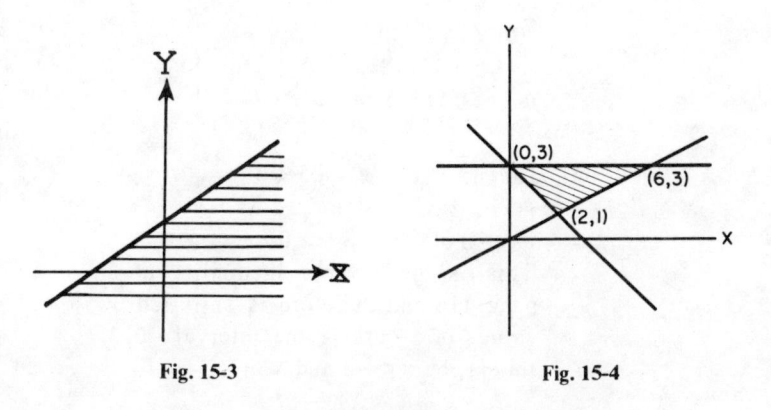

Fig. 15-3

Fig. 15-4

SOLUTIONS TO EXERCISES

I

i. Given: $\quad 7x - \sqrt{48} > 3x - 4\sqrt{3}$
   Add $-3x$: $\quad 4x - \sqrt{48} > -4\sqrt{3}$
   Transpose $\sqrt{48}$: $\quad 4x > \sqrt{48} - 4\sqrt{3}$
   Simplify: $\quad 4x > 4\sqrt{3} - 4\sqrt{3}$
   $\quad 4x > 0,$
   $\quad x > 0$

ii. Given: $\quad 6x + 3i^2 - 5i^3 + 7i < 3i - 9i^3$
    Simplify: $\quad 6x - 3 + 5i + 7i < 3i + 9i$
    $\quad 6x - 3 + 12i < 12i$
    $\quad 6x < 3,$
    $\quad x < \frac{1}{2}$

iii. Given: $\quad ax - c \geq 3ax + c$
     Transpose: $\quad ax - 3ax \geq c + c$
     Simplify: $\quad -2ax \geq 2c$
     Divide by $-2a$: $\quad x \leq \frac{c}{a}$

iv. Given: $6x^2 - 13x + 6 < 0$
    $(3x - 2)(2x - 3) = 0; x = \frac{2}{3}$ and $\frac{3}{2}$
    At $x = 1$, $6 - 13 + 6 = -1 < 0$.
    $\therefore 6x^2 - 13x + 6 < 0$ at $\frac{2}{3} < x < \frac{3}{2}$.

v. Given: $x^2 - 2ax + a^2 < 0$ or $(x - a)^2 < 0$
   Since the square of a real number is always positive, $(x - a)^2 > 0$ for all x; and no values of x satisfy the inequality.

vi. Given: $x^2 - 2ax - a^2 < 0$
    $f(x) = x^2 - 2ax - a^2 = 0$, $x = a \pm a\sqrt{2}$
    $f(0) = -a^2 < 0$
    $\therefore f(x) = x^2 - 2ax - a^2 < 0$ for
    $a(1 - \sqrt{2}) < x < a(1 + \sqrt{2})$

vii. Given: $i^2x^2 + x - 6i^2 < 0$
     $f(x) = i^2x^2 + x - 6i^2 = -x^2 + x + 6 < 0$
     or $g(x) = x^2 - x - 6 > 0$
     This is the reverse inequality of 3, page 116 and therefore is satisfied by values of x outside the interval $-2, 3$; that is, by $x < -2$ and $x > 3$.

II

Fig. 15-5: $2x + 2y - 5 < 0$
$2x - 2y - 1 < 0$
$10x - 2y - 1 > 0$

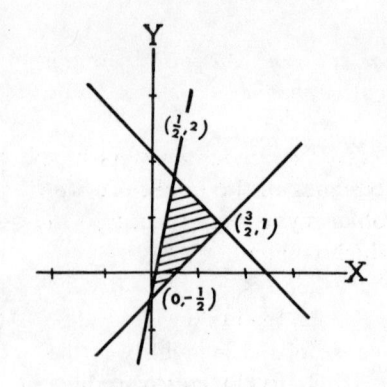

III

i. Given: $\quad a^2 > b^2$
   Transpose: $\quad a^2 - b^2 > 0$
   Factor: $\quad (a + b)(a - b) > 0$
   Since $a > 0$ and $b > 0$,
   divide by $a + b$: $a - b > 0$ or $a > b$.

ii. Given $c > a$, then $c - a > 0$.
    Since $k < 0$, then $k(c - a) < 0$.

IV

i. If $c > 0$ and $x > y$, then $y - c < x - c$. This statement is *true* since $x > y$ means $y < x$ and $\therefore y - c < x - c$ for all c.

ii. If $c < 0$ and $x > y$, then $cx > cy$. *false*

iii. If $a = b > 0$ and $c > d$, then $ac > bd$. *true*

iv. If x is negative, then $x^3 < 0$. This statement is *true* since the cube of a negative number is a negative number.

v. If $a > b$, $c > d$, and $x > y$, then $2a + cx > b + dy$. *true*

21

EXPONENTIAL AND LOGARITHMIC FUNCTIONS

SELF-TEST

DIRECTIONS: Write your answers in the numbered regions to the right or use the graph paper in the back of the book. To check your answers, turn the page. Study the solutions to the problems you missed, and do the exercises following any problem or group of problems in which you had errors.

1. Evaluate y if $y = 16^{-\frac{1}{2}} \; 16^{\frac{3}{4}}$.

2. Evaluate y if $y = \log_3 27$.

3. Express $\log_3 \left(\frac{1}{81}\right) = -4$ in exponential form.

4. Express $16^{\frac{3}{4}} = 8$ in logarithmic form.

5. Express $\log_4 7 + \frac{3}{2} \log_4 9 - 2 \log_4 10$ as the logarithm of a single number.

6. Express $\log \sqrt[3]{\dfrac{8.65 \times 32.54}{0.06}}$ in terms of the logarithms of its factors.

7. Find the solution set of $9^{x-3} = 27^{x+4}$ without using logarithms.

8. Find, correct to the nearest tenth, the value of x which satisfies $3^{2x} = 64.6$.

9. Find the solution set of the following equation.

$$\log(x^2 - 4) - \log(x + 2) = \log 16$$

In **10–13**, let $f: x \to a^x$ and $f(3) = 8$, and let $g: x \to b^x$ and $g(-2) = \frac{1}{9}$.

10. Find the values of a and b.
12. Find the product $(fg)\left(\frac{1}{3}\right)$.

11. Find the sum $(f + g)(3)$.
13. Find the composite $(f \circ g)(2)$.

14. Sketch the graphs of $y = 3^x$ and $y = \log_3 x$ on the same set of axes.

1
2
3
4
5
6
7
8
9
10 $a =$
    $b =$
11
12
13
14 Use graph paper page 181.

SOLUTIONS

1. Apply the definitions for rational exponents.
$$y = 16^{-\frac{1}{2}} \; 16^{\frac{3}{4}} = \frac{1}{\sqrt{16}} \cdot \sqrt[4]{16^3}$$
$$= \frac{1}{4} \cdot 2^3 = \frac{1}{4} \cdot 8 = 2$$

2. $(y = \log_3 27) \leftrightarrow (3^y = 27), \therefore y = 3$

3. The exponential form of $\log_3\left(\frac{1}{81}\right) = -4$ is $3^{-4} = \frac{1}{81}$.

4. The logarithmic form of $16^{\frac{3}{4}} = 8$ is $\log_{16} 8 = \frac{3}{4}$.

EXERCISES I

In **i–viii**, find the value of y.

i. $y = 125^{\frac{2}{3}} 25^{-\frac{1}{2}}$ **ii.** $y = \ln e^{\frac{1}{2}}$
iii. $y = \log 0.0001$ **iv.** $y = \log_3 81$
v. $y = \log_5 \sqrt[3]{5}$ **vi.** $y = \ln e$
vii. $y = \log_a a$ **viii.** $y = 4^{\log_2 8}$

ix. Express $\log_{125} 5 = \frac{1}{3}$ in exponential form.

x. Express $3^{-5} = \frac{1}{405}$ in logarithmic form.

---

5. $\log_4 7 + \left(\frac{3}{2}\right)\log_4 9 - 2\log_4 10$
$= \log_4 7 + \log_4 9^{\frac{3}{2}} - \log_4 10^2$
$= \log_4 7 + \log_4 27 - \log_4 100$
$= \log_4 (7)(27) - \log_4 100$
$= \log_4 \frac{(7)(27)}{100}$
$= \log_4 1.89$

6. $\log \sqrt[3]{\frac{8.65 \times 32.54}{0.06}} = \left(\frac{8.65 \times 32.54}{0.06}\right)^{\frac{1}{3}}$
$= \frac{1}{3} \log \frac{8.65 \times 32.54}{0.06}$
$= \frac{1}{3}(\log 8.65 \times 32.54 - \log 0.06)$
$= \frac{1}{3}(\log 8.65 + \log 32.54 - \log 0.06)$
$= \frac{1}{3}\log 8.65 + \frac{1}{3}\log 32.54 - \frac{1}{3}\log 0.06$

EXERCISES II

i. Express
$$\frac{1}{3}\log_b 27 - \left(\frac{2}{5}\log_b 32 + \log_b 7\right)$$
as the logarithm of a single number.

Match **ii–iv** with **a–c**.

ii. $\log x_1^2 x_2$ **a.** $\frac{1}{2}(\log x_1 - \log x_2)$
iii. $\log \sqrt{x_1/x_2}$ **b.** $\log x_1 + \frac{1}{2}\log x_2$
iv. $\log x_1 \sqrt{x_2}$ **c.** $2\log x_1 + \log x_2$

---

7. Express both members $9^{x-3} = 27^{x+4}$
as powers of 3. $(3^2)^{x-3} = (3^3)^{x+4}$
Multiply exponents. $3^{2x-6} = 3^{3x+12}$
Equate exponents. $2x - 6 = 3x + 12$
Solve for x. $x = -18$

8. $3^{2x} = 64.6$
$\log 3^{2x} = \log 64.6$
$2x \log 3 = \log 64.6$
$x = \frac{1}{2}\left(\frac{\log 64.6}{\log 3}\right)$
$x = \frac{1}{2}\left(\frac{1.8102}{0.4771}\right) = 1.9$, to the nearest tenth

9. $\log(x^2 - 4) - \log(x + 2) = \log 16$
$\log \frac{x^2 - 4}{x + 2} = \log 16$
$\frac{x^2 - 4}{x + 2} = 16$
$\frac{(x + 2)(x - 2)}{(x + 2)} = 16$
$x - 2 = 16$
$x = 18$

EXERCISES III

In **i–vi**, find the solution set.

i. $16^{2x} = 8^{\frac{1}{2}}$
ii. $10^{2x} = 0.001$
iii. $1.23^x = 4$
iv. $\log(2x - 2) - \log(x + 4) = \log 12$
v. $2\log(x - 6) = \log 16$
vi. $\log(x + 2) + \log(x - 1) = 1$

Exponential and Logarithmic Functions 123

10. $f: x \to a^x$ and $f(3) = 8$
$f(3) = a^3, 8 = a^3, a = 2$
$g: x \to b^x$ and $g(-2) = \frac{1}{9}$
$g(-2) = b^{-2}, \frac{1}{9} = b^{-2}, \therefore b = 3$

11. $(f + g)(3) = f(3) + g(3) = 2^3 + 3^3 = 35$

12. $(fg)\left(\frac{1}{3}\right) = f\left(\frac{1}{3}\right)g\left(\frac{1}{3}\right) = 2^{\frac{1}{3}} 3^{\frac{1}{3}}$
$= (2 \cdot 3)^{\frac{1}{3}} = 6^{\frac{1}{3}}$

13. $(f \circ g)(2) = f(g(2)) = f(3^2)$
$= f(9) = 2^9$

14. $y = 3^x$

x	-3	-2	-1	0	1	2	3
y	$\frac{1}{27}$	$\frac{1}{9}$	$\frac{1}{3}$	1	3	9	27

Since $y = 3^x$ and $y = \log_3 x$ are inverse functions, the graph of $y = \log_3 x$ can be obtained by reflecting the graph of $y = 3^x$ in the line $y = x$. The graph can also be obtained by constructing a table of values. In this case it is easier to assign values to y and compute the corresponding values for x.

$y = \log_3 x \longleftrightarrow 3^y = x$

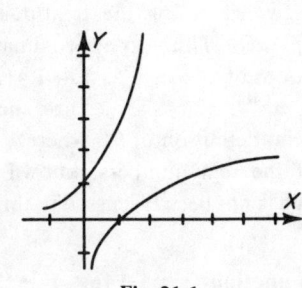

x	$\frac{1}{27}$	$\frac{1}{9}$	$\frac{1}{3}$	1	3	9	27
y	-3	-2	-1	0	1	2	3

Fig. 21-1

Solutions to the exercises appear on page 126.

ANSWERS

1. $y = 2$
2. $y = 3$
3. $3^{-4} = \frac{1}{81}$
4. $\log_{16} 8 = \frac{3}{4}$
5. $\log_4 1.89$
6. $\frac{1}{3} \log 8.65 + \frac{1}{3} \log 32.54 - \frac{1}{3} \log 0.06$
7. $\{-18\}$
8. $x = 1.9$
9. $\{18\}$
10. $a = 2$
    $b = 3$
11. 35
12. $6^{\frac{1}{3}}$
13. 2^9
14. Figure 21-1

1
2
3
4
5
6
7
8
9
10 $a =$
   $b =$
11
12
13
14 Use graph paper, page 181.

BASIC FACTS

Functions of the form
$$f: x \rightarrow b^x, b > 0$$
are called **exponential functions** with base b. The domain of f is the set of real numbers, and the range is the set of positive real numbers.

All exponential functions have the following properties.

$$b^{x_1} = b^{x_2} \leftrightarrow x_1 = x_2, b \neq 1$$
$$b^{x_1} b^{x_2} = b^{x_1 + x_2}$$
$$(b^{x_1})^{x_2} = b^{x_1 x_2}$$
$$(ab)^{x_1} = a^{x_1} b^{x_1}$$

Where x_1 and x_2 are real numbers.

Graphs of exponential functions, defined by $y = b^x$, $b > 0$, vary with the value of b.

If $b = 1$, the function degenerates into the constant function $f: x \rightarrow 1$, since $1^x = 1$ for all x.

If $b > 1$, the function is one-to-one and increasing since b^x increases as x increases. As x becomes smaller, b^x becomes smaller and approaches but never attains the value zero.

If $b < 1$, the function is one-to-one and decreasing since b^x decreases as x increases. As x becomes large, b^x approaches but never reaches 0.

For any $b > 0$, the graph of the function has a y-intercept at 1, since $f(0) = 1$. The graph lies above the X-axis since $b^x > 0$ for all x. Hence the function has no zeros.

These graphs are given in Figure 21-2.

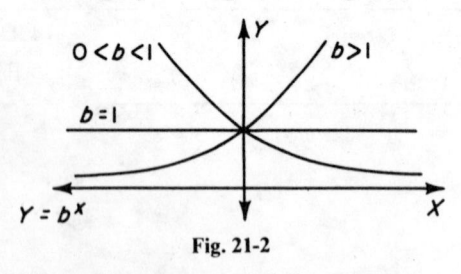

Fig. 21-2

Since the exponential function f defined by $y = b^x$, $b > 0$, $b \neq 1$, is one-to-one, it has an inverse function f^{-1} defined by $x = b^y$, $b > 0$, $b \neq 1$. The solu-

ADDITIONAL INFORMATION

The Number e

A number of great interest and importance in mathematics is an irrational number, denoted e, and approximately equal to 2.71828. The use of e as a base for the exponential function is of such consequence in theoretical mathematics that the function
$$f: x \rightarrow e^x$$
is referred to as *the* **exponential function** and the base not mentioned. The graph of $y = e^x$ has the property that the slope of the line tangent to the curve at any point (x,y) is e^x. Tables for values of e^x and e^{-x} are given in the appendix, page 169.

The exponential functions are used to define a set of functions called the **hyperbolic functions**, the hyperbolic sin (sinh x), the hyperbolic cos (cosh x), and the hyperbolic tan (tanh x).

$$\sinh x = \frac{e^x - e^{-x}}{2} \quad \cosh x = \frac{e^x + e^{-x}}{2} \quad \tanh x = \frac{e^x - e^{-x}}{e^x + e^{-x}}$$

Evaluating the Exponential Function $f: x \rightarrow b^x$

The domain of definition for the exponential function with base b is the set of real numbers. If x is a rational number, the functional value, b^x, is obtained from the elementary definitions for exponents. Thus, if m and n are positive integers,

$$b^n = b \cdot b \cdot b \cdot \ldots \cdot b \ (n \text{ factors})$$
$$b^0 = 1, b \neq 0$$
$$b^{-n} = \frac{1}{b^n}, b \neq 0$$
$$b^{\frac{1}{n}} = \sqrt[n]{b}$$
$$b^{\frac{m}{n}} = \left(b^{\frac{1}{n}}\right)^m = (\sqrt[n]{b})^m = \sqrt[n]{b^m}.$$

If x is an irrational number, the functional value b^x can be obtained to any approximation by expanding the irrational to a sufficient number of decimal places. Thus, to approximate $b^{\sqrt{2}}$, consider successive approximations of $\sqrt{2}$ such as 1.4, 1.41, 1.414, 1.4142, ... Then $b^{1.4}$, $b^{1.41}$, $b^{1.414}$, $b^{1.4142}$, ..., are successive approximations of $b^{\sqrt{2}}$. A formal definition of b^x, where x is irrational, requires a property of the real numbers, known as the Axiom of Completeness, which has not been discussed in this book.

Evaluating the Logarithmic Function $f: x \rightarrow \log_b x$

When e is used as a base for the logarithmic function, the logarithms are called **natural logarithms** and denoted ln x with the base omitted. When 10 is used as a base, the logarithms are called **common logarithms** and denoted by log x, again omitting the base. Tables for the logarithms of numbers to bases e and 10 are given in the appendix, page 166.

For bases other than 10 and e, the functional value $\log_b x$ is obtion of f^{-1} for y is denoted $y = \log_b x$, and y is called the **logarithm** of x to the base b.

Functions of the form

$$f: x \to \log_b x, b > 0, b \neq 1$$

where $y = \log_b x \leftrightarrow x = b^y$, are called **logarithmic functions** to the base b. The domain of f is the set of positive real numbers and the range is the set of real numbers.

Since the exponential function with base b and the logarithmic function to the base b are inverse functions, the graph of f^{-1} is the reflection of the graph of f in the line $y = x$. (See Figure 21-3.)

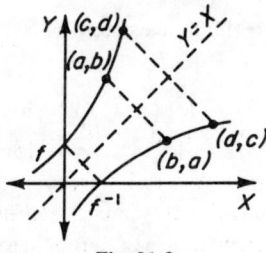

Fig. 21-3

All logarithmic functions have the following properties.

$$\log_b x_1 = \log_b x_2 \leftrightarrow x_1 = x_2$$

$$\log_b x_1 x_2 = \log_b x_1 + \log_b x_2$$

$$\log_b 1/x_1 = -\log_b x_1$$

$$\log_b x_1/x_2 = \log_b x_1 - \log_b x_2$$

$$\log_b x_1^p = p \log_b x_1, p \text{ rational}$$

where x_1 and x_2 are positive real numbers.

Graphs of logarithmic functions defined by $y = \log_b x$, $b > 0$, $b \neq 1$, vary with the value of b. If $b > 1$, the graph is increasing. If $b < 1$, the graph is decreasing. (See Figure 21-4.)

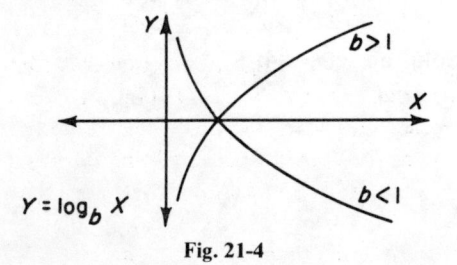

Fig. 21-4

tained by changing from base b to either base 10 or base e and then using the tables. For example, if $f: x \to \log_4 x$, then $f(5) = \log_4 5$. To evaluate $\log_4 5$, convert to base 10 and use the tables for common logarithms.

$$\log_b x = \frac{\log_a x}{\log_a b} \; ; \quad \log_4 5 = \frac{\log_{10} 5}{\log_{10} 4} = \frac{0.6990}{0.6021} = 1.1443$$

Exponential and Logarithmic Equations

An **exponential equation** is an equation in which the unknown appears in an exponent. The solution set for some exponential equations is very difficult to obtain; but the simpler ones may be solved in either of the following ways.

METHOD I: 1. Express both members of the equation as powers of the same base.
2. Apply the first property for exponential functions (see Basic Facts), and equate the exponents.
3. Solve the resulting equation for x.

METHOD II: 1. Apply the first property for logarithmic functions (see Basic Facts), and equate the logarithims of both members.
2. Solve the resulting equation for x.

By Method I, $3^{2x} = 27$
$3^{2x} = 3^3$
$2x = 3$
$x = 1\frac{1}{2}$.

By Method II, $5^{2x+4} = 120$
$\log(5^{2x+4}) = \log 120$
$(2x + 4) \log 5 = \log 120$
$2x + 4 = \frac{\log 120}{\log 5}$
$2x + 4 = \frac{2.0792}{0.6990}$
$2x + 4 = 29.74$
$2x = 25.74$
$x = 12.87$.

A **logarithmic equation** is an equation in which the unknown appears in an expression whose logarithm is indicated. Log $(x + 3) = \log 9$ is a logarithmic equation, but $x + \log 10 = \log 3$ is not. Some logarithmic equations can be solved by applying the definition and properties of the logarithmic function.

$$\log (x + 3) = \log 9$$
$$x + 3 = 9$$
$$x = 6$$

It is important to remember that the domain of the logarithmic function does not include negative numbers and, therefore, that the solution set of a logarithmic equation can include no numbers which, when substituted into the original equation, result in a logarithm of a negative number.

SOLUTIONS TO EXERCISES

I

i. $125^{\frac{2}{3}} \, 25^{-\frac{1}{2}} = \sqrt[3]{125^2} \cdot \frac{1}{\sqrt{25}} = \frac{5^2}{5} = 5$

ii. $y = \ln e^{\frac{1}{2}} \leftrightarrow e^y = e^{\frac{1}{2}}, \therefore y = \frac{1}{2}$

iii. $(y = \log 0.0001) \leftrightarrow (10^y = 0.0001 = 10^{-4}) \therefore y = -4$

iv. $(y = \log_3 81) \leftrightarrow (3^y = 81), \therefore y = 4$

v. $(y = \log_5 \sqrt[3]{5}) \leftrightarrow (5^y = 5^{\frac{1}{3}}), \therefore y = \frac{1}{3}$

vi. $(y = \ln e) \leftrightarrow (e^y = e), \therefore y = 1$

vii. $(y = \log_a a) \leftrightarrow (a^y = a), \therefore y = 1$

viii. $y = 4^{\log_2 8} = 4^3 = 64$

ix. $\log_{125} 5 = \frac{1}{3}$

In exponential form, $125^{\frac{1}{3}} = 5$.

x. $3^{-5} = \frac{1}{405}$

In logarithmic form, $\log_3 \frac{1}{405} = -5$.

II

i. $\frac{1}{3} \log_b 27 - \left(\frac{2}{5} \log_b 32 + \log_b 7 \right)$

$= \log_b 27^{\frac{1}{3}} - (\log_b 32^{\frac{2}{5}} + \log_b 7)$
$= \log_b 3 - (\log_b 4 + \log_b 7)$
$= \log_b 3 - \log_b (4)(7)$
$= \log_b 3 - \log_b 28$
$= \log_b \frac{3}{28}$

ii. $\log x_1^2 x_2 = \log x_1^2 + \log x_2$
$= 2 \log x_1 + \log x_2$

iii. $\log \sqrt{\frac{x_1}{x_2}} = \log \left(\frac{x_1}{x_2} \right)^{\frac{1}{2}} = \frac{1}{2} \log \frac{x_1}{x_2}$
$= \frac{1}{2} (\log x_1 - \log x_2)$

iv. $\log x_1 \sqrt{x_2} = \log x_1 + \log \sqrt{x_2}$
$= \log x_1 + \log x_2^{\frac{1}{2}}$
$= \log x_1 + \frac{1}{2} \log x_2$

III

i. $16^{2x} = 8^{\frac{1}{2}}$
$(2^4)^{2x} = (2^3)^{\frac{1}{2}}$
$2^{8x} = 2^{\frac{3}{2}}$
$8x = \frac{3}{2}$
$x = \frac{3}{16}$

Solution set $= \left\{ \frac{3}{16} \right\}$

ii. $10^{2x} = 0.001$
$10^{2x} = 10^{-3}$
$2x = -3$
$x = -\frac{3}{2}$

Solution set $= \left\{ -\frac{3}{2} \right\}$

iii. $1.23^x = 4$
$\log(1.23^x) = \log 4$
$x \log 1.23 = \log 4$
$x = \frac{\log 4}{\log 1.23} = \frac{0.6021}{0.0899}$
$x = 6.7$, to the nearest tenth
Solution set $= \{6.7\}$

iv. $\log(2x - 2) - \log(x + 4) = \log 12$
$\log \frac{2x - 2}{x + 4} = \log 12$
$\frac{2x - 2}{x + 4} = 12$
$2x - 2 = 12x + 48$
$x = -5$

Since the logarithmic function is defined only for positive numbers, -5 is not acceptable and the solution set is ϕ.

v. $2 \log(x - 6) = \log 16$
$\log(x - 6)^2 = \log 16$
$(x - 6)^2 = 16$
$x^2 - 12x + 20 = 0$
$(x - 10)(x - 2) = 0$
$x = 10 \lor x = 2$

$x = 2$ is not acceptable since it results in the logarithm of a negative number. The solution set is $\{10\}$.

vi. $\log(x + 2) + \log(x - 1) = 1$
$\log[(x + 2)(x - 1)] = \log 10$
$x^2 + x - 12 = 0$
$(x + 3)(x - 4) = 0$
$x = -3 \lor x = 4$

Solution set $= \{4\}$

22 SEQUENCES, SERIES, LIMITS

SELF-TEST

DIRECTIONS: Write your answers in the numbered regions to the right. To check your answers, turn the page. Study the solutions to the problems you missed, and do the exercises following any problem or group or problems in which you had errors.

1. Write the first five terms of the sequence defined by
$$a_n = \frac{1}{n^2 + 4}.$$

2. Expand $\sum_{n=1}^{\infty} (-1)^n + n$ to five terms.

3. Write the first five terms of an arithmetic sequence in which the first term is a and the third term is b.

4. The first term of a geometric sequence is $\frac{1}{16}$ and the common ratio is 2. Find the tenth term.

5. Find the sum of the following series.
$$\sum_{n=1}^{100} 3n + 4$$

6. Prove by mathematical induction that the sum of the first n terms of a geometric series is $S_n = a_1 \frac{(r^n - 1)}{r - 1}$, where a_1 is the first term and r is the common ratio, $r \neq 1$.

7. What four terms (arithmetic means) can be inserted between $2\frac{1}{2}$ and 10 so that the six terms are in arithmetic progression?

8. If the general term of a sequence is $\frac{3n^4 + n^3 + 2n - 4}{5n^4 + 2n^3}$, find the limit of the sequence as n increases without bound.

9. If $f: x \to \frac{6x^3 - 4x^2 + 8}{x^2 - x}$, evaluate $\lim_{x \to -1} f(x)$.

10. If $f(x) = \frac{4x^2 - 4}{x + 1}$ for all real x where $x \neq -1$, define $f(-1)$ so that $f(x)$ will be continuous at $x = -1$.

SOLUTIONS

1. $a_n = \dfrac{1}{(n^2 + 4)}$

$a_1 = \dfrac{1}{1+4} = \dfrac{1}{5}, a_2 = \dfrac{1}{8}, a_3 = \dfrac{1}{13}, a_4 = \dfrac{1}{20},$

$a_5 = \dfrac{1}{29}$

2. $\displaystyle\sum_{n=1}^{\infty} (-1)^n + n =$

$[(-1)^1 + 1] + [(-1)^2 + 2] +$
$[(-1)^3 + 3] + [(-1)^4 + 4] +$
$[(-1)^5 + 5] = 0 + 3 + 2 + 5 + 4,$
to five terms

EXERCISES I

In i–iv, write the first five terms of the given sequence or series.

i. $a_n = \dfrac{n}{2^n}$ **iii.** $\displaystyle\sum_{n=1}^{\infty} \dfrac{n}{2n+4}$

ii. $a_n = \log 2n$ **iv.** $\displaystyle\sum_{n=2}^{\infty} \dfrac{(-n)^3}{n-1}$

v. Write a series of twelve terms in solution to the following problem.

Starting with a single pair of rabbits, how many pairs will be produced in a year if every month each pair produces a new pair which become productive from the second month on?

---

3. $a_1 = a$ and $a_3 = b$. The common difference is found by using $a_n = a_1 + (n-1)d$ where $n = 3$. Hence $b = a + 2d$ and $d = \dfrac{(b-a)}{2}$.

$a_2 = a_1 + d = a + \dfrac{b-a}{2} = \dfrac{a+b}{2}$

$a_4 = a_3 + d = b + \dfrac{b-a}{2} = \dfrac{3b-a}{2}$

$a_5 = a_4 + d = \dfrac{3b-a}{2} + \dfrac{b-a}{2} = \dfrac{4b-2a}{2}$

$= 2b - a$

4. Since $a_n = a_1 r^{n-1}, a_{10} = (1/16)(2)^9 = 32.$

5. $\displaystyle\sum_{n=1}^{100} 3n + 4 = 7 + 10 + 13 + 16 + \cdots$

to 100 terms. This is an arithmetic series with a common difference of 3.

$a_n = a_1 + (n-1)d$
$a_{100} = 7 + (99)(3) = 304$
$S_n = \dfrac{n}{2}(a_1 + a_n)$
$S_{100} = \dfrac{100}{2}(7 + 304) = 15{,}550$

6. $p_n: S_n = \dfrac{a_n(r^n - 1)}{(r-1)}$

p_n is true for $n = 1$ since $S_1 = \dfrac{a_1(r-1)}{r-1}$

$= a_1.$

Assume that p_k is true and use this to prove that p_{k+1} is true.

$p_k: S_k = \dfrac{a_1(r^k - 1)}{-1}$

$p_{k+1}: S_{k+1} - S_k + a_{k+1}$

$= \dfrac{a_1(r^k - 1)}{r - 1} + a_1 r^k$

$= \dfrac{a_1(r^k - 1) + a_1 r^k(r - 1)}{r - 1}$

$= \dfrac{a_1 r^k - a_1 + a_1 r^{k+1} - a_1 r^k}{r - 1}$

$= \dfrac{a_1 r^{k+1} - a_1}{r - 1}$

$= \dfrac{a_1(r^{k+1} - 1)}{r - 1}$

Since the result of operating on p_k is the same as the result obtained by substituting $n = k + 1$ in p_n, $p_k \rightarrow p_{k+1}$.

Since p_n satisfies both conditions of the Axiom of Mathematical Induction, it is true for all positive integral n.

7. Let $a_1 = 2\tfrac{1}{2}$ and $a_6 = 10$. To find the common difference, use $a_n = a_1 + (n-1)d$ where $n = 6$.

$10 = 2\tfrac{1}{2} + 5d$ and $d = 1\tfrac{1}{2}$

Therefore, $a_2 = 4$, $a_3 = 5\tfrac{1}{2}$, $a_4 = 7$,

$a_5 = 8\tfrac{1}{2}$.

Sequences, Series, Limits

EXERCISES II

i. Find the 25th term of the arithmetic sequence $-8, -5, -2, \ldots$.

ii. Find the sum of the first ten powers of ten, starting with 10^1.

8. $\dfrac{3n^4 + n^3 + 2n - 4}{5n^4 + 2n^3} = \dfrac{3 + \dfrac{1}{n} + \dfrac{2}{n^3} - \dfrac{4}{n^4}}{5 + \dfrac{2}{n}}$

$\lim\limits_{n \to \infty} \left(\dfrac{3 + \dfrac{1}{n} + \dfrac{2}{n^3} - \dfrac{4}{n^4}}{5 + \dfrac{2}{n}} \right) = \dfrac{3}{5}$

EXERCISES III

Find the limit of each sequence, if the limit exists.

i. $a_n = 7$ iii. $a_n = 3n^2 + n$

ii. $a_n = \dfrac{3}{n^3} + 6$ iv. $a_n = \dfrac{3 - n}{n}$

9. $\lim\limits_{x \to -1} \dfrac{(6x^3 - 4x^2 + 8)}{(x^2 - x)}$

$= \dfrac{\lim\limits_{x \to -1} 6x^3 - 4x^2 + 8}{\lim\limits_{x \to -1} x^2 - x} = \dfrac{-2}{2} = -1$

EXERCISES IV

Find the following limits.

i. $\lim\limits_{x \to -2} f(x)$ where $f(x) = \begin{cases} x + 2 \text{ if } x \geq 0 \\ 3 \text{ if } x < 0 \end{cases}$

ii. $\lim\limits_{x \to 2} \dfrac{|2x - 5|}{4}$ iii. $\lim\limits_{x \to -4} \dfrac{x^3(x + 4)}{x + 4}$

10. $f(x) = \dfrac{4x^2 - 4}{x + 1}, x \neq 1$

$\dfrac{4x^2 - 4}{x + 1} = \dfrac{4(x + 1)(x - 1)}{x + 1} = 4(x - 1)$

$\lim\limits_{x \to -1} 4(x - 1) = -8$

$f(-1) = -8$ makes $f(x)$ continuous at $x = -1$.

ANSWERS

1. $\dfrac{1}{5}, \dfrac{1}{8}, \dfrac{1}{13}, \dfrac{1}{20}, \dfrac{1}{29}$

2. $0 + 3 + 2 + 5 + 4$

3. $a, \dfrac{a + b}{2}, b, \dfrac{3b - a}{2}, 2b - a$

4. 32

5. 15,550

6. See solutions.

7. $4, 5\dfrac{1}{2}, 7, 8\dfrac{1}{2}$

8. $\dfrac{3}{5}$

9. -1

10. $f(-1) = -8$

EXERCISE V

Determine the point of discontinuity for each function. Define the function at this point so as to make it continuous.

i. $f(x) = \dfrac{3x - 6}{x - 2}$

ii. $f(x) = \dfrac{5x(x - 3)^2}{x^2 - 6x + 9}$

Solutions to the exercises appear on page 132.

BASIC FACTS

A **sequence** of real numbers is a function $a(n)$ from the set of positive integers to the set of real numbers. The functional values or **terms** of the sequence, $a(1), a(2), a(3), \ldots, a(n), \ldots$, are usually denoted $a_1, a_2, a_3, \ldots, a_n, \ldots$ to distinguish the sequence from other functions. a_1 is the first term, a_2 the second term, and a_n is the nth or **general term**.

The sequence
$$a = \{(1,a_1), (2,a_2), (3,a_3), \ldots, (n,a_n), \ldots\}$$
is denoted in subscript notation as
$$a = a_1, a_2, a_3, \ldots, a_n, \ldots$$
or in set notation as
$$a = \{a_n\}_{n=1}^{\infty}.$$

A **series** is the indicated sum of a sequence. The series
$$a_1 + a_2 + a_3 + \cdots + a_n + \cdots$$
can be expressed in summation notation as
$$\sum_{n=1}^{\infty} a_n$$
which is read "the summation of a_n from $n = 1$ to ∞."

Consider the sequences $a = 1, \frac{1}{2}, \frac{1}{3}, \frac{1}{4}, \ldots, \frac{1}{n}, \ldots$ and $b = 1, 8, 27, 64, \ldots, n^3, \ldots$. These sequences differ according to the way in which the terms behave as n becomes large. In sequence a, $\frac{1}{n}$ gets close to the fixed number zero as n becomes large. Zero is said to be the limit of the terms of the sequence or the limit of the sequence.

In general, if the values of a_n for all sufficiently large n are close to the fixed number A, then A is the **limit of the sequence**. This is written
$$\lim_{n \to \infty} a_n = A$$
and read "the limit of a_n as n becomes infinite is A." Since it is not possible for a_n to be close to each of two different numbers for all n sufficiently large, a sequence cannot have two limits. A

ADDITIONAL INFORMATION

The sequence $a = \{(1,a_1), (2,a_2), (3,a_3), \ldots, (n,a_n), \ldots\}$ can be written $a = a_1, a_2, a_3, \ldots, a_n, \ldots$ since the natural order for the terms indicates the correspondence between the domain and range. In such a listing, the general term or some other rule of correspondence must be given. Tacit assumption of the values of the unlisted terms is not sufficient. For example, $0, 1, 0, 1, \ldots$ is not a sequence since no rule for the construction of the remaining terms is given and many possibilities exist. Two such possibilities are

$$a = 0, 1, 0, 1, 0, 1, 0, \ldots, a_n, \ldots \text{ where } a_n = \begin{cases} 0 \text{ if } n \text{ is odd} \\ 1 \text{ if } n \text{ is even} \end{cases}$$

$$b = 0, 1, 0, 1, 4, 9, 16, \ldots, b_n, \ldots \text{ where } b_n = \begin{cases} 0 \text{ if } n = 1 \\ (n-3)^2 \text{ if } n > 1. \end{cases}$$

Two special sequences which occur frequently in mathematics are the arithmetic and geometric sequences. An **arithmetic sequence** is a sequence in which every term after the first is obtained by adding a constant (the common difference) to the preceding term. The nth term of any arithmetic sequence is given by the formula $a_n = a_1 + (n - 1)d$, where d is the common difference. The terms of an arithmetic sequence are said to be in arithmetic progression.

An **arithmetic series** is the indicated sum of the terms of an arithmetic sequence. The actual sum of the first n terms of such a series is given by the formula $S_n = \frac{n}{2}(a_1 + a_n)$.

A **geometric sequence** is a sequence in which every term after the first is obtained by multiplying the preceding term by a constant (the common ratio). The nth term of a geometric sequence is given by the formula $a_n = a_1 r^{n-1}$, where r is the common ratio. The terms of a geometric sequence are said to be in geometric progression.

A **geometric series** is the indicated sum of the terms of a geometric sequence. The actual sum of the first n terms is given by the formula $S_n = \dfrac{a_1(r^n - 1)}{r - 1}$, where r is the common ratio and $r \neq 1$. If $r = 1, S_n = na_1$.

The formal definition of the limit of a sequence clarifies the somewhat ambiguous terminology ('close to' and 'sufficiently large') of the more intuitive explanation given in the Basic Facts.

The **limit of a sequence** $a_1, a_2, a_3, \ldots, a_n, \ldots$ is the number A if, given any positive number ϵ, there exists a real number N such that $|a_n - A| < \epsilon$ for all $n > N$.

The given number ϵ specifies how close n must be to A, and the number N specifies how large n should be to insure this degree of closeness. In general, N depends on ϵ because as ϵ is made smaller, N becomes larger.

This concept can be interpreted geometrically. Figure 22-1 is

sequence which has a limit is said to be **convergent**.

In the sequence b, as n becomes large so does n^3. Since there is no fixed number that n^3 approaches, the sequence has no limit. A sequence which has no limit is said to be **divergent**.

The limits of many sequences can be found by applying the following theorem.

THEOREM 22-1: Let $\lim_{n \to \infty} a_n = A$, and $\lim_{n \to \infty} b_n = B$, then

$\lim_{n \to \infty} c = c$, c the constant sequence

$\lim_{n \to \infty} (a_n \pm b_n) = A \pm B$

$\lim_{n \to \infty} a_n b_n = AB$

$\lim_{n \to \infty} ka_n = kA$, k a constant

$\lim_{n \to \infty} \frac{a_n}{b_n} = \frac{A}{B}$, $B \neq 0$.

Consider the function f defined by $f(x) = x + 4$. As x gets close to 2, $f(x)$ gets close to 6; and 6 is said to be the limit of f at 2.

In general, if a is a point in an open interval of real numbers, if f is a function defined on this interval (except perhaps at the point $x = a$), and if $f(x)$ gets close to the number F as x gets close to a, then F is the **limit of the function** at a. This is written

$$\lim_{x \to a} f(x) = F.$$

Limits of functions may be found by applying the following theorem.

THEOREM 22-2: Let $\lim_{x \to a} f(x) = F$, and $\lim_{x \to a} g(x) = G$, then

$\lim_{x \to a} c(x) = c$, c the constant function

$\lim_{x \to a} (f(x) \pm g(x)) = F \pm G$

$\lim_{x \to a} f(x) \cdot g(x) = FG$

$\lim_{x \to a} k f(x) = kF$, k a constant

$\lim_{x \to a} \frac{f(x)}{g(x)} = \frac{F}{G}$, $G \neq 0$.

the graph of a sequence whose limit is A. An ϵ is chosen. Then there must be a point N on the X-axis such that for all n to the right of N the graph of the sequence lies in the region 2ϵ about the line A. As ϵ is made smaller (Figure 22-2), 2ϵ becomes narrower and one must go further out on the X-axis to find an appropriate N. If this N could not be found for any $\epsilon > 0$, the sequence would have no limit.

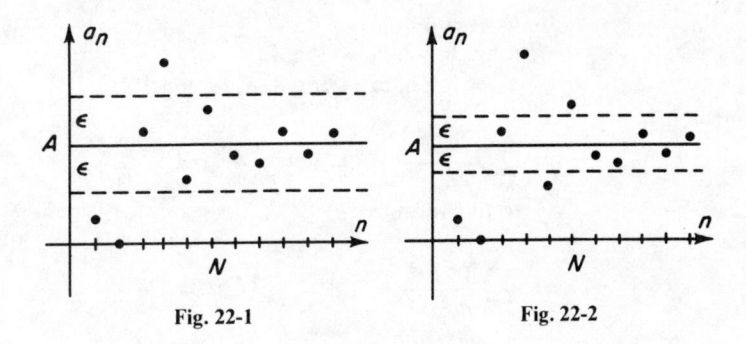

Fig. 22-1

Fig. 22-2

The concept of the limit of a function defined on an interval of real numbers is also made more precise by a formal definition.

If a function f is defined at every point in an open interval containing a, except perhaps at $x = a$, then the **limit of the function** at a is the number F, if for every positive number ϵ there exists a positive number δ such that $|f(x) - F| < \epsilon$ for all x having the property that $0 < |x - a| < \delta$.

The given number ϵ specifies how close $f(x)$ must be to F. The number δ specifies how close x must be to a to insure that $f(x)$ is within distance ϵ of F.

In Figure 22-3, $\lim_{x \to a} f(x) = F$ since, for any $\epsilon > 0$, as x ranges from $a - \delta$ to $a + \delta$, $x \neq a$, $f(x)$ is within the interval $(F - \epsilon, F + \epsilon)$. The situation at $x = a$ is irrelevant to the existence of a limit at a since $0 < |x - a|$. The limit of a function at a need not equal the value of the function at a. If $\lim_{x \to a} f(x) = f(a)$, f is said to be continuous at the point a (Figure 22-4). If f is continuous at every point in an interval, it is continuous in that interval; if not, it is discontinuous. All polynomial functions are everywhere continuous.

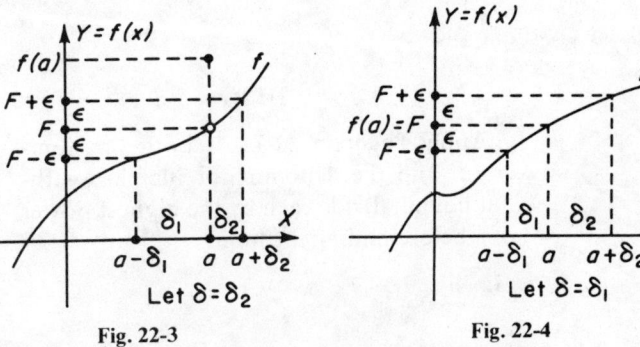

Fig. 22-3

Fig. 22-4

SOLUTIONS TO EXERCISES

I

i. $a_n = \dfrac{n}{2^n}$

$a_1 = \frac{1}{2}, a_2 = \frac{1}{2}, a_3 = \frac{3}{8}, a_4 = \frac{1}{4}, a_5 = \frac{5}{32}$

ii. $a_n = \log 2n$

$a_1 = \log 2, a_2 = \log 4,$
$a_3 = \log 6, a_4 = \log 8, a_5 = \log 10$

iii. $\sum\limits_{n=1}^{\infty} \dfrac{n}{2n+4} = \frac{1}{6} + \frac{1}{4} + \frac{3}{10} + \frac{1}{3} + \frac{5}{14}$, to five terms

iv. $\sum\limits_{n=2}^{\infty} \dfrac{(-n)^3}{n-1} = -8 - \frac{27}{2} - \frac{64}{3} - \frac{125}{4} - \frac{216}{5}$, to five terms

v. The series is $1 + 1 + 2 + 3 + 5 + 8 + 13 + 21 + 34 + 56 + 90 + 146$ to twelve terms. The terms of this series form an interesting sequence known as the *Fibonacci Sequence*.

II

i. $-8, -5, -2, \ldots$ is an arithmetic sequence.

$a_n = a_1 + (n-1)d, d = a_2 - a_1 = 3$
$a_{25} = -8 + (25-1)3 = 64$

ii. The first ten powers of ten are in geometric progression.

$S_n = \dfrac{a_1(r^{n-1} - 1)}{(r-1)}, r = \dfrac{a_2}{a_1} = 10$

$S_{10} = \dfrac{10(10^9 - 1)}{(10-1)} = 1{,}111{,}111{,}110$

III

Apply Theorem 22-1. If both the numerator and the denominator increase without bound, divide each by the highest power of n before finding the limit.

i. $\lim\limits_{n\to\infty} 7 = 7$

ii. $\lim\limits_{n\to\infty} \dfrac{3}{n^3} + 6 = \lim\limits_{n\to\infty} \dfrac{3}{n^3} + \lim\limits_{n\to\infty} 6 = 0 + 6 = 6$

iii. $\lim\limits_{n\to\infty} 3n^2 + n$ does not exist since, as n grows large, so does $3n^2 + n$.

iv. $\lim\limits_{n\to\infty} \dfrac{3-n}{n} = \lim\limits_{n\to\infty} \dfrac{\frac{3}{n} - 1}{1} = \dfrac{0-1}{1} = -1$

IV

Apply Theorem 22-2. If $\lim = \dfrac{0}{0}$, divide out common factors.

i. $\lim\limits_{x\to -2} f(x)$ where $f(x) = \begin{cases} x + 2 \text{ if } x \geq 0 \\ 3 \text{ if } x < 0 \end{cases}$

$= \lim\limits_{x\to -2} 3 = 3$

ii. $\lim\limits_{x\to 2} \dfrac{|2x - 5|}{4} = \dfrac{|4-5|}{4} = \dfrac{1}{4}$

iii. $\lim\limits_{x\to -4} \dfrac{x^3(x+4)}{x+4} = \lim\limits_{x\to -4} x^3 = -64$

V

i. $f(x) = \dfrac{3x-6}{x-2}, f(2) = \dfrac{0}{0}$ is a point of discontinuity.

$\lim\limits_{x\to 2} \dfrac{3x-6}{x-2} = \lim\limits_{x\to 2} \dfrac{3(x-2)}{x-2} = \lim\limits_{x\to 2} 3 = 3$

If $f(2) = 3$, the function will be continuous.

ii. $f(x) = \dfrac{5x(x-3)^2}{x^2 - 6x + 9} = \dfrac{5x(x-3)^2}{(x-3)^2}$

$f(3)$ is not defined and is therefore a point of discontinuity.

$\lim\limits_{x\to 3} 5x = 15$

If $f(3) = 15$, the function will be continuous.

23 PERMUTATIONS AND COMBINATIONS

SELF-TEST

DIRECTIONS: Write your answers in the numbered regions to the right. To check your answers, turn the page. Study the solutions to the problems you missed, and do the exercises following any problem or group of problems in which you had errors.

Tables for $n!$ from $n = 1$ to $n = 15$ appear in the appendix, page 174.

1. The membership of a math-science club includes one more math major than science major. The club wishes to admit six new members one at a time so that at all times there will be more math than science members. If three of the new members are math majors and three are science majors, draw a tree diagram to illustrate the number of different ways the applicants can be admitted.

2. Evaluate $_6P_2$ and $_{12}P_{12}$.

3. $0! = 1$ by definition. Is this definition consistent with the meaning of $_nP_n$ and $_nP_r$?

4. Evaluate $\binom{6}{4}$ and $\binom{6}{2}$. Is $\binom{n}{r}$ always equal to $\binom{n}{n-r}$?

5. If $_nP_2 = 110$, find n.

6. If there are ten different airlines operating flights from New York to San Francisco, in how many ways could a round trip be made if the return flight is with the same airline as the departure flight? a different airline?

7. How many four-digit numerals are there? How many of the numbers they represent are not divisible by five? How many of the numerals have repetitions of digits?

8. How many different eight-letter nonsense words can be formed using each of *tttrrrdd* only once?

9. Ten people apply for positions with a firm. In how many ways can the employer choose at least four but not more than six of these applicants?

10. What is the middle term in the expansion of $\left(\dfrac{a}{2} + \dfrac{b}{3}\right)^6$? What is the numerical coefficient of the term involving $b^4 a^2$?

| 1 |
| 2 |
| 3 |
| 4 |
| 5 |
| 6 |
| 7 |
| 8 |
| 9 |
| 10 |

133

SOLUTIONS

1. Let M indicate a math major and S indicate a science major. There are five arrangements.

```
                M—S—S—S
                 M—S—S
              M—S
  O—M           S—M—S
              S—M—S
        S—M
              S—M—S
```

> **EXERCISE I**
>
> A die is rolled. If the number facing up is even, the die is rolled again; if the number is odd, a coin is tossed. Draw a tree diagram indicating all possible outcomes of this experiment.

2. $_6P_2 = \dfrac{6!}{4!} = \dfrac{6\cdot 5\cdot 4!}{4!} = 30$

$_{12}P_{12} = 12! = 479{,}001{,}600$

3. $0! = 1$ is consistent with the meaning of $_nP_r$ and $_nP_n$. $_nP_n = n!$ and

$_nP_r = \dfrac{n!}{(n-r)!} = \dfrac{n!}{0!}$ if $r = n$.

$n! = \dfrac{n!}{0!}$ only if $0! = 1$.

4. $\binom{6}{4} = \dfrac{6!}{4!(6-4)!} = \dfrac{6\cdot 5\cdot 4!}{4!2!} = 15$

$\binom{6}{2} = \dfrac{6!}{2!(6-2)!} = \dfrac{6\cdot 5\cdot 4!}{2!4!} = 15$

$\binom{n}{r} = \binom{n}{n-r}$

$\binom{n}{r} = \dfrac{n!}{r!(n-r)!}$

$\binom{n}{n-r} = \dfrac{n!}{(n-r)!(n-(n-r))!}$

$\qquad\quad = \dfrac{n!}{r!(n-r)!}$

5. $_nP_2 = \dfrac{n!}{(n-2)!} = \dfrac{n(n-1)(n-2)!}{(n-2)!}$

$\qquad = n(n-1)$

$n(n-1) = 110$

$n^2 - n - 110 = 0 \rightarrow n = 11 \text{ or } -10$

Since n must be positive, $n = 11$.

> **EXERCISES II**
>
> **i.** Evaluate $_5P_2$, $_{15}P_{13}$, $_{14}P_{14}$, and $_9P_1$.
>
> **ii.** Evaluate $_7C_3$, $\binom{8}{4}$, $_{20}C_{20}$, and $_{20}C_1$.
>
> **iii.** $\binom{n}{0} = 1$ by definition. Show that this definition is consistent with the meaning of $\binom{n}{r}$.
>
> **iv.** If $_nC_3 = 56$, find $_nP_3$.

6. Apply the multiplication principle. The flight going can be made in ten ways. The flight returning can be made in one way if the airline is the same, and in nine ways if the airline is different.

$10\cdot 1 = 10$

$10\cdot 9 = 90$

7. There are ten digits: 0,1,2,3,4,5,6,7,8,9. In constructing a four-digit numeral, there are four places to fill. Since zero must be excluded from the first place on the left, that place can be filled in nine ways. Each of the other three places can be filled in ten ways, repetitions permitted. $9\cdot 10\cdot 10\cdot 10 = 9000$ four-digit numerals.

Zero and five must be excluded from the fourth place of the numeral for a number not divisible by five. $9\cdot 10\cdot 10\cdot 8 = 7200$ numbers which are not divisible by five.

To find the number of numerals that have repetitions, find the number which don't and subtract. $9\cdot 9\cdot 8\cdot 7 = 4536$ without repetitions. There are 4464 with repetitions.

8. The number of different nonsense words that can be formed from *tttrrrdd* is

$$\frac{8!}{3!3!2!} = \frac{8 \cdot 7 \cdot 6 \cdot 5 \cdot 4 \cdot 3!}{3! 3 \cdot 2 \cdot 2} = 560.$$

9. From the ten people applying for positions, the employer may choose four, five, or six of the applicants. These choices are mutually exclusive. Find the sum of the number of ways he may make each choice. (Order is not important.)

$$_{10}C_4 + {_{10}C_5} + {_{10}C_6} = 210 + 252 + 210 = 672$$

EXERCISES III

i. How many five-letter nonsense words can be formed from the letters a,b,c,d,f if repetitions are not to occur and the letters c and d do not occur together?

ii. How many different ways can a man tip a porter if the man has a nickel, a dime, a quarter, and a half-dollar?

iii. How many different license plates can be made bearing two letters of the English alphabet (O excluded) and then a four-digit numeral?

iv. In how many ways can five different keys be put on a circular key ring?

v. In how many ways can six people, two of whom drive, be seated for a trip in a six-seat automobile?

vi. In how many ways can a professor assign two A's and two F's to a class of ten students?

10. $\left(\frac{a}{2} + \frac{b}{3}\right)^6$ has seven terms in its expansion. The fourth or middle term is:

$$\binom{6}{3}\left(\frac{a}{2}\right)^3\left(\frac{b}{3}\right)^3 = \frac{6 \cdot 5 \cdot 4}{1 \cdot 2 \cdot 3}\left(\frac{a^3}{8}\right)\left(\frac{b^3}{27}\right) = \frac{5}{54}a^3b^3.$$

The coefficient of the term involving b^4a^2 is the coefficient of the fifth term, or

$$\binom{6}{4} = \frac{6 \cdot 5}{1 \cdot 2} = 15 \text{ times } \left(\frac{1}{3}\right)^4\left(\frac{1}{2}\right)^2 = \frac{5}{108}.$$

ANSWERS

1. five ways	1
2. 30 479,001,600	2
3. yes	3
4. 15 15 yes	4
5. $n = 11$	5
6. 10 90	6
7. 9,000 7,200 4,464	7
8. 560	8
9. 672	9
10. $\frac{5}{54}a^3b^3$ $\frac{5}{108}$	10

EXERCISES IV

i. Express the expansion of $(a + b)^n$, n a positive integer, in summation notation.

ii. Write the expansion of $(3a - 2b)^4$.

iii. Find the sixth term of the expansion of $(x^2 + y)^8$.

iv. Find the term involving x^7 in the expansion of $(1 + x)^{12}$.

Solutions to the exercises are on page 138.

BASIC FACTS

A **permutation** of a set of elements is any arrangement of certain of these elements in a definite order. Order is essential in such arrangements; a change in order yields a different permutation. ABC, CAB and BCA are three different permutations of the letters A, B, C.

One method for exhibiting the different permutations of a set of elements is the **tree diagram**. In such a diagram, each path represents a possible permutation, and the total number of paths gives the number of distinct permutations. The tree diagram below exhibits all possible permutations of the letters A, B, C when these letters are considered three at a time.

1st Position	2nd Position	3rd Position	Permutations
A	B	C	ABC
A	C	B	ACB
B	A	C	BAC
B	C	A	BCA
C	A	B	CAB
C	B	A	CBA

A tree diagram is advantageous in that it gives not only the number of permutations but also the actual arrangements. However, if the number of elements in a set is large, the construction of a tree diagram is laborious.

A general method for determining the number of permutations of a set is given by the **multiplication principle** which is as follows.

> If a task can be performed in r different ways, and if, for each of these r ways, a second task can be performed in s different ways, and if, for each of the first two, a third task can be performed in t different ways, and so on, then the sequence

ADDITIONAL INFORMATION

When constructing a **tree diagram** to determine the number of permutations of a set, first determine the number of elements in each permutation. Each branch in the first set of branches (those from the origin O) represents a possible first element in one of the permutations. For each of these branches there is a second set of branches. Each branch in any one of these second sets represents a possible second element in the permutation whose first element is the branch from which this second branch is drawn. This process is continued until the number of sets of branches equals the number of elements in each permutation. The number of different paths that can be traced without backtracking from the origin to the last branch is the number of different permutations that can be formed.

The **multiplication principle** is actually a short-cut method for counting these paths. Each task corresponds to filling in an element of one of the permutations. If the first task can be done in r ways, the second in s ways, and the third in t ways, the tree diagram would have r branches from the starting point. From each of these r branches there would be s new branches, and from each of these s branches there would be t new branches. The total number of paths would then be $r \cdot s \cdot t$.

The multiplication principle applies to situations where a succession of tasks is performed. First one is performed and then the next. In certain situations, tasks are **mutually exclusive**: they cannot occur simultaneously. One is performed or the other, but not both. In such situations, the total number of permutations or combinations is determined by the **addition principle**.

> If two tasks are mutually exclusive and the first can be performed in r different ways and the second can be performed in s different ways, then one task or the other can be performed in $r + s$ ways.

Consider the number of ways of selecting r objects from a set of $n + 1$ objects. This is denoted $\binom{n+1}{r}$. In making the selection, a particular element in the set of $n + 1$ elements is either chosen or not chosen. Since these choices are mutually exclusive, the total number of ways of making the selection will be the sum of the number of ways of making the selection if the particular element is included and of the number of ways of making the selection if the particular element is not included. If the element is included in the selection, the remaining $r - 1$ elements can be chosen from the remaining n elements in $\binom{n}{r-1}$ ways; if it is not included, the r elements can be chosen from the remaining n elements in $\binom{n}{r}$ ways. Therefore, $\binom{n}{r-1} + \binom{n}{r} = \binom{n+1}{r}$. This result, known as **Pascal's Rule**, is useful in constructing a table of values for $\binom{n}{r}$.

of tasks can be performed in $r \cdot s \cdot t \cdot \ldots$ ways.

In applying the multiplication principle, a task which must be performed in a special way should be considered first.

In certain cases, the method suggested by the multiplication principle can be shortened by formulas for determining the number of permutations.

The symbol $n!$, read "n factorial," denotes the product of all whole numbers from 1 to n.

$$n! = n(n-1)(n-2)\ldots 3 \cdot 2 \cdot 1$$

The symbol $_nP_n$ or $P_{n,n}$ denotes the total number of permutations of a set of n different elements considered n at a time, without repetitions.

$$_nP_n = n!$$

The symbol $_nP_r$ or $P_{n,r}$ denotes the total number of permutations of a set of n different elements considered r at a time, without repetitions.

$$_nP_r = \frac{n!}{(n-r)!}$$

The symbol $^{s_1,s_2,\ldots}{}_nP_n$ or $P_{n,n}^{s_1,s_2,\ldots}$, where $s_1 + s_2 + \cdots = n$, denotes the total number of permutations of a set of n elements considered n at a time where there are s_1 elements of one kind, s_2 elements of another kind, and so on.

$$^{s_1,s_2,\ldots}{}_nP_n = \frac{n!}{s_1!s_2!\ldots}$$

A **combination** of a set of elements is any selection of certain of these elements without regard to their order. AB and BA represent the same combination.

The symbol $_nC_r$, $C_{n,r}$, or $\binom{n}{r}$, denotes the number of combinations of a set of n distinct elements considered r at a time.

$$_nC_r = \binom{n}{r} = \frac{n!}{r!(n-r)!}, r \leq n$$

r \ n	0	1	2	3	4	5
0	1					
1	1	1				
2	1	2	1			
3	1	3	3	1		
4	1	4	6	4	1	
5	1	5	10	10	5	1

```
             1
            1 1
           1 2 1
          1 3 3 1
         1 4 6 4 1
        1 5 10 10 5 1
```

The row headings (the vertical column on the left) correspond to the values of n; the column headings (the horizontal row on the top) correspond to the values of r. Since $\binom{n}{0} = \binom{n}{n} = 1$, the first and last entry in each row is 1. All other entries are computed by applying Pascal's Rule. For example, the entry for $n = 5, r = 3$ is the sum of the entries for $n = 4, r = 2$ and $n = 4, r = 3$. Thus, $\binom{5}{3} = \binom{4}{2} + \binom{4}{3}$ or $\binom{5}{3} = 6 + 4 = 10$. The elements of this table are often written as **Pascal's Triangle**, the array to the right of the table.

One application of Pascal's Triangle is in the expansion of a power of a binomial. Consider the first few integral powers of the binomial $a + b$.

$$
\begin{aligned}
(a+b)^0 &= 1 \\
(a+b)^1 &= 1a^1 + 1b^1 \\
(a+b)^2 &= 1a^2 + 2a^1b^1 + 1b^2 \\
(a+b)^3 &= 1a^3 + 3a^2b^1 + 3a^1b^2 + 1b^3 \\
(a+b)^4 &= 1a^4 + 4a^3b^1 + 6a^2b^2 + 4a^1b^3 + 1b^4
\end{aligned}
$$

A pattern is evident. The coefficients of these expansions are the same as the values of Pascal's Triangle and may therefore be expressed in terms of $\binom{n}{r}$. Each expansion begins with a^n and ends with b^n, where n is the exponent of the binomial. In each successive intermediate term, the exponents of a decrease by one and the exponents of b increase by one. Thus, in each term the sum of the exponents is n. The following theorem generalizes this pattern.

THEOREM 23-1: **The Binomial Theorem**

If n is any positive integer, then

$$(a+b)^n = \binom{n}{0}a^n + \binom{n}{1}a^{n-1}b + \binom{n}{2}a^{n-2}b^2 + \cdots + \binom{n}{r}a^{n-r}b^r + \cdots + \binom{n}{n}b^n.$$

The $(r+1)$st term of this expansion is $\binom{n}{r}a^{n-r}b^r$. Using this general term, the value of any term of the expansion of a positive integral power of a binomial can be found.

SOLUTIONS TO EXERCISES

I

II

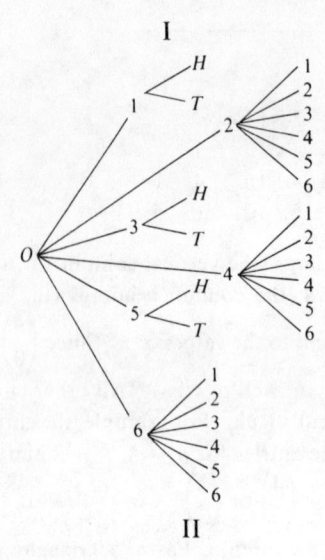

i. $_5P_2 = \frac{5!}{3!} = 5 \cdot 4 = 20$

$_{15}P_{13} = \frac{15!}{2!} + \frac{1,307,674,368,000}{2}$
$= 653,837,184,000$

$_{14}P_{14} = 14! = 87,178,291,200$

$_9P_1 = \frac{9!}{8!} = 9$

ii. $_7C_3 = \frac{7!}{3!4!} = 35$

$\binom{8}{4} = \frac{8!}{4!4!} = \frac{8 \cdot 7 \cdot 6 \cdot 5 \cdot 4!}{4 \cdot 3 \cdot 2 \cdot 1 \cdot 4!} = 70$

$_{20}C_{20} = 1$

$_{20}C_1 = 20$

iii. $\binom{n}{r} = \frac{n!}{r!(n-r)!}$

If $r = 0$, $\binom{n}{0} = \frac{n!}{0!n!} = \frac{n!}{1 \cdot n!} = 1$.

The definition of $\binom{n}{0}$ as 1 is consistent.

iv. $_nC_3 = 56$. To find $_nP_3$, note that $_nP_r = r!(_nC_r) \therefore {}_nP_3 = 3! \, 56 = 336$.

III

i. Find $_5P_5$, the number of words with no restrictions. Consider d and c as one letter and find $2 \cdot {}_4P_4$, the number of words with d and c together, as dc and cd. $_5P_5 - 2 \cdot {}_4P_4 = 120 - 48 = 72$ is the number of words in which d and c are not together.

ii. If a man has a nickel, a dime, a quarter, and a half-dollar, he has four coins and can give one, two, three, or four of them. These operations are mutually exclusive. $_4C_1 + {}_4C_2 + {}_4C_3 + {}_4C_4 = 15$

iii. Each license plate will have six elements: the first two from the alphabet (O excluded) and the last four from the digits. Since repetitions are not restricted, the first two places can be filled in 25 ways each. Since the digits must form a four-digit number, 0 is excluded from the first position on the left which can then be filled in nine ways. The other three positions can be filled in ten ways each.

$25 \cdot 25 \cdot 9 \cdot 10 \cdot 10 \cdot 10 = 5,625,000$.

iv. There are $_5P_5 = 120$ linear arrangements of the keys. Since for every five linear arrangements there will be one circular arrangement, $\frac{120}{5} = 24$ is the number of ways to arrange the keys.

v. The driver's seat can be filled in two ways. The next seat can be filled in five ways, the next in four ways, and so on. $2 \cdot 5 \cdot 4 \cdot 3 \cdot 2 \cdot 1 = 240$ is the number of possible seating arrangements.

vi. 2 A's and 2 F's can be assigned to a class of ten in $10 \cdot 9 \cdot 8 \cdot 7 = 5040$ ways.

IV

i. $(a + b)^n = \sum_{r=0}^{n} \binom{n}{r} a^{n-r} b^r$

ii. $(3a + (-2b))^4 = 81a^4 - 216a^3b + 216a^2b^2 - 96ab^3 + 16b^4$

iii. The sixth term of $(x^2 + y)^8$ is $\binom{8}{5}(x^2)^3(y)^5 = 56x^6y^5$.

iv. x^7 is in the eighth term of $(1 + x)^{12}$. This term is $\binom{12}{7}(1)^5(x)^7 = 792x^7$

24 PROBABILITY

SELF-TEST

DIRECTIONS: Write your answers in the numbered regions to the right. To check your answers, turn the page. Study the solutions to the problems you missed, and do the exercises following any problem or group of problems in which you had errors.

1. An experiment consists of drawing one coin from a bag containing four nickels, six dimes, and two quarters.

    a. Set up a sample space U. Assume that coins of the same denomination are not distinguishable.
    b. Assign weights to each element of U.
    c. Find the probability measure for each subset of U.

2. Two experiments are performed in succession. First a coin is tossed. If h (head) shows, then a marble is drawn from a container in which there is one yellow, one green, and one black marble. If t (tail) shows, a card is chosen from a deck containing a king, queen, jack, and ace.

    a. Construct a tree diagram for this sequence indicating the tree measure.
    b. What is the probability of a green marble?
    c. What is the probability of a head and a jack?
    d. What is the probability of a tail or a green marble?
    e. What is the probability of not getting a yellow marble?

3. A pair of dice, one white and one black, is tossed. Let w be the number on the top face of the white die and b be the number on the top face of the black die.

    $p: w \leq 4$ $q: b > 3$ $r: b + w \geq 8$

    Find the probability of the following statements.

    a. p **c.** r **e.** $p \rightarrow q$
    b. q **d.** $p \vee q$
    f. What is the conditional probability of p given r?

4. A coin is tossed four times. What is the probability that the number of heads is greater than the number of tails?

139

SOLUTIONS

1. a. A sample space for drawing one coin from a bag containing four nickels, six dimes, and two quarters is $U = \{n, d, q\}$.

b. The possibility of drawing n is twice that of drawing q and the possibility of drawing d is three times that of drawing q. The sum of the weights of U must be 1. Let the weight of $q = x$.
Then $x + 2x + 3x = 1 \rightarrow x = \frac{1}{6}$.
Hence, weight $q = \frac{1}{6}$, weight $n = \frac{1}{3}$, and weight $d = \frac{1}{2}$.

c. The probability measure of a subset of U is the sum of the weights of its elements.
$m(\phi) = 0, \quad m(n) = \frac{1}{3}, \quad m(d) = \frac{1}{2},$
$m(q) = \frac{1}{6}, \quad m(nd) = \frac{1}{3} + \frac{1}{2} = \frac{5}{6},$
$m(nq) = \frac{1}{2}, \quad m(dq) = \frac{2}{3},$
$m(U) = 1$

EXERCISES I

i. Set up a sample space U for an experiment in which digits from 0 to 9 are drawn.

ii. Refer to i. Find $m(X)$ if $X \subset U$ and X contains only prime numbers.

---

2. a. TREE MEASURE

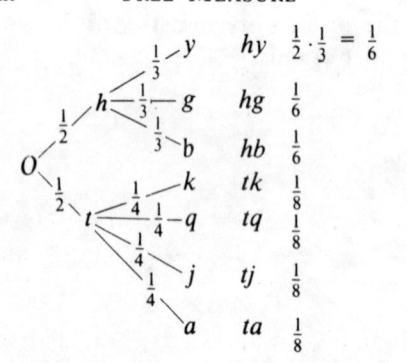

$$
\begin{array}{l}
hy \quad \frac{1}{2} \cdot \frac{1}{3} = \frac{1}{6} \\
hg \quad \frac{1}{6} \\
hb \quad \frac{1}{6} \\
tk \quad \frac{1}{8} \\
tq \quad \frac{1}{8} \\
tj \quad \frac{1}{8} \\
ta \quad \frac{1}{8}
\end{array}
$$

b. The probability of a green marble is $\frac{1}{6}$.

c. The probability of a head and a jack is 0.

d. To find the probability of a tail or a green marble, let p be 'A tail is drawn' and q be 'A green marble is drawn,' then $\text{PR}[p \vee q] = m(P) + m(Q) - m(P \cap Q)$. P, the truth set of p is $\{tk, tq, tj, ta\}$, Q, the truth set of q is $\{hg\}$, $P \cap Q = \phi$, $m(P) = \frac{1}{2}$, $m(Q) = \frac{1}{6}$, and $m(\phi) = 0$. Hence, $\text{PR}[p \vee q] = \frac{4}{6} = \frac{2}{3}$.

e. Let p be 'A yellow marble is drawn.' $\text{PR}[\sim p] = 1 - \text{PR}[p] = 1 - m(P) = 1 - \frac{1}{6} = \frac{5}{6}$

3. An appropriate sample space is composed of 36 elements each of which is an ordered pair (w, b). A graphic representation (Figure 24-1) is the set of lattice points. Each point has weight $\frac{1}{36}$.

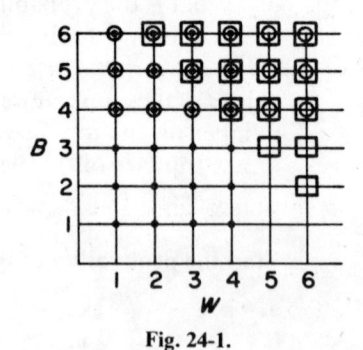

Fig. 24-1.

p: $w \leq 4$, P has 24 elements (dots)
q: $b > 3$, Q has 18 elements (circles)
r: $b + w \geq 8$, R has 15 elements (boxes)

a. $\text{PR}[p] = m(P) = \frac{24}{36} = \frac{2}{3}$

b. $\text{PR}[q] = m(Q) = \frac{18}{36} = \frac{1}{2}$

c. $\text{PR}[r] = m(R) = \frac{15}{36} = \frac{5}{12}$

d. $\text{PR}[p \lor q]$
$= \text{PR}[p] + \text{PR}[q] - \text{PR}[p \land q]$
$= \frac{2}{3} + \frac{1}{2} - \frac{1}{3} = \frac{5}{6}$

e. $\text{PR}[p \to q] = \text{PR}[\sim p \lor q]$
$= \frac{1}{3} + \frac{1}{2} - \frac{1}{6} = \frac{2}{3}$

f. $\text{PR}[p \mid r] = \frac{m(P \cap R)}{m(R)} = \frac{6}{36} \div \frac{5}{12} = \frac{2}{5}$

EXERCISES II

i. Three urns marked I, II, and III, contain respectively one black and two red marbles, two black and two red marbles, and three red marbles. An urn is chosen and one marble withdrawn.

a. Set up a tree diagram to illustrate the possible outcomes. Indicate the tree measure.

b. If p is 'A red marble is chosen,' q is 'A black marble is chosen,' and r, s, and t are respectively 'Urns I, II, and III are chosen,' find $\text{PR}[p]$, $\text{PR}[q]$, $\text{PR}[r]$, $\text{PR}[s]$, $\text{PR}[t]$, $\text{PR}[p \land q]$, $\text{PR}[p \mid q]$, $\text{PR}[p \land t]$, and $\text{PR}[p \mid t]$.

ii. If the probability of A winning a contest is $\frac{1}{5}$ and the probability of B winning is $\frac{1}{7}$, 'A wins' and 'B wins' are independent statements. Find

a. the probability of both A and B winning

b. the probability of either A or B winning

c. the probability of A winning and B losing

d. the probability of both losing

---

4. Each toss of the coin is an independent trial with two outcomes. Applying the binomial expansion, $(h + t)^4 = h^4 + 4h^3t + 6h^2t^2 + 4ht^3 + t^4$. h^4 and h^3t represent outcomes in which the number of heads is greater than tails. Since they occur five out of sixteen times, the probability is $\frac{5}{16}$.

ANSWERS

1. a. $U = \{n, d, q\}$

b. weight of $n = \frac{1}{3}$; $d = \frac{1}{2}$; $q = \frac{1}{6}$

c. $m(\phi) = 0$, $m(n) = \frac{1}{3}$, $m(d) = \frac{1}{2}$,
$m(q) = \frac{1}{6}$, $m(nd) = \frac{5}{6}$, $m(nq) = \frac{1}{2}$,
$m(dq) = \frac{2}{3}$, $m(U) = 1$

---

2. a. See solutions. **d.** $\frac{2}{3}$

b. $\frac{1}{6}$

c. 0 **e.** $\frac{5}{6}$

---

3. a. $\frac{2}{3}$ **d.** $\frac{5}{6}$

b. $\frac{1}{2}$ **e.** $\frac{2}{3}$

c. $\frac{5}{12}$ **f.** $\frac{2}{5}$

---

4. $\frac{5}{16}$

EXERCISES III

i. A coin is biased so that the probability of tossing a head is $\frac{1}{3}$. A second coin is also biased so that the probability of tossing a tail is $\frac{1}{4}$. If both coins are tossed, find the probability of

a. two tails **b.** exactly one head

c. no tails **d.** at least one head

Solutions to the exercises are on page 144.

BASIC FACTS

Consider an experiment in which the outcome is unknown. Let p be a statement relative to a logically possible outcome of this experiment. The **truth set** of p, denoted P, is that subset of logical possibilities for which statement p is true.

To determine the **probability** of statement p, denoted $\text{PR}[p]$, two things must be done. First, a sample space, or set of logically possible outcomes of the experiment, denoted U, must be determined. Secondly, to each subset X of U, a number, or probability measure, denoted $m(X)$, must be assigned. The probability of p is defined as the probability measure of its truth set.

$$\text{PR}[p] = m(P)$$

A set of logical outcomes of an experiment is a **sample space**, U, for that experiment if and only if each of its elements is the result of exactly one outcome of the experiment. The elements of U are sometimes called simple events and any subset of U is called an **event**. Two events, X_1 and X_2, are mutually exclusive if no element of U belongs to both X_1 and X_2, that is, if $X_1 \cap X_2 = \phi$.

A sample space for a given experiment is not necessarily unique. It may contain few or many elements depending on how thoroughly one analyzes the logical possibilities. However, if it is to be of use in determining the probabilities of a set of statements, it must be large enough to contain the truth set of each statement.

To determine the measure of a subset X of U, a positive number or **weight** is assigned to each element of U so that the sum of these weights is one.

If a sample space contains n elements and if each element is assumed to be an equally likely outcome, each element of U is assigned weight $\frac{1}{n}$. Thus, if a die is tossed and a sample space $\{1,2,3,4,5,6\}$ is considered, the six outcomes seem equally likely and weight $\frac{1}{6}$ can reason-

ADDITIONAL INFORMATION

An experiment consists of drawing a card from a standard deck. Some **sample spaces** for this experiment are $U_1 = \{\text{red, black}\}$, $U_2 = \{\text{red, club, spade}\}$, $U_3 = \{\text{heart, diamond, club, spade}\}$, $U_4 = \{x \mid x \text{ is the denomination}\}$, $U_5 = \{(x,y) \mid x \text{ is the denomination}, y \text{ is the suit}\}$, and $U_6 = \{\text{marked, unmarked}\}$. The set $\{\text{red, club}\}$ is not a sample space. All logical possibilities must be included, but drawing a spade does not correspond to any of these elements. Similarly, the set $\{\text{red, club, black}\}$ does not constitute a sample space. Each logical possibility must be assigned to only one element but drawing a club corresponds to 'club' and to 'black.'

To determine the **probability measure** of a subset of a sample space, **weights** are assigned to each element of the sample space so that the sum of the weights of all of the elements is 1. The weight of each element in U_1 is $\frac{1}{2}$ since each of 'red' and 'black' is an equally likely outcome. In U_2 a red card will occur twice as often as a club or a spade. Hence 'red' is assigned weight $\frac{1}{2}$, while 'club' and 'spade' are each assigned weight $\frac{1}{4}$. Each element in U_3 is equally likely and is assigned weight $\frac{1}{4}$. Similarly, each element of U_4 is given weight $\frac{1}{13}$ and each element of U_5 is given weight $\frac{1}{52}$. There is no natural way to assign weights to the elements of U_6. If the experiment were performed 100 times and unmarked occurred 87 times and marked occurred 13 times, weight $\frac{87}{100}$ could be assigned to unmarked and weight $\frac{13}{100}$ to marked.

If a sample space is used to determine the **probability of a statement**, it must contain the truth set of that statement. To determine the probability of p: 'A red card is chosen,' U_1, U_2, U_3, or U_5 could be used. U_1 is the least refined and U_5 the most refined of these sample spaces. If U_1 is used, P, the truth set of p, = $\{\text{red}\}$. $\text{PR}[p] = m(P) = \frac{1}{2}$. If U_3 is used, $P = \{\text{heart, diamond}\}$. $\text{PR}[p] = m(P) = \frac{1}{4} + \frac{1}{4} = \frac{1}{2}$. If U_5 is used, $P = \{(x,y) \mid y \text{ is a heart or a diamond}\}$. P contains 26 elements. $\text{PR}[p] = m(P) = 26\left(\frac{1}{52}\right) = \frac{1}{2}$. The sample space which offers the least information is the simplest one to use.

In many instances concern is centered upon finding the probability of p given additional information q. This is called the **conditional probability** of p given q and denoted $\text{PR}[p \mid q]$. By definition,

$$\text{PR}[p \mid q] = \frac{m(P \cap Q)}{m(Q)} = \frac{\text{PR}[p \wedge q]}{\text{PR}[q]}.$$

The following example illustrates this definition. Suppose a die is tossed. What is the probability that a three shows? This is an unconditional probability. Since $U = \{1,2,3,4,5,6\}$, the probability is $\frac{1}{6}$. Now, suppose

ably be assigned to each element. If another sample space {3, not −3} is considered, the outcomes are not equally likely. However, it seems natural to assign a weight of $\frac{1}{6}$ to 3 and a weight of $\frac{5}{6}$ to not −3. If the weights cannot be assigned naturally, the experiment is performed a number of times and the weights assigned according to the number of times each outcome occurs.

$m(X)$, the **probability measure** of X is the sum of the weights of the elements of X. This definition implies the following properties.

$$m(X) = 0 \text{ if and only if } X = \phi$$

$$0 \le m(X) \le 1 \text{ for all } X$$

For all X_1 and X_2,
$$m(X_1 \cup X_2) = m(X_1) + m(X_2) - m(X_1 \cap X_2).$$

If X_1 and X_2 are mutually exclusive, that is, $X_1 \cap X_2 = \phi$, then
$$m(X_1 \cup X_2) = m(X_1) + m(X_2).$$

If X' is the complement of X, then
$$m(X') = 1 - m(X).$$

Since the **probability of a statement** is the probability measure of its truth set, properties for the probabilities of statements can be obtained from the properties of the probability measure.

$\text{PR}[p] = 0$ if and only if $P = \phi$; that is, if and only if p is logically false.

$0 \le \text{PR}[p] \le 1$ for any p.

$\text{PR}[p \vee q] = \text{PR}[p] + \text{PR}[q] - \text{PR}[p \wedge q]$ for any p and q.

$\text{PR}[p \vee q] = \text{PR}[p] + \text{PR}[q]$ if and only if $P \cap Q = \phi$; that is, if and only if p and q are inconsistent.

$$\text{PR}[\sim p] = 1 - \text{PR}[p].$$

the die is tossed and the information that the number is odd is offered. This information has the effect of reducing the sample space; that is, of limiting the logical possibilities. U (reduced) = {1,3,5} and the probability is $\frac{1}{3}$. Rather than reduce the sample space and reassign measures to the subsets, the definition for $\text{PR}[p \mid q]$ can be applied, where p is 'A three shows' and q is 'The number is odd.' $P = \{3\}$, $Q = \{1,3,5\}$, $P \cap Q = \{3\}$, $m(P \cap Q) = \frac{1}{6}$, and $m(Q) = \frac{3}{6}$. Hence, $\text{PR}[p \mid q] = \frac{1}{6} \div \frac{3}{6} = \frac{1}{3}$.

If $\text{PR}[p \mid q] = \text{PR}[p]$, that is, if $\text{PR}[p \wedge q] = \text{PR}[p] \cdot \text{PR}[q]$, then p and q are **independent statements**.

The tree diagram at the right represents a finite sequence of experiments each of which has a finite number of outcomes. $\{a,b,c,d,e\}$ is the set of all possible outcomes for the experiments. Each path represents a possible outcome for the sequence of experiments and is assigned a probability measure, called a **tree measure** as follows. Each branch is assigned a probability. Thus, p_1 is the probability that a will occur on the first experiment; p_8 is the probability that b will occur on the second experiment given that c has occurred on the first experiment, and so on. The tree measure is the product of the probabilities along any path.

	OUT-COMES	TREE MEASURES
	ab	$p_1 p_4$
	ad	$p_1 p_5$
	ae	$p_1 p_6$
	ba	$p_2 p_7$
	cb	$p_3 p_8$
	cc	$p_3 p_9$

If a sequence of experiments consists in the repetition of a single experiment in such a manner that the result of any one experiment in no way affects the result of any other experiment, each experiment is called an **independent trial**. If an experiment has two outcomes, a and b, and it is repeated n times, the outcomes for the sequence of experiments, disregarding order, are given by the terms of the binomial expansion $(a + b)^n$. The coefficient of each term indicates the number of times the variables occur in that pattern and 2^n represents the number of outcomes for the sequence of experiments taking order into consideration. If a and b are equally likely, the probability that $a^{n-r}b^r$ will occur is $\binom{n}{r} \div 2^n$.

The tree diagram at the right represents the tossing of a coin three times. Since this is three independent trials with two outcomes, h (heads) and t (tails), the binomial expansion $(h + t)^3 = h^3 + 3h^2t + 3ht^2 + t^3$ indicates that h^3 and t^3 each occur once while ht^2 and h^2t each occur three times.

	OUT-COMES	TREE MEASURES
	h^3	$\frac{1}{8}$
	h^2t	$\frac{1}{8}$
	h^2t	$\frac{1}{8}$
	ht^2	$\frac{1}{8}$
	h^2t	$\frac{1}{8}$
	ht^2	$\frac{1}{8}$
	ht^2	$\frac{1}{8}$
	t^3	$\frac{1}{8}$

SOLUTIONS TO EXERCISES

I

i. A sample space for an experiment in which digits from 0 to 9 are drawn is $U = \{0,1,2,3,4,5,6,7,8,9\}$.

ii. If X contains only primes, $X = \{2,3,5,7\}$. Since each element in U is equally likely, each has weight $\frac{1}{10}$.

$m(X) = \frac{1}{10} + \frac{1}{10} + \frac{1}{10} + \frac{1}{10} = \frac{2}{5}$.

II

i. a.

TREE MEASURE

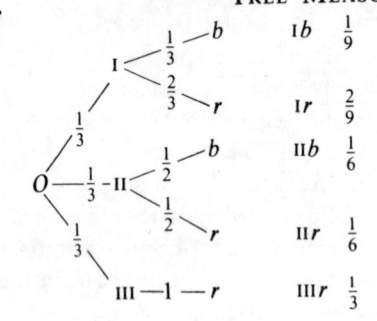

(tree diagram)
- I, b: $\frac{1}{3} \cdot \frac{1}{3}$, $\text{I}b$ $\frac{1}{9}$
- I, r: $\frac{2}{3}$, $\text{I}r$ $\frac{2}{9}$
- II, b: $\frac{1}{2}$, $\text{II}b$ $\frac{1}{6}$
- II, r: $\frac{1}{2}$, $\text{II}r$ $\frac{1}{6}$
- III, r: 1, $\text{III}r$ $\frac{1}{3}$

b. $P = \{\text{I}r, \text{II}r, \text{III}r\}$,

$m(P) = \frac{2}{9} + \frac{1}{6} + \frac{1}{3} = \frac{13}{18}$

$Q = \{\text{I}b, \text{II}b\}, m(Q) = \frac{1}{9} + \frac{1}{6} = \frac{5}{18}$

$R = \{\text{I}b, \text{I}r\}, m(R) = \frac{1}{9} + \frac{2}{9} = \frac{1}{3}$

$S = \{\text{II}b, \text{II}r\}, m(S) = \frac{1}{6} + \frac{1}{6} = \frac{1}{3}$

$T = \{\text{III}r\}, m(T) = \frac{1}{3}$

$\text{PR}[p] = m(P) = \frac{13}{18}$

$\text{PR}[q] = m(Q) = \frac{5}{18}$

$\text{PR}[r] = m(R) = \frac{1}{3}$

$\text{PR}[s] = m(S) = \frac{1}{3}$

$\text{PR}[t] = m(T) = \frac{1}{3}$

$\text{PR}[p \wedge q] = m(P \cap Q)$
$= m(\phi) = 0$

$\text{PR}[p \mid q] = \frac{\text{PR}[p \wedge q]}{\text{PR}[q]} = 0 \div \frac{1}{3} = 0$

$\text{PR}[p \wedge t] = m(P \cap T) = \frac{1}{3}$

$\text{PR}[p \mid t] = \frac{\text{PR}[p \wedge t]}{\text{PR}[t]} = \frac{1}{3} \div \frac{1}{3} = 1$

ii. Let p be 'A wins,' then $\text{PR}[p] = \frac{1}{5}$.

Let q be 'B wins,' then $\text{PR}[q] = \frac{1}{7}$.

a. The probability of both A and B winning is

$\text{PR}[p \wedge q] = \text{PR}[p] \cdot \text{PR}[q] = \frac{1}{35}$.

b. The probability of either A or B winning is $\text{PR}[p \vee q] = \text{PR}[p] + \text{PR}[q] - \text{PR}[p \wedge q] = \frac{11}{35}$.

c. The probability of A winning and B losing is $\text{PR}[p \wedge \sim q]$
$= \text{PR}[p] \cdot \text{PR}[\sim q] = \frac{6}{35}$.

d. The probability of both losing is $\text{PR}[\sim p \wedge \sim q] = \text{PR}[\sim p] \cdot \text{PR}[\sim q]$
$= \frac{24}{35}$.

III

i. Let h_1 and t_1 represent head and tail of the first coin. Since the probability of h_1 is $\frac{1}{3}$, the probability of t_1 is $\frac{2}{3}$. Let h_2 and t_2 represent head and tail of the second coin. Since the probability of t_2 is $\frac{1}{4}$, the probability of h_2 is $\frac{3}{4}$. Since the outcome on the toss of two coins simultaneously is the same as the outcome on the toss of first one and then the other, a tree diagram summarizes the possible outcomes.

TREE MEASURE

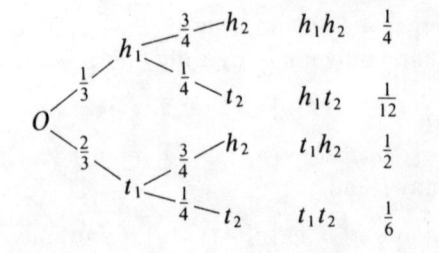

- h_1, h_2: $\frac{1}{3} \cdot \frac{3}{4}$, $h_1 h_2$ $\frac{1}{4}$
- h_1, t_2: $\frac{1}{4}$, $h_1 t_2$ $\frac{1}{12}$
- t_1, h_2: $\frac{2}{3} \cdot \frac{3}{4}$, $t_1 h_2$ $\frac{1}{2}$
- t_1, t_2: $\frac{1}{4}$, $t_1 t_2$ $\frac{1}{6}$

a. The probability of two tails is $\frac{1}{6}$.

b. The probability of exactly one head is $\frac{1}{2} + \frac{1}{12} = \frac{7}{12}$.

c. The probability of no tails is $\frac{1}{4}$.

d. The probability of at least one head is $\frac{7}{12} + \frac{1}{4} = \frac{5}{6}$.

25 ALGEBRA OF ORDERED PAIRS, VECTORS

SELF-TEST

DIRECTIONS: Write your answers in the numbered regions to the right. To check your answers, turn the page. Study the solutions to the problems you missed, and do the exercises following any problem or group of problems in which you had errors.

In **1–5** perform the indicated operations.

1. $(4, 6) + (-2, 3)$

2. $(1, -3) - (-2, 1)$

3. $(2, 3) \cdot (3, -2)$

4. $\dfrac{(3, 1) - 2(3, 4)}{(-1, -2)}$

5. Show that addition for ordered pairs is commutative.

6. Find the norm of $3\mathbf{v} - 4\mathbf{w}$ if $\mathbf{v} = (2, 5)$ and $\mathbf{w} = (3, -2)$.

7. Find the direction angle of $(1, 1)$.

8. Find $(3\mathbf{i} + 4\mathbf{j}) \cdot (2\mathbf{i} - 3\mathbf{j})$.

9. Find the angle between $\mathbf{v} + \mathbf{w}$ and $\mathbf{v} - \mathbf{w}$ if
$\mathbf{v} = (3, 5)$ and $\mathbf{w} = (1, 2)$.

10. Find the unit vector along $\mathbf{v} = (3, 4)$.

11. Express the vector $\mathbf{v}$ as an ordered pair if
$|\mathbf{v}| = 4$ and the direction angle $\theta = 30°$.

12. Find the value of y for which $\mathbf{v} = (3, 2)$ is perpendicular to $\mathbf{w} = (4, y)$.

13. Resolve the vector $\mathbf{u} = 2\mathbf{v} - 3\mathbf{w}$ along $\mathbf{i}$ and $\mathbf{j}$ if
$\mathbf{v} = (13, 2)$ and $\mathbf{w} = (6, -2)$.

14. Find $(2\mathbf{v} - 3\mathbf{w}) \cdot (4\mathbf{v} + 2\mathbf{w})$ if $\mathbf{v} = (-1, -2)$ and $\mathbf{w} = (2, -3)$.

15. Find $(6, -1, -3) \cdot (2, -3, 4)$.

16. Find $|\mathbf{v}|$ if $\mathbf{v} = (2, -1, 5)$.

17. Find $\cos \phi$ if ϕ is the angle between $3\mathbf{v}$ and $2\mathbf{w}$ and $\mathbf{v} = (1, 3, 2)$ and $\mathbf{w} = (5, -2, -3)$.

SOLUTIONS

1. $(4,6) + (-2,3) = (4-2, 6+3)$
   $= (2,9)$

2. $(1,-3) - (-2,1) = (1+2, -3-1)$
   $= (3,-4)$

3. $(2,3)(3,-2) = (6+6, -4+9)$
   $= (12,5)$

4. $\dfrac{(3,1) - 2(3,4)}{(-1,-2)} = \dfrac{(-3,-7)}{(-1,-2)}$
   $= \left(\dfrac{3+14}{1+4}, \dfrac{7-6}{1+4}\right) = \left(\dfrac{17}{5}, \dfrac{1}{5}\right)$

EXERCISES I

Find the value of (x, y).

i. $(x,y) = 3(5,-4) + 2(-3,5)$
ii. $(x,y) = 3(2,1)(7,-5)$
iii. $(4,2) = (7,3) + (x,y)$
iv. $2(x,y) = (3,4) - (-1,3)$
v. $(3,6)(x,y) = (7,2)$

---

5. $(x_1, y_1) + (x_2, y_2) = (x_2, y_2) + (x_1, y_1)$
   By definition
   $(x_1, y_1) + (x_2, y_2) = (x_1 + x_2, y_1 + y_2)$
   $(x_2, y_2) + (x_1, y_1) = (x_2 + x_1, y_2 + y_1)$
   Since x_1, y_1, x_2, and $y_2 \in R$ and the commutative law holds for addition,
   $x_1 + x_2 = x_2 + x_1$ and
   $y_1 + y_2 = y_2 + y_1$
   Then by definition of equality
   $(x_1 + x_2, y_1 + y_2) = (x_2 + x_1, y_2 + y_1)$

6. $\mathbf{v} = (2,5)$ and $\mathbf{w} = (3,-2)$
   $3\mathbf{v} - 4\mathbf{w} = (6,15) - (12,-8)$
   $= (-6, 23)$
   $|3\mathbf{v} - 4\mathbf{w}| = \sqrt{36 + 529} = \sqrt{565}$

7. $\mathbf{v} = (1,1)$
   $\cos \theta = \dfrac{x}{|\mathbf{v}|} = \dfrac{1}{\sqrt{1+1}} = \dfrac{1}{\sqrt{2}}$;
   $\theta = 45°$

8. $(3\mathbf{i} + 4\mathbf{j}) \cdot (2\mathbf{i} - 3\mathbf{j}) = 6 - 12 = -6$

EXERCISES II

Show that

i. $(x,y) \cdot (a,b) = (a,b) \cdot (x,y)$
ii. $c(x,y) \cdot (a,b) = (x,y) \cdot c(a,b)$
iii. $(x,y) \cdot \left(\dfrac{x}{x^2+y^2}, \dfrac{-y}{x^2+y^2}\right) = (1,0)$

---

9. $\mathbf{v} = (3,5)$ and $\mathbf{w} = (1,2)$
   $\mathbf{v} + \mathbf{w} = (4,7)$ and $\mathbf{v} - \mathbf{w} = (2,3)$
   $\cos \phi = \dfrac{ac+bd}{|\mathbf{s}| |\mathbf{d}|} = \dfrac{8+21}{\sqrt{65}\sqrt{13}}$
   $= \dfrac{29}{\sqrt{5}\sqrt{13}\sqrt{13}} = \dfrac{29}{13\sqrt{5}}$
   $\phi = \arccos\left(\dfrac{29}{13\sqrt{5}}\right)$

10. $\mathbf{v} = (3,4)$; $|\mathbf{v}| = \sqrt{9+16} = 5$
    $\mathbf{u} = \dfrac{\mathbf{v}}{|\mathbf{v}|} = \left(\dfrac{3}{5}, \dfrac{4}{5}\right)$

11. By definition $\sin \theta = \dfrac{y}{|\mathbf{v}|}$
    $\sin 30° = \dfrac{1}{2} = \dfrac{y}{4}, \; y = 2$
    $|\mathbf{v}| = 4 = \sqrt{x^2 + y^2} = \sqrt{x^2 + 4}$
    $x^2 + 4 = 16; \; x^2 = 12, \; x = 2\sqrt{3}$
    $[x = -2\sqrt{3}$ is extraneous since $\theta = 30°]$
    $\mathbf{v} = (x,y) = (2\sqrt{3}, 2)$

EXERCISES III

Find $\mathbf{v}$.

i. $\mathbf{v} = (2\mathbf{i} - 3\mathbf{j}) + (-\mathbf{i} + \mathbf{j})$
ii. If $\mathbf{v} \perp (4,-3)$ and $\mathbf{v} = (x, 8)$
iii. $|\mathbf{v}| = 9$ and $\theta = 45°$

---

12. If $\mathbf{v} \perp \mathbf{w}$, then $\cos \phi = 0$ and $\mathbf{v} \cdot \mathbf{w} = 0$.
    $\mathbf{v} \cdot \mathbf{w} = ac + bd = 12 + 2y = 0$
    $2y = -12, \; y = -6$.

13. $\mathbf{v} = (13,2)$ and $\mathbf{w} = (6,-2)$
    $2\mathbf{v} = (26,4)$ and $3\mathbf{w} = (18,-6)$
    $\mathbf{u} = (26-18, 4+6) = (8,10)$
    $= 8\mathbf{i} + 10\mathbf{j}$

Algebra of Ordered Pairs, Vectors   147

14. $\mathbf{v} = (-1, -2)$ and $\mathbf{w} = (2, -3)$
   $2\mathbf{v} - 3\mathbf{w} = (-2 - 6, -4 + 9)$
   $\qquad = (-8, 5) = \mathbf{u}$
   $4\mathbf{v} + 2\mathbf{w} = (-4 + 4, -8 - 6)$
   $\qquad = (0, -14) = \mathbf{r}$
   $\mathbf{u} \cdot \mathbf{r} = 0 - 70 = -70$

15. $(6, -1, -3) \cdot (2, -3, 4)$
   $\qquad = 6(2) + (-1)(-3) + (-3)(4)$
   $\qquad = 12 + 3 - 12 = 3$

16. $\mathbf{v} = (2, -1, 5)$
   $|\mathbf{v}| = \sqrt{4 + 1 + 25} = \sqrt{30}$

17. $\mathbf{v} = (1, 3, 2)$ and $\mathbf{w} = (5, -2, -3)$
   $3\mathbf{v} = (3, 9, 6); \ 2\mathbf{w} = (10, -4, -6)$
   $\cos \phi = \dfrac{3(10) + 9(-4) + 6(-6)}{|3\mathbf{v}| |2\mathbf{w}|}$
   $\qquad = \dfrac{30 - 36 - 36}{\sqrt{9 + 81 + 36}\sqrt{100 + 16 + 36}}$
   $\qquad = -\dfrac{42}{\sqrt{126}\sqrt{152}} = -\dfrac{7}{6\sqrt{133}}$

EXERCISES IV

i. Find $(3\mathbf{i} - 2\mathbf{j}) \cdot (2\mathbf{i} - 3\mathbf{j})$.

ii. Find $3\mathbf{v} \cdot 2\mathbf{w}$ if
   $\mathbf{v} = 2\mathbf{i} - \mathbf{j}$ and $\mathbf{w} = -3\mathbf{i} + \mathbf{j}$.

iii. Find x such that
   $|\mathbf{v}| = 5$ and $\mathbf{v} = 2x\mathbf{i} + 3\mathbf{j}$.

iv. Find x such that
   $(x\mathbf{i} - 4\mathbf{j}) \perp (x\mathbf{i} + 9\mathbf{j})$.

v. Find the angle between $2\mathbf{v}$ and $\mathbf{w}$ if
   $\mathbf{v} = 3\mathbf{i} - \mathbf{j}$ and $\mathbf{w} = 4\mathbf{i} + 2\mathbf{j}$.

Solutions to exercises are on page 150.

ANSWERS

1. $(2, 9)$	1
2. $(3, 4)$	2
3. $(12, 5)$	3
4. $\left(\frac{17}{5}, \frac{1}{5}\right)$	4
5. Commutative law holds	5
6. $\sqrt{565}$	6
7. $45°$	7
8. -6	8
9. $\text{Arccos}\left(\dfrac{29}{13\sqrt{5}}\right)$	9
10. $\left(\frac{3}{5}, \frac{4}{5}\right)$	10
11. $\mathbf{v} = (2\sqrt{3}, 2)$	11
12. $y = -6$	12
13. $8\mathbf{i} + 10\mathbf{j}$	13
14. -70	14
15. 3	15
16. $\sqrt{30}$	16
17. $-\dfrac{7}{6\sqrt{133}}$	17

BASIC FACTS

Let A be the set of **ordered pairs** of real numbers (x, y) defined by
$$A = \{(x, y) \mid x, y \in R\}$$

Definitions

$(x_1, y_1) = (x_2, y_2)$ if and only if $x_1 = x_2$ and $y_1 = y_2$.

$(x_1, y_1) + (x_2, y_2) = (x_1 + x_2, y_1 + y_2)$

$(0, 0)$ is the identity element for addition: $(x, y) + (0, 0) = (x, y)$

If $c \in R$ (called a scalar), then
$$c(x, y) = (cx, cy)$$

If $c = -1$, $-1(x, y) = (-x, -y)$; $(-x, -y)$ is the inverse element for addition: $(x, y) + (-x, -y) = (0, 0)$

$(x_1, y_1) - (x_2, y_2) = (x_1 - x_2, y_1 - y_2)$

Properties. The set A forms a commutative group with respect to addition.

1) A is closed under addition
2) The addition of ordered pairs in A is commutative
3) The associative property for addition holds
4) The identity element exists $(0, 0)$
5) The inverse element exists $(-x, -y)$
6) If c and d are scalars, then $(cd)(x, y) = c(d(x, y))$
7) $(c + d)(x, y) = c(x, y) + d(x, y)$
8) $c[(x_1, y_1) + (x_2, y_2)]$ $= c(x_1, y_1) + c(x_2, y_2)$
9) $1(x, y) = (x, y)$

A set satisfying these nine properties forms a **linear space over the real numbers**. The algebraic system is also called a **vector space**.

Definition. The product of two ordered pairs (x_1, y_1) and (x_2, y_2) is the ordered pair $(x_1 x_2 - y_1 y_2, x_1 y_2 + x_2 y_1)$. The ordered pair $(1, 0)$ is the identity element for multiplication.

The quotient of two ordered pairs is:
$$\frac{(x_1, y_1)}{(x_2, y_2)} = \frac{x_1 x_2 + y_1 y_2}{x_2^2 + y_2^2}, \frac{x_2 y_1 - x_1 y_2}{x_2^2 + y_2^2}$$

ADDITIONAL INFORMATION

The addition and subtraction operations of **ordered pairs** are usually clear. Most of the difficulty arises from the operations of multiplication and division. Since the complete theory of complex variables can be based on the ordered pair definition, these two operations coincide with the results obtained in the corresponding operations with complex numbers. To illustrate this let us consider multiplication.

Ordered Pair Definition	$i^2 = -1$ Definition
$z_1 = (x_1, y_1)$ and $z_2 = (x_2, y_2)$	$z_1 = x_1 + y_1 i$ and $z_2 = x_2 + y_2 i$
$z_1 z_2 = (x_1, y_1)(x_2, y_2)$ $= (x_1 x_2 - y_1 y_2,$ $\quad x_1 y_2 + x_2 y_1)$	$z_1 z_2 = (x_1 + y_1 i)(x_2 + y_2 i)$ $= x_1 x_2 - y_1 y_2$ $\quad + (y_1 x_2 + x_1 y_2)i$

The same comparison can be made for division.

When we compare the **vector** defined by an ordered pair and the complex number plotted in the Argand diagram, we can see the consistency of the geometric interpretation. In the following table do not confuse the unit vector **i** with $i^2 = -1$.

Vector	Complex Number
$\mathbf{v} = (x, y) = x\mathbf{i} + y\mathbf{j}$	$z = (x, y) = x + yi$

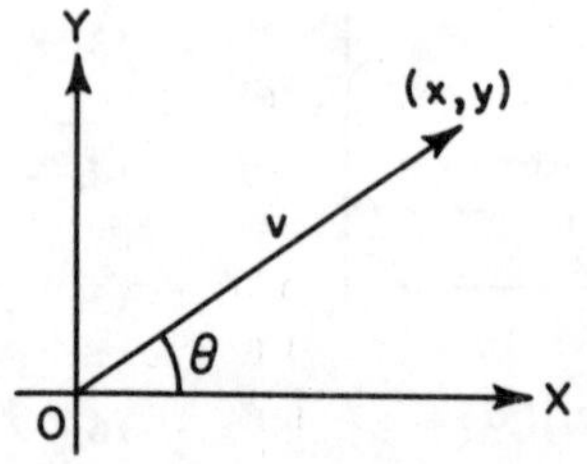

Fig. 25-1   Fig. 25-2

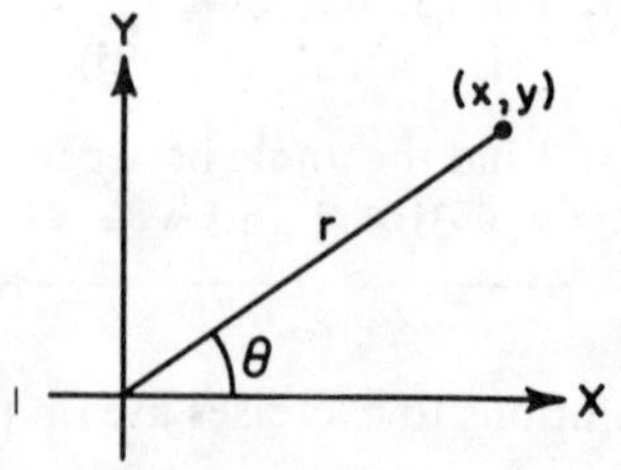

$|\mathbf{v}| = \sqrt{x^2 + y^2}$ $|z| = r = \sqrt{x^2 + y^2}$

$\cos \theta = \dfrac{x}{|\mathbf{v}|}$; $\sin \theta = \dfrac{y}{|\mathbf{v}|}$ $\cos \theta = \dfrac{x}{r}$; $\sin \theta = \dfrac{y}{r}$

$\mathbf{v} = |\mathbf{v}|(\cos \theta \mathbf{i} + \sin \theta \mathbf{j})$ $z = r(\cos \theta + i \sin \theta)$

Sum of two vectors: Sum of two complex numbers:

$\mathbf{v}_1 + \mathbf{v}_2 = (x_1 + x_2, y_1 + y_2)$ $z_1 + z_2 = (x_1 + y_1 i)$
$= (x_1 + x_2)\mathbf{i} + (y_1 + y_2)\mathbf{j}$ $\quad + (x_2 + y_2 i)$
   $= (x_1 + x_2) + (y_1 + y_2)i$

Vectors

A **complex number** z is an ordered pair of real numbers (x, y) where x is called the **real** part and y is called the **imaginary** part. It is also defined by $z = x + yi$ where x and $y \in R$ and $i^2 = -1$. A two dimensional **vector** $\mathbf{v}$ is an ordered pair of real numbers (x, y). A **vector** $\mathbf{v}$ in a plane is a directed line segment of fixed length. The numbers x and y are called **components** of $\mathbf{v}$.

Properties. If $\mathbf{v} = (x, y)$, then

1. The **norm** of $\mathbf{v} = |\mathbf{v}| = \sqrt{x^2 + y^2}$
2. The **length** of $\mathbf{v} = |\mathbf{v}|$
3. The direction angle θ of $\mathbf{v}$
   $$\cos\theta = \frac{x}{|\mathbf{v}|}, \quad \sin\theta = \frac{y}{|\mathbf{v}|}$$
   $0° \leq \theta < 360°$
4. The resultant of two vectors $\mathbf{v}_1 + \mathbf{v}_2 = (x_1 + x_2, y_1 + y_2)$ is the diagonal of a parallelogram formed with its two sides coincident with $\mathbf{v}_1$ and $\mathbf{v}_2$.

A **unit vector** is defined by

$$\mathbf{u} = \frac{\mathbf{v}}{|\mathbf{v}|}$$

Let $\mathbf{i} = (1, 0)$ be the unit vector along the positive X-axis and $\mathbf{j} = (0, 1)$ be the unit vector along the positive Y-axis, then we can write $\mathbf{v} = x\mathbf{i} + y\mathbf{j}$.

If $\mathbf{v} = a\mathbf{i} + b\mathbf{j}$ and $\mathbf{w} = c\mathbf{i} + d\mathbf{j}$, then the angle between $\mathbf{v}$ and $\mathbf{w}$ is found by

$$\cos\phi = \frac{ac + bd}{|\mathbf{v}||\mathbf{w}|}$$

$$\mathbf{v} + \mathbf{w} = (a + c)\mathbf{i} + (b + d)\mathbf{j}$$

The **dot (inner) product** of two vectors $\mathbf{v} \cdot \mathbf{w}$ is a scalar

$$\mathbf{v} \cdot \mathbf{w} = ac + bd$$

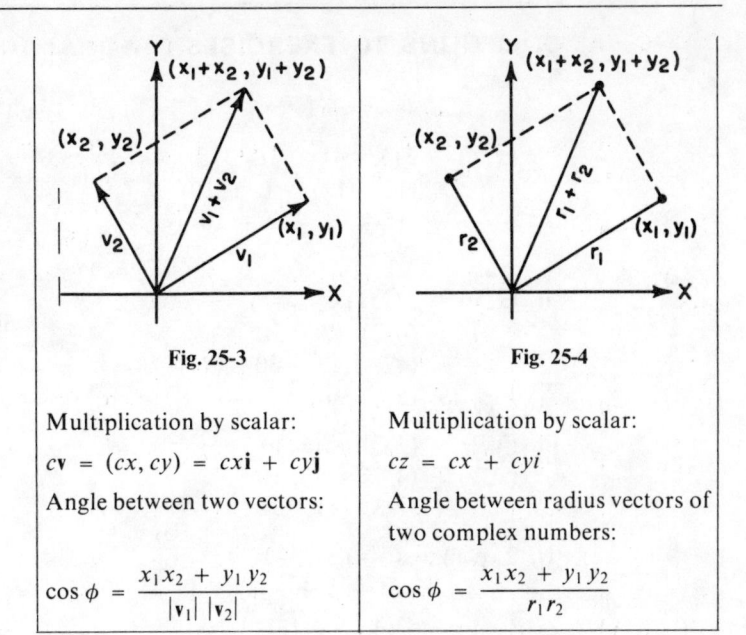

Fig. 25-3

Multiplication by scalar:
$$c\mathbf{v} = (cx, cy) = cx\mathbf{i} + cy\mathbf{j}$$
Angle between two vectors:
$$\cos\phi = \frac{x_1 x_2 + y_1 y_2}{|\mathbf{v}_1||\mathbf{v}_2|}$$

Fig. 25-4

Multiplication by scalar:
$$cz = cx + cyi$$
Angle between radius vectors of two complex numbers:
$$\cos\phi = \frac{x_1 x_2 + y_1 y_2}{r_1 r_2}$$

The last formulas may both be derived by using the law of cosines.

Three-Dimensional Vectors

This comparison of vectors and complex numbers stops as we expand the concept of vectors beyond ordered pairs to ordered triples and in general to n dimensions. We now write the vector as a row or column matrix and introduce a new multiplication to coincide with the concept of multiplying matrices. Thus we define the dot (or inner) product of two vectors to be a scalar (real number) and equal to the sum of the products of corresponding elements.

$$(a_1, b_1, c_1) \cdot (a_2, b_2, c_2) = a_1 a_2 + b_1 b_2 + c_1 c_2$$

Note that the result is the same as that obtained by multiplying a row matrix (a_1, b_1, c_1) by a column matrix composed of a_2, b_2, and c_2.

Vectors defined by ordered triples can be written in terms of their components using unit vectors $\mathbf{i} = (1, 0, 0)$, $\mathbf{j} = (0, 1, 0)$, and $\mathbf{k} = (0, 0, 1)$ directed along the positive (X, Y, Z)-axes of three dimensional geometry:

$$\mathbf{v} = (a, b, c) = a\mathbf{i} + b\mathbf{j} + c\mathbf{k}$$

The properties given for a two dimensional vector also hold for a three-dimensional vector.

150 Algebra of Ordered Pairs, Vectors

SOLUTIONS TO EXERCISES

I

i. $(x, y) = 3(5, -4) + 2(-3, 5)$
$\qquad = (15, -12) + (-6, 10)$
$\qquad = (15 - 6, -12 + 10)$
$\qquad = (9, -2)$

ii. $(x, y) = 3(2, 1) \cdot (7, -5)$
$\qquad = (6, 3) \cdot (7, -5)$
$\qquad = (42 + 15, -30 + 21)$
$\qquad = (57, -9)$

iii. $(4, 2) = (7, 3) + (x, y)$
$(x, y) = (4, 2) - (7, 3)$
$\qquad = (-3, -1)$

iv. $2(x, y) = (3, 4) - (-1, 3)$
$\qquad = (3 + 1, 4 - 3) = (4, 1)$
$(x, y) = \frac{1}{2}(4, 1) = \left(2, \frac{1}{2}\right)$

v. $(3, 6) \cdot (x, y) = (7, 2)$
$(x, y) = \frac{(7, 2)}{(3, 6)} = \left(\frac{21 + 12}{9 + 36}, \frac{6 - 42}{9 + 36}\right)$
$\qquad = \left(\frac{33}{45}, -\frac{36}{45}\right) = \left(\frac{11}{15}, -\frac{4}{5}\right)$

II

i. $(x, y) \cdot (a, b) = (a, b) \cdot (x, y)$
By definition
$(x, y) \cdot (a, b) = (ax - by, bx + ay)$
$(a, b) \cdot (x, y) = (ax - by, ay + bx)$
Since $ax - by = ax - by$ and
$bx + ay = ay + bx$ we have
$(x, y) \cdot (a, b) = (a, b) \cdot (x, y)$
by definition of equality

ii. $c(x, y) \cdot (a, b) = (x, y) \cdot c(a, b)$
$c(x, y) \cdot (a, b) = (cx, cy) \cdot (a, b)$
$\qquad = (acx - bcy, bcx + acy)$
$(x, y) \cdot c(a, b) = (x, y) \cdot (ac, bc)$
$\qquad = (acx - bcy, bcx + acy)$
The statement is true by the definition of equality.

iii. $(x, y) \cdot \left(\frac{x}{x^2 + y^2}, \frac{-y}{x^2 + y^2}\right) = (1, 0)$
$\left(\frac{x^2}{x^2 + y^2} - \frac{-y^2}{x^2 + y^2}, -\frac{xy}{x^2 + y^2} + \frac{xy}{x^2 + y^2}\right)$
$\qquad = \left(\frac{x^2 + y^2}{x^2 + y^2}, \frac{-xy + xy}{x^2 + y^2}\right) = (1, 0)$

III

i. $\mathbf{v} = (2\mathbf{i} - 3\mathbf{j}) + (-\mathbf{i} + \mathbf{j})$
$\qquad = (2 - 1)\mathbf{i} + (-3 + 1)\mathbf{j}$
$\qquad = \mathbf{i} - 2\mathbf{j}$

ii. Let $\mathbf{v} \equiv (x, 8)$
If $\mathbf{v} \perp (4, -3)$, then $4x - 24 = 0$
and $x = 6$; $\mathbf{v} = (6, 8)$

iii. $|\mathbf{v}| = 9$ and $\theta = 45°$
$\cos 45° = \frac{1}{\sqrt{2}} = \frac{x}{9}; \ x = \frac{9}{\sqrt{2}}$
$\sin 45° = \frac{1}{\sqrt{2}} = \frac{y}{9}; \ y = \frac{9}{\sqrt{2}}$
$\mathbf{v} = \left(\frac{9}{\sqrt{2}}, \frac{9}{\sqrt{2}}\right)$

IV

i. $(3\mathbf{i} - 2\mathbf{j})(2\mathbf{i} - 3\mathbf{j})$
$\qquad = (3)(2) + (-2)(-3) = 12$

ii. $\mathbf{v} = 2\mathbf{i} - \mathbf{j}$ and $\mathbf{w} = -3\mathbf{i} + \mathbf{j}$
$3\mathbf{v} = 6\mathbf{i} - 3\mathbf{j}; \ 2\mathbf{w} = -6\mathbf{i} + 2\mathbf{j}$
$3\mathbf{v} \cdot 2\mathbf{w} = (6)(-6) + (-3)(2) = -42$

iii. $\mathbf{v} = 2x\mathbf{i} + 3\mathbf{j}; \ |\mathbf{v}| = 5$
$|\mathbf{v}|^2 = (2x)^2 + 3^2 = 4x^2 + 9 = 25$
$4x^2 = 25 - 9 = 16, \ x^2 = 4, \ x = \pm 2$

iv. If $(x\mathbf{i} - 4\mathbf{j}) \perp (x\mathbf{i} + 9\mathbf{j})$, then
$x^2 - 36 = 0$ and $x = \pm 6$

v. $\mathbf{v} = 3\mathbf{i} - \mathbf{j}, \ 2\mathbf{v} = 6\mathbf{i} - 2\mathbf{j}, \ \mathbf{w} = 4\mathbf{i} + 2\mathbf{j}$
$\cos \theta = \frac{(6)(4) + (-2)(2)}{\sqrt{36 + 4}\sqrt{16 + 4}}$
$\qquad = \frac{24 - 4}{\sqrt{40}\sqrt{20}} = \frac{20}{\sqrt{2}\sqrt{20}\sqrt{20}}$
$\qquad = \frac{20}{20\sqrt{2}} = \frac{1}{\sqrt{2}}$

The acute angle is $45°$.

FINAL EXAMINATION

Answers are given on page 157.

1. Let $A = \{2, 5, 12, 20\}$, $B = \{9, 12, 15, 23\}$, and $C = \{2, 5\}$. Are the following statements true or false?
   a. $2 \in A$
   b. $2 \subset A$
   c. $C \subset A$
   d. $C \subset C$
   e. $\emptyset \subseteq A$
   f. $A = B$
   g. $A \equiv B$
   h. $n(A) = 4$
   i. A has fifteen proper subsets.
   j. A one-to-one correspondence can be established between A and C.

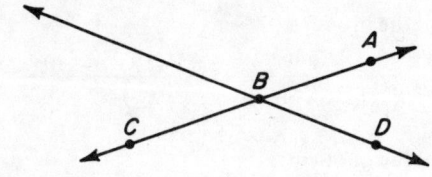

2.

   On the basis of the above figure determine the truth or falsity of each of the following statements.
   a. $\overrightarrow{AB} \subset \overleftrightarrow{AB}$
   b. $\{B\} \subset \overrightarrow{BD}$
   c. $\overrightarrow{CB} = \overrightarrow{BA}$
   d. $\angle ABD = \overrightarrow{BA} \cup \overrightarrow{BD}$
   e. $\overleftrightarrow{AC} \cap \overrightarrow{CB}$ is a finite point set.

3. If A is the real interval $(-3, 7]$ and B is the real interval $(2, 10)$, find $A \cap B$.

4. Let $A = \{1, 2\}$ and $B = \{3, 4, 5\}$. Which of the following statements are true?
   a. $A \times B = B \times A$
   b. $(2, 3) \notin A \times B$
   c. $A \times B$ has 2^6 subsets.
   d. $A \times B$ and $B \times A$ are disjoint sets.

5. Let $U = \{x \text{(integer)} \mid 0 < x < 10\}$, $P = \{\text{single digit prime numbers}\}$, $E = \{2, 4, 6, 8\}$, and $O = \{1, 3, 5, 7, 9\}$.
   a. List the elements in P'.
   b. List the elements in $E \cap P$.
   c. List the elements in $O' \cap (P \cup E')$.
   d. Is $O \cap E = U'$?

6. Which of the following statements represents the shaded region of the Venn diagram?
   a. $B' \cap (A \cup C)$
   b. $C \cap (A \cup B')$
   c. $C \cup (A \cap B')$
   d. $(B' \cup C)$

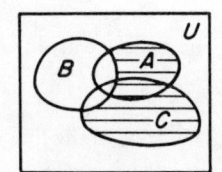

7. Is the following statement true or false?
   $$[(A \cup B') \cap B]' = A' \cup B'$$

8. A survey of 100 students entering a university indicated that each student had at least one sequence, 54 students had sequences in mathematics, 45 had sequences in science, 40 had sequences in mathematics and science, 12 had sequences in mathematics and language, 15 had sequences in science and language, and 10 had sequences in all three areas. Draw a Venn diagram to illustrate the given data and use the diagram to determine the number of students that had exactly two sequences.

9. p_x: x is a number such that $3x - 6 = 0$
   q_x: x is a number such that $2x - 8 > 0$
   If the domain of x is $\{0, 1, 2, 3, 4, 5, 6, 7\}$, list the elements in the truth set of $p_x \vee q_x$.

10. Graph set S where $S = \{-2, 0, 3, 7\}$.

11. Graph the interval $(-5, 7]$.

12. Graph the set of ordered pairs $\{(-2, 1), (3, 2), (-1, -1), (2, -3)\}$.

13. What is an axiomatic system?

14. Consider the relation '—is a subset of—' in the universe of sets. Which of the following statements are true?
    a. This relation is a dyadic relation.
    b. This relation is symmetric.
    c. This relation is transitive.
    d. This relation is an equivalence relation.

15. Let $A = \{0, 1, 2\}$ and let $*$ be a rule defined such that $a * b$ is the remainder when $a + b$ is divided by 3.
    a. Evaluate $a * b$ for $(0, 0)$, $(1, 2)$, $(2, 0)$ and $(1, 0)$.
    b. Is the result of $*$ on the set A always an element of A?
    c. Is $*$ an operation?
    d. What is the identity element for $*$?
    e. What, if any, is the inverse of 1?

16. What, if any, is the identity element for the set of even integers with respect to the operation of addition?

151

17. Let b be any non-zero element in the set of rational numbers. Does b have an inverse with respect to the operation of multiplication? If so, give the formula for the inverse.

18. If n' is the successor of n, what is the successor of $(3 \cdot 4')'$?

19. If the nth term of a sequence is $3n^{n+2}$, what is the $(k + 1)$st term?

20. State the Axiom of Mathematical Induction.

21. Can the method of mathematical induction be used to prove the following true statement?

    For all natural numbers n, $x^n - y^n$ is divisible by $x - y$.

22. a. Find the prime factorization of 90 and of 300.
    b. What is the greatest common divisor of 90 and 300?
    c. Are 90 and 300 relatively prime?

23. Express $0.2\overline{34}$ as a number in the form $\frac{a}{b}$ where a and b are integers.

24. Evaluate: $|-3| - |7| + \sqrt[3]{64^2}$

25. The conjugate of $a - bi$ is _____.

26. Change $-3 + 3i$ to its trigonometric form.

27. Perform the indicated operations. Express each result as a complex number in standard form.
    a. $(2 - 3i) + (6 + 4i)$
    b. $\left(5 + \frac{1}{2}i\right) - \left(2 + \frac{1}{4}i\right) + (2i)$
    c. $(3 + 2i)(4 - i)$
    d. $i^3(-1 + 2i)$
    e. $(2 - 3i)(2 + 3i)$

28. Determine the truth or falsity of each of the following statements.
    a. The set of rational numbers and the set of irrational numbers are disjoint sets.
    b. The union of the set of positive real numbers and the set of negative real numbers is the set of real numbers.
    c. π is a member of the set of complex numbers.
    d. The sum of two conjugate complex numbers could be an imaginary number.

29. Simplify:
    $$3x\{x^2 - [4 + 7x + (x^2 - 3)] + 10\}$$

30. Express in lowest terms:
    $$\frac{x^3 + 4x^2 - 21x}{x^3 + 5x^2 - 24x}$$

31. For what values of x is the following expression *not* defined?
    $$\frac{x + 4}{x^3 - 16x}$$

32. Find the sum:
    $$\frac{x^2 - 4}{x} + \frac{x^2 - 7}{x + 3}$$

33. Find the product:
    $$\frac{x^2 - 9}{x^2 + 13x + 30} \times \frac{x^3 + 3x^2 - 70x}{x^2 - 3x}$$

34. Find the quotient:
    $$\frac{3x^2 + 19x + 20}{x^2 - 1} \div \frac{x^2 + 5x}{x^2 - 6x - 7}$$

35. Find the quotient and remainder when $3x^3 + 5x^2 + 28x + 7$ is divided by $x + 4$

36. Let $R = \{(-2, 4), (-1, 8), (2, -2), (4, 4)\}$.
    a. Does R define a function?
    b. What is the domain of R?

37. If x and $f(x)$ are real numbers, find the domain and range of the following function.
    $$f : x \rightarrow \frac{2}{x} - 4$$

38. Let f be a function defined by the equation $f(x) = 3x - 7$. Write an equation which defines f^{-1}, the inverse function of f.

39. f is a constant function and $f(9) = \frac{1}{2}$. Find the value of $f(-6)$.

40. The abscissa of the midpoint of a line between $P_1(x_1, y_1)$ and $P_2(x_2, y_2)$ is _____.

41. The distance from $P_1(x_1, y_1)$ to $ax + by + c = 0$ is _____.

42. The slope of the line $72x + 18y = 36$ is _____.

43. In the equation $y = mx + b$, the number b is the _____ of the line.

44. Find the value of x in the expression.
$$\sqrt{x-1} + 1 = 0$$

In problems **45–47**, determine if the statements are true or false.

45. If $9C = 5(F - 32)$, then $F = 212$ when $C = 100$.

46. The lines $3x - 5 = 0$ and $4y + 3 = 0$ are perpendicular.

47. The lines represented by the equations $2x - 3y = 5$ and $2y - 3x = 5$ are parallel.

Given the points $P_1(4, 2)$, $P_2(2, 6)$, $P_3(-2, 2)$, and $P_4(2, 0)$:

48. Find the midpoint of $\overline{P_2 P_3}$.

49. Find the slope of the line $\overline{P_1 P_4}$.

50. Find the equation of the line through P_2 and P_3.

51. Is $\overline{P_1 P_4}$ parallel to $\overline{P_2 P_3}$?

52. Find the equation of a line through P_1 and perpendicular to $\overline{P_2 P_3}$?

53. Find the distance from P_1 to $\overline{P_2 P_3}$.

54. Let A and B be matrices such that
$$A = \begin{pmatrix} 4 & -7 \\ 6 & 2 \end{pmatrix} \text{ and } B = \begin{pmatrix} 7 & 9 \\ 2 & 1 \end{pmatrix}.$$
   a. Find the matrix $A + B$.
   b. Find the matrix AB.

55. Evaluate the following determinant.
$$\begin{vmatrix} -2 & 2 & 5 \\ 2 & 7 & 0 \\ -3 & -1 & 2 \end{vmatrix}$$

56. Which of the following statements are true?
   a. The additive identity for the set of 2 by 2 matrices is the following matrix.
$$\begin{pmatrix} 1 & 0 \\ 0 & 1 \end{pmatrix}$$
   b. If A is a 2 by 4 matrix and B is a 4 by 3 matrix, the product matrix AB is a 2 by 3 matrix.
   c. If any two rows of a square matrix are identical, its determinant is zero.
   d. Cramer's Rule may be used to solve any system of linear equations.

57. Can matrix A be transformed into matrix B by row operations if
$$A = \begin{pmatrix} \frac{1}{2} & -1 & 2 \\ 1 & 2 & -1 \\ 4 & -6 & 2 \end{pmatrix} \text{ and } B = \begin{pmatrix} 1 & 0 & 0 \\ 0 & 1 & 0 \\ 0 & 0 & 1 \end{pmatrix}?$$

Complete the following statements.

58. The solutions of $ax^2 + bx + c = 0$ are _____.

59. The equations of a circle with center at (h, k) is _____.

60. The formula for the remainder when $f(x)$ is divided by $x + r$ is _____.

61. The discriminant of $ax^2 + bx + c = 0$ is _____.

62. An ellipse is the locus of a point which moves so that the _____ of its distances from two fixed points is a constant.

63. If $b^2 - 4ac$ is a perfect square, the roots of $ax^2 + bx + c = 0$ are _____.

64. The graph of $s = t - 16t^2$ is a(n) _____.

65. The number of negative real roots of $8x^4 - 18x^3 + 2x^2 + 7x - 6 = 0$ cannot exceed _____.

66. The equation in Problem 65 has _____ roots.

67. Dividend = Divisor × Quotient + _____.

Find the values of x and/or y which satisfy the following conditions.

68. $3x - 5 > 0$

69. $x^2 - x - 6 = 0$

70. $x + y = 7$ and $x - y = 5$

71. $x^2 - x - 6 < 0$

72. $2x - 3y = -1$ and $3x - 2y = -9$

73. $2ix^2 + 5x - 2i = 0$ and $i^2 = -1$

74. $x^2 - 7x + 12 < 0$

75. $x^4 + x^3 - 3x^2 - 3x = 0$

76. The coordinates of the center $C(x, y)$ of the circle $x^2 + y^2 - 6x - 8y = 0$

77. $2x^2 - 10x + 17 = 0$

Given the ellipse $9x^2 + 25y^2 - 18x - 150y + 9 = 0$:

78. Find the coordinates of the center.

79. Find the semimajor axis.

80. Find the eccentricity.

81. Find the coordinates of the foci.

82. Find the equations of the directrices.

In problems **83–98**, determine if the statements are true or false.

83. The sum of the roots of $ax^2 + bx + c = 0$ is c/a.

84. One solution of $6x^4 + x^3 - 26x^2 - 4x + 8 = 0$ is $x = -\frac{1}{2}$.

85. If $x - 2 > 0$ and $y - 3 < 0$, then $xy - 2y - 3x + 6 < 0$.

86. If $c > 0$ and $x > y$, then $cy < cx$.

87. If a and b are positive and $a^2 > b^2$, then $a > b$.

88. If $x = 2$ and $a > b$, then $\frac{b}{x} < \frac{a}{x}$.

89. If $7x - 3 > 2x + 7$, then $x > 2$.

90. The graph of $xy = 5$ is a hyperbola.

91. The only root of $x^3 = 1$ is $x = 1$.

92. If $e = 1$, the conic is a parabola.

93. The following equations are consistent $x + 2y = 4$ and $3x + 6y = 12$.

94. If r is a zero of the polynomial $f(x)$, then $(x + r)$ is a factor of $f(x)$.

95. $X = \begin{pmatrix} 2 \\ -1 \end{pmatrix}$ is a solution of $AX = C$ if
$$A = \begin{pmatrix} 2 & -3 \\ 3 & 1 \end{pmatrix} \text{ and } C = \begin{pmatrix} 7 \\ 5 \end{pmatrix}$$

96. If $a + \sqrt{b}$ is a root of a rational integral equation, then $a - \sqrt{b}$ is a root.

97. The length of the latus rectum of the parabola $2x^2 = 9y$ is $\frac{9}{2}$.

98. The sum of the roots of the equation $2x^2 - 10x + 7 = 0$ is 5.

99. An exponential function is defined by the equation $f(x) = 2^x$.
   a. What is the domain of this function?
   b. What is the range of this function?
   c. Write an expression for the inverse of f.
   d. Sketch the graph of f on a set of coordinate axes.
   e. Using the same set of axes, sketch the graph of the inverse function of f.

100. Express $\log_5 125 = 3$ in exponential form.

101. Find the solution set of the exponential equation $3^{x+2} = \left(\frac{1}{27}\right)^{2x}$. Do not use logarithms.

102. Find the solution set of the following logarithmic equation.
$$\log(x - 2) + \log(x + 3) = \log 6$$

103. Write the tenth term of the sequence defined by $a_n = \frac{n + 3}{n!}$.

104. Find the sum of the following series.
$$\sum_{1}^{4} 3n - 2$$

105. Which of the following statements is true?
    a. The terms 1, −1, 1, −1, 1 are in arithmetic progression.
    b. If the first term of an arithmetic sequence is −4 and the common difference is 12, then the twenty-first term is 192.
    c. A sequence can have two different limits.
    d. Every polynomial function is continuous.

106. Find the limit of the sequence whose general term is $\frac{5n^3 - 3n^2 + 7}{2n^3 - 4}$.

107. If $f(x) = \frac{2x^3 - 3x + 10}{x - 5}$, evaluate $\lim_{x \to 2} f(x)$.

108. The function $f: x \to \frac{x^2 - x - 6}{x - 3}$ is defined for all real x except $x = 3$.
    a. What can be said about the graph of f at $x = 3$?
    b. Define $f(3)$ so that f will be continuous at $x = 3$.

109. How many different batting orders can be formed once the nine members of a baseball team have been chosen?

110. Six people enter a subway car. In how many ways can three unoccupied seats in the car be filled?

111. A student has two identical algebra books, two identical geometry books, and one calculus book.
    a. In how many ways can he line up these books on a shelf?
    b. How many of the arrangements in (a) have the identical books next to one another?
    c. How many of the arrangements in (a) have no algebra books in the three middle positions?

112. How many committees consisting of six members can be formed from a set of twenty people?

113. How many committees consisting of two Democrats and two Republicans can be formed from a set of five Democrats and seven Republicans?

114. The opening player in a scrabble game has a set of seven different letters. If he wishes to form at least a six letter word, how many different arrangements of the letters could he consider?

115. What is the seventh term of the expansion of $(x + y)^{12}$?

116. In the expansion of $(x + 2)^7$, what is the coefficient of the term involving x^6?

117. An experiment consists of drawing one coin from a bag in which there is a nickel and a dime. Is {dime, nickel, head, tail} an acceptable sample space for this experiment?

118. An experiment consists of tossing five coins and recording the number of heads and tails.
    a. How many different outcomes has the experiment if the coins are not distinguishable?
    b. How many different outcomes result if the coins are distinguishable?
    c. What is the probability of exactly two heads if the coins are distinguishable?
    d. What is the probability of at least three tails if the coins are distinguishable?
    e. What is the probability of three heads and at least three tails?

119. A drawer contains four black socks and two pink socks. Two socks are taken from the drawer in the dark.
    a. What is the probability that the socks are the same color?
    b. What is the probability that the socks are a different color?
    c. What is the probability that one of the socks is black?
    d. If one of the socks is known to be pink, what is the probability that the other sock is pink?

120. An integer is chosen at random from the first fifty positive integers. What is the probability that the integer is divisible by 4 or 6?

121. Two dice with faces labeled 1, 2, 3, 4, 5, 6 are tossed. Let p be 'the sum of the numbers on the upper faces is greater than seven' and q be 'the number on each upper face is greater than four.'
    a. Find PR$[p]$.
    b. Find PR$[q]$.
    c. Find PR$[p \wedge q]$.

122. What can be said about statement p if (1) the probability of p is 1, or if (2) the probability of p is 0?

Determine if the following statements are true or false.

123. $|\mathbf{v}| = 13$ if $\mathbf{v} = (5, -12)$

124. $\mathbf{v} = (3, 2)$ is perpendicular to $\mathbf{w} = (4, -6)$

125. If $\mathbf{v} = (3, -1)$ and $\mathbf{w} = (2, 3)$, then $\mathbf{u} = 3\mathbf{v} - 2\mathbf{w} = 5\mathbf{i} - 9\mathbf{j}$

126. The unit vector along $\mathbf{v} = (3, 4)$ is $\mathbf{u} = \left(\frac{3}{7}, \frac{4}{7}\right)$

ANSWERS

1. a. True f. False
   b. False g. True
   c. True h. True
   d. False i. True
   e. True j. False
2. a. True c. True
   b. False d. True
   e. False
3. $(2, 7]$
4. c, d
5. a. $P' = \{1, 4, 6, 8, 9\}$
   b. $E \cap P = \{2\}$
   c. $O' \cap (P \cup E') = \{2\}$
   d. Yes
6. a
7. True
8.

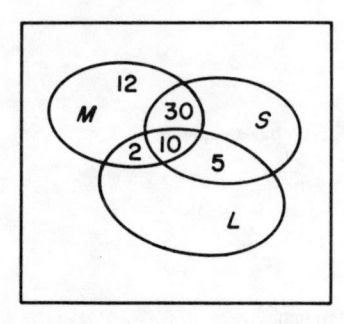

37 students had exactly two sequences.

9. $\{2, 5, 6, 7\}$
10.

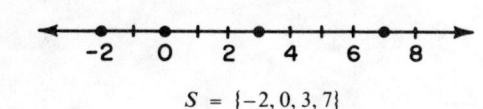

$S = \{-2, 0, 3, 7\}$

11.

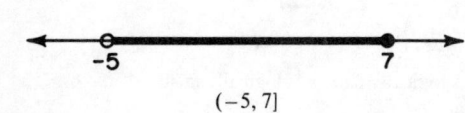

$(-5, 7]$

12.

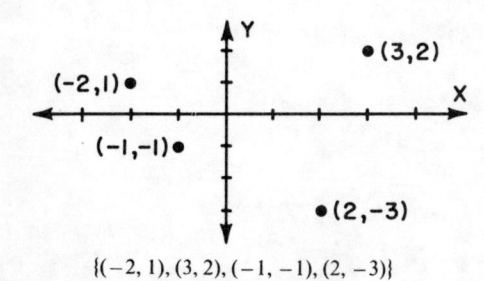

$\{(-2, 1), (3, 2), (-1, -1), (2, -3)\}$

13. An axiomatic system is any logical system having primitive terms and postulates.
14. a, c
15. a. 0, 0, 2, 1
    b. yes
    c. yes
    d. 0
    e. 2
16. 0
17. The inverse of b is $\frac{1}{b}$.
18. 17
19. $3(k + 1)^{k+3}$
20. If S is a set of natural numbers such that $1 \in S$ and $K \in S \to K + 1 \in S$, then all natural numbers belong to S.
21. yes
22. a. $90 = 2 \cdot 3^2 \cdot 5, 300 = 2^2 \cdot 3 \cdot 5^2$
    b. 30
    c. no
23. $\frac{232}{990}$ or $\frac{116}{495}$
24. 12
25. $a + bi$
26. $3\sqrt{2}(\cos 135° + i \sin 135°)$
27. a. $8 + i$ c. $14 + 5i$
    b. $3 + 2\frac{1}{4}i$ d. $2 + i$
    e. $13 + 0i$
28. a. True c. True
    b. False d. False
29. $27x - 21x^2$
30. $\frac{x + 7}{x + 8}$
31. $0, 4, -4$
32. $\frac{2x^3 + 3x^2 - 11x - 12}{x(x + 3)}$
33. $x - 7$
34. $\frac{(3x + 4)(x - 7)}{x(x - 1)}$
35. Quotient: $3x^2 - 7x + 28$
    Remainder: -105
36. a. yes
    b. $\{-2, -1, 2, 4\}$
37. domain: $\{x \text{ (real)} \mid x \neq 0\}$
    range: $\{y \text{ (real)} \mid y \neq -4\}$
38. $f^{-1}(x) = \frac{x + 7}{3}$
4. a. $f + g = \{(3, 11), (5, -1), (9, 3)\}$
    $f \circ g = \{(3, 6), (6, 0), (9, 4)\}$
   b. 6
39. $\frac{1}{2}$
40. $\frac{1}{2}(x_1 + x_2)$
41. $\frac{ax_1 + by_1 + c}{\sqrt{a^2 + b^2}}$
42. -4
43. y-intercept
44. no solution
45. True 46. True

Final Examination Answers

47. False
48. (0, 4)
49. 1
50. $y = x + 4$
51. Yes
52. $y + x = 6$
53. $3\sqrt{2}$
54. a. $\begin{pmatrix} 11 & 2 \\ 8 & 3 \end{pmatrix}$ b. $\begin{pmatrix} 14 & 29 \\ 46 & 56 \end{pmatrix}$
55. 59
56. b, c
57. Yes
58. $\dfrac{-b \pm \sqrt{b^2 - 4ac}}{2a}$
59. $(x - h)^2 + (y - k)^2 = r^2$
60. $f(-r)$
61. $b^2 - 4ac$
62. sum
63. rational
64. parabola
65. one
66. four
67. Remainder
68. $x > \dfrac{5}{3}$
69. $-2, 3$
70. $(6, 1)$
71. $-2 < x < 3$
72. $(-5, -3)$
73. $2i, \dfrac{i}{2}$
74. $3 < x < 4$
75. $0, -1, +\sqrt{3}$
76. $(3, 4)$
77. $\dfrac{5}{2} + \dfrac{3}{2}i$
78. $(1, 3)$
79. 5
80. $\dfrac{4}{5}$
81. $(5, 3), (-3, 3)$
82. $(6, 3), (-4, 3)$
83. False
84. False
85. True
86. True
87. True
88. True
89. True
90. True
91. False
92. True
93. False
94. False
95. True
96. True
97. True
98. True
99. a. {reals}
    b. {non-negative reals}

c. $f^{-1}(x) = \log_2 x$
d–e.

100. $5^3 = 125$
101. $\left\{-\dfrac{2}{7}\right\}$
102. {3}
103. $\dfrac{13}{10!}$
104. 22
105. d
106. $\dfrac{5}{2}$
107. $-\dfrac{20}{3}$
108. a. The graph has a gap at $x = 3$.
     b. $f(3) = 5$
109. $9! = 362{,}880$
110. 120
111. a. 30; b. 6; c. 3
112. 38,760
113. 210
114. $2(7!) = 10{,}080$
115. $924 x^6 y^6$
116. 14
117. no
118. a. 6   c. $\dfrac{5}{16}$
     b. 32 d. $\dfrac{1}{2}$
     e. 0
119. a. $\dfrac{7}{15}$; b. $\dfrac{8}{15}$; c. $\dfrac{14}{15}$; d. $\dfrac{1}{5}$
120. $\dfrac{8}{25}$
121. a. $\dfrac{5}{12}$; b. $\dfrac{1}{9}$; c. $\dfrac{1}{9}$
122. p is logically true; p is logically false.
123. True
124. True
125. True
126. False

APPENDICES

Appendix A Trigonometric Functions

In this appendix we shall review the trigonometric functions and some of their properties which are used in this book. Basic to this review is the general angle which may be measured in sexagesimal measure or circular (radian) measure.

$$\begin{aligned} \text{one degree} &= \tfrac{1}{360} \text{ of one revolution} = 1° \\ \text{one minute} &= \tfrac{1}{60} \text{ of one degree} = 1' \\ \text{one second} &= \tfrac{1}{60} \text{ of one minute} = 1'' \end{aligned}$$

A radian is defined to be the central angle subtended by an arc of a circle equal in length to the radius of the circle. Since a radian is an arc length equal to a radius, and since there are 2π radii on the circumference of a circle ($C = 2\pi R$), there are 2π radians in one revolution. Thus, we may formulate degree-radian conversion factors.

$$\begin{aligned} 360 \text{ degrees} &= 2\pi \text{ radians} \\ 1 \text{ degree} &= \tfrac{\pi}{180} \text{ radians} \\ &= 0.017453 \text{ radians} \\ 2 \text{ radians} &= \tfrac{360}{\pi} \text{ degrees} \\ 1 \text{ radian} &= \tfrac{180}{\pi} \text{ degrees} \\ &= 57.2958° \end{aligned}$$

We shall give three sets of definitions for the trigonometric functions. These definitions are *identical* over their common domain of definition.

A. Trigonometric Functions of a Real Number

Basic to the idea of trigonometric functions is the circular coordinate systems. Consider a circle with its center at the origin, and let A be the point of intersection of the circle and the positive X-axis. Starting at A, measure a distance of t units along the circumference of the circle. The value, t, is positive (negative) if

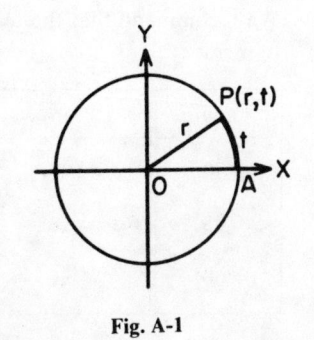

Fig. A-1

we measure the distance in a counterclockwise (clockwise) direction. The point P has the circular coordinates (r, t). The same point has more than one pair of circular coordinates. Thus $(r, 5)$ and $(r, 5 + 2\pi r)$ locate the same point.

If we use the circular coordinates and consider the unit circle ($r = 1$), then a point T on the circumference can be described by the coordinates $(1, t)$, and the distance t is equal to the radian measure of the angle formed by the X-axis and the line from the origin to T. (See Figure A-2.) The point T also has coordinates (x, y). The trigonometric functions of the real number t are defined as follows.

$$\begin{aligned} \sin t &= y \\ \cos t &= x \\ \tan t &= \tfrac{y}{x} \\ \cot t &= \tfrac{x}{y} \\ \sec t &= \tfrac{1}{x} \\ \csc t &= \tfrac{1}{y} \end{aligned}$$

Fig. A-2

Since these definitions are given in terms of ratios, and division by zero is not permissible, it is necessary to limit the values of x and y for specific functions. Thus if $x = 0$, which happens when $t = \pi/2$ or $3\pi/2$, then $\tan t$ and $\sec t$ are undefined. If $y = 0$, then $\cot t$ and $\csc t$ are undefined.

B. Trigonometric Functions of a General Angle

Let θ be any angle placed in its standard position. Let P be any point (not the origin) having coordinates (x, y) and lying on the terminal side of θ. Let the radius vector of P be denoted by r. See Figure A-3. Then the trigonometric functions of θ are defined as follows,

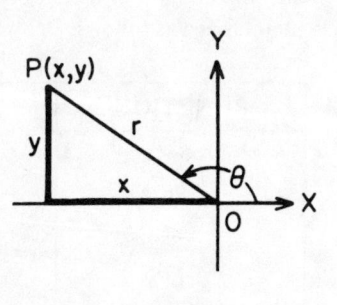

Fig. A-3

$$\sin \theta = \frac{\text{ordinate}}{\text{radius vector}} = \frac{y}{r}$$

$$\cos\theta = \frac{\text{abscissa}}{\text{radius vector}} = \frac{x}{r}$$

$$\tan\theta = \frac{\text{ordinate}}{\text{abscissa}} = \frac{y}{x}$$

$$\cot\theta = \frac{\text{abscissa}}{\text{ordinate}} = \frac{x}{y}$$

$$\sec\theta = \frac{\text{radius vector}}{\text{abscissa}} = \frac{r}{x}$$

$$\csc\theta = \frac{\text{radius vector}}{\text{ordinate}} = \frac{r}{y}$$

We note that the same restrictions must be placed on x and y as those for the trigonometric functions of the real numbers. On the unit circle the circular coordinate t is equal to the radian measure of the central angle θ. Consequently, *the trigonometric functions of the real number* t *are identical to the trigonometric functions of the angle* θ *when* θ *is measured in terms of radians.*

C. Trigonometric Functions of the Acute Angles

Let the angle θ be restricted to $0 < \theta < 90°$. Then the point P falls in the first quadrant, and a right triangle is formed with r = hypotenuse, y = side opposite θ, and x = side adjacent to θ. The trigonometric functions of θ for $0 < \theta < 90°$ can then be defined as follows.

$$\sin\theta = \frac{\text{opposite side}}{\text{hypotenuse}} \qquad \cot\theta = \frac{\text{adjacent side}}{\text{opposite side}}$$

$$\cos\theta = \frac{\text{adjacent side}}{\text{hypotenuse}} \qquad \sec\theta = \frac{\text{hypotenuse}}{\text{adjacent side}}$$

$$\tan\theta = \frac{\text{opposite side}}{\text{adjacent side}} \qquad \csc\theta = \frac{\text{hypotenuse}}{\text{opposite side}}$$

From the above general definitions, we see that the domain of definition of the trigonometric functions can be taken to be the real number system with certain exceptions. These exceptions are also referred to as *inadmissible* values of the independent variable. Let R denote the set of real numbers and $n = 0, 1, 2, \ldots$, then the domain of definition for each function may be summarized as follows.

FUNCTION	DOMAIN
sin	R
cos	R
tan	R except $\left\{\frac{\pi}{2} \pm n\pi\right\}$
cot	R except $\{\pm n\pi\}$
sec	R except $\left\{\frac{\pi}{2} \pm n\pi\right\}$
csc	R except $\{\pm n\pi\}$

D. Signs of the Trigonometric Functions

Since the definitions of the trigonometric functions are given in terms of x and y and since these variables can be positive or negative depending on the location of P, we see that the trigonometric functions can be positive or negative depending on the size of the argument. The signs can be determined from the quadrant in which P falls, for this determines the signs of x and y.

We recommend that this diagram become a part of the reader's memory system for rapid recall.

$$\left.\begin{matrix}\sin\\\csc\end{matrix}\right\}+ \qquad \text{all}\}+$$

$$\left.\begin{matrix}\text{all}\\\text{others}\end{matrix}\right\}-$$

$$\left.\begin{matrix}\tan\\\cot\end{matrix}\right\}+ \qquad \left.\begin{matrix}\cos\\\sec\end{matrix}\right\}+$$

$$\left.\begin{matrix}\text{all}\\\text{others}\end{matrix}\right\}- \qquad \left.\begin{matrix}\text{all}\\\text{others}\end{matrix}\right\}-$$

E. Functions of Special Angles

The functions of the quadrantal angles (0°, 90°, 180°, 270°) are determined directly from the definitions. These are also the trigonometric functions of the real numbers $0, \frac{\pi}{2}, \pi,$ and $\frac{3\pi}{2}$.

angle	sin	cos	tan
0°	0	1	0
90°	1	0	...
180°	0	−1	0
270°	−1	0	...

angle	cot	sec	csc
0°	...	1	...
90°	0	...	1
180°	...	−1	...
270°	0	...	−1

The trigonometric functions of 30°, 45°, and 60° $\left(\frac{\pi}{6}, \frac{\pi}{4}, \frac{\pi}{3}\right)$ can be expressed in terms of exact numbers and, consequently, are used in many illustrations. We recommend that the following values be committed to memory.

angle	sin	cos	tan
30°	$\frac{1}{2}$	$\frac{1}{2}\sqrt{3}$	$\frac{1}{3}\sqrt{3}$
45°	$\frac{1}{2}\sqrt{2}$	$\frac{1}{2}\sqrt{2}$	1
60°	$\frac{1}{2}\sqrt{3}$	$\frac{1}{2}$	$\sqrt{3}$

angle	cot	sec	csc
30°	$\sqrt{3}$	$\frac{2}{3}\sqrt{3}$	2
45°	1	$\sqrt{2}$	$\sqrt{2}$
60°	$\frac{1}{3}\sqrt{3}$	2	$\frac{2}{3}\sqrt{3}$

F. Functions of Negative Arguments

To obtain the trigonometric functions of a negative argument, employ the following theorem.

THEOREM. *If θ is any positive number, then*

$$\sin(-\theta) = -\sin\theta \qquad \cot(-\theta) = -\cot\theta$$
$$\cos(-\theta) = \cos\theta \qquad \sec(-\theta) = \sec\theta$$
$$\tan(-\theta) = -\tan\theta \qquad \csc(-\theta) = -\csc\theta$$

for all admissible values of θ.

G. Reduction to $0 \leq t \leq \frac{\pi}{2}$

The trigonometric functions of any real number may be reduced to the trigonometric functions of a number in the interval $0 \leq t \leq \frac{\pi}{2}$. We shall again develop this concept by considering the argument to be an angle and define a **reference angle** as the acute angle α between the terminal side and the x-axis. Thus the reference angle for different intervals of θ can be found by using the following table.

θ	α
$0 < \theta < 90°$	$\alpha = \theta$
$90° < \theta < 180°$	$\alpha = 180° - \theta$
$180° < \theta < 270°$	$\alpha = \theta - 180°$
$270° < \theta < 360°$	$\alpha = 360° - \theta$

If the angle is given in radians, the equivalent radian measure is used in the above table; i.e., $180° = \pi$, $360° = 2\pi$.

H. Inverse Trigonometric Functions

We discussed the concept of inverse functions in Chapter 11. Since the trigonometric functions satisfy the definitions of a general function in their domain of definition, it is of interest to see if we can find the inverse of these functions. Let us recall that given a function f, we can find an inverse function f^{-1} by interchanging the independent and dependent variables and then solving for the new dependent variable in terms of the new independent variable. If the resulting relation satisfies the definition of a function, we have found the inverse function. Consider the function y given by

$$y = \sin x$$

and interchange x and y to obtain

$$x = \sin y$$

In order to solve this last equation for y, we need to introduce new terminology and symbols;

$$y = \arcsin x$$

is an expression for the number whose sine is x. For the trigonometric functions of angles, the expression arcsin x represents an *angle* (i.e., arcsin $x = \theta$) whose sine is x (i.e., $\sin\theta = x$). The expressions arcsin x and sin y are **inverse relations**.

In order to obtain the inverse function we restrict the range of the inverse relation and call this the *principal value*.

DEFINITION. *The* **principal value** *of arcsin x, arccsc x, arctan x, and arccot x is the smallest numerical value; of arccos x and arcsec x, it is the smallest positive value of the angle.*

The principal value is usually indicated by capitalizing the "A" in "arc." These values can be summarized.

$$-\frac{\pi}{2} \leq \text{Arcsin } x \leq \frac{\pi}{2} \qquad -\frac{\pi}{2} \leq \text{Arccot } x \leq \frac{\pi}{2}$$
$$0 \leq \text{Arccos } x \leq \pi \qquad 0 \leq \text{Arcsec } x \leq \pi$$
$$-\frac{\pi}{2} \leq \text{Arctan } x \leq \frac{\pi}{2} \qquad -\frac{\pi}{2} \leq \text{Arccsc } x \leq \frac{\pi}{2}$$

Appendix B Logarithmic Tables

Mathematical tables or tables which exhibit values that remain constant for specified arguments form a body of knowledge which is becoming more and more important. The subject of logarithms provides an excellent introduction to the use of tables. The entries in the tables are approximate numbers expressed in decimal form correct to a given number of places. There are many such tables;* here we shall use an abbreviated four-place table.

Before turning to the tables, let us first consider the matter of expressing a number correct to a certain number of places. Since, in general, these numbers are continuing decimals, we must "round off" the numbers to four places. This is accomplished by discarding all the digits to the right of the fourth place. If the digit

---

*For a complete discussion of trigonometric tables see K. L. Nielsen and J. H. Vanlonkhuyzen, *Plane and Spherical Trigonometry* (New York: Barnes and Noble, 1954) pp. 8–12.

in the 5th place is
  (a) greater than half a unit in the 4th place, increase the digit in that place by 1;
  (b) less than half a unit in the 4th place, leave the digit in that place unaltered;
  (c) exactly half a unit in the 4th place,
      (i) increase an *odd* digit in the 4th place by 1,
      (ii) leave an *even* digit in the 4th place unaltered.

ILLUSTRATIONS Express the following numbers to four places.
1. 64.347. Answer: 64.35.
2. 13.342. Answer: 13.34.
3. 7.2345. Answer: 7.234.
4. 7.2375. Answer: 7.238.

A. Common Logarithms (See Chapter 21)

It is easy to write down a table of numbers which are integral powers of 10 and the corresponding common logarithms.

Exponential Form	Logarithmic Form
...	...
$10^3 = 1000$	$\log 1000 = 3$
$10^2 = 100$	$\log 100 = 2$
$10^1 = 10$	$\log 10 = 1$
$10^0 = 1$	$\log 1 = 0$
$10^{-1} = 0.1$	$\log 0.1 = -1$
$10^{-2} = 0.01$	$\log 0.01 = -2$
$10^{-3} = 0.001$	$\log 0.001 = -3$
...	...

To find the logarithm of a number which is not an integral power of 10 can best be explained by considering an example such as $N = 110$. Since 110 is between 100 and 1000, it is natural to suppose, from the above table, that log 110 is between 2 and 3 or, in other words, the logarithm is 2 + (a proper fraction). Since we can express the proper fraction in decimal form, we have, in general,

$$\log N = \text{(an integer)} + (0 \leq \text{decimal fraction} < 1)$$

The integral part is called the **characteristic**. The decimal fraction is called the **mantissa**. Thus

$$\log N = \text{characteristic} + \text{mantissa}$$

Since the mantissas may be nonrepeating infinite decimals, they are approximated to as many places as desired. The approximations have been tabulated in four-place, five-place, or higher-place tables which are called logarithmic tables. Thus the mantissas, or decimal parts, are found from tables, and the values in the tables are **always positive**.

The characteristic is determined according to the following two rules.

RULE I. *If the number* N *is greater than* 1, *the characteristic of its logarithm is one less than the number of digits to the left of the decimal point.*

RULE II. *If the number* N *is less than* 1, *the characteristic of its logarithm is negative; if the first digit which is not zero occurs in the* k*th decimal place, the characteristic is* $-k$.

Since the characteristic and mantissa are combined to give the complete logarithm,

$$\log N = \text{characteristic} + \text{mantissa}$$

and, further, since the mantissa is always positive, it is best to write a negative characteristic, $-k$, as

$$(10 - k) - 10.$$

ILLUSTRATIONS
1. Write the logarithms given a mantissa of .3942 and characteristics 1, 0, -1, and -2.

SOLUTION

Characteristic	Logarithm
1	1.3942
0	0.3942
-1	$9.3942 - 10$
-2	$8.3942 - 10$

2. Find the characteristics of the logarithms of the numbers given in the left column.

SOLUTION

Number	Characteristic
197.3	2
81.72	1
6.291	0
0.3962	$9 - 10$
0.0815	$8 - 10$
0.000073	$5 - 10$

If the logarithm is given, the problem becomes one of finding the number corresponding to this logarithm. The number is called the **antilogarithm**. The characteristic of the given logarithm determines the position of the decimal point in the antilogarithm. In placing the decimal point we use the same two rules given for determining the characteristic of a number. However, we must remember it is a reverse problem.

ILLUSTRATION. The digits of an antilogarithm are 7329. Place the decimal point if the characteristic is 1, 2, 8 − 10.
SOLUTION

Characteristic	Number
1	73.29
2	732.9
8 − 10 = −2	0.07329

B. Logarithmic Tables and Their Use

To find the mantissa of a logarithm, we shall use the Table of the Common Logarithms on page 166. This table is a four-place table, and, although it will not yield as accurate results as a table of more places, it will serve to illustrate the methods.* The numbers .04021, .0421, 402100.0 are said to have the *same sequence of digits*. The *mantissa* of the logarithm for each of these numbers is the *same*; the characteristics are, of course, different.

I. *To find the logarithm of a given number.*
ILLUSTRATIONS
1. Find log 32.4.
SOLUTION. By Rule 1 the characteristic is 1. To find the mantissa turn to the table and locate the first two digits (32) in the left column headed by "N." In the "32" row and in the column headed by "4" (the third digit of the number) find 5105. This number is the mantissa. Thus log 32.4 = 1.5105.

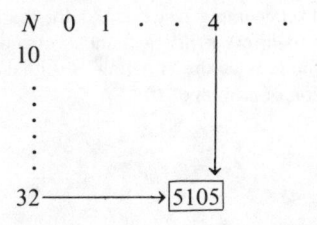

```
N  0  1  . . 4 . . .
10
  .
  .
  .
  .
32 ─────────→ 5105
```

2. Find log .06732.
SOLUTION. By Rule 2 the characteristic is 8 − 10. To find the mantissa it is necessary to interpolate. We shall use simple linear interpolation. Locate "67" in left column. In the "67" row and column headed by "3," find 8280, and in the column headed by "4," find 8287. The difference 8287 − 8280 = 7 is now multiplied by .2 (2 being the fourth significant digit of given number); (7)(.2) = 1.4, which is rounded off to the nearest whole

---
*For a five-place table see Kaj L. Nielsen, *Logarithmic and Trigonometric Tables* (2nd ed.; New York: Barnes and Noble, 1961), pp. 2–21.

number, 1. This number is added to the smaller mantissa; 8280 + 1 = 8281. The answer is given by

$$\log .06732 = 8.8281 - 10$$

II. *To find the antilogarithm of a given logarithm.*
ILLUSTRATIONS
1. Find N if log N = 7.6503 − 10.
SOLUTION. First find the mantissa 6503 in the table; it appears in the column headed by "7" in the row which has "44" at the left under "N." Thus the sequence of digits is 447.

Now the characteristic is 7 − 10 = −3. Using Rule 2 for the characteristic, the first significant digit after the decimal point should occur in the *third* place. The answer is N = .00447.

2. Find N if log N = 1.5952.
SOLUTION. First seek the mantissa in the table. We find 5944 and 5955 corresponding to the numbers 393 and 394. Consequently, our number is between these two numbers. Form the quotient.

$$\left(\frac{\text{given mantissa} - \text{smaller table mantissa}}{\text{larger table mantissa} - \text{smaller table mantissa}}\right)10$$

$$10\left(\frac{5952 - 5944}{5955 - 5944}\right) = \frac{8}{11}(10) = \frac{80}{11} = 7+$$

This number is rounded off to an integer which is "attached to" the smaller number N resulting from the table. Thus 7 is attached to 393 to give us the sequence of numbers 3937. The given characteristic is 1; so by Rule 1, we have N = 39.37.

C. Computations Using Logarithms

In carrying out computations using logarithms it is desirable to have a systematic form in which to display the work. We recommend the form given in the illustrations. The student should check each step carefully.
ILLUSTRATIONS
1. Find N if $N = \dfrac{5.367}{(12.93)(0.06321)}$

SOLUTION. We employ logarithms and, using the three properties of logarithms, we get

$$\log N = \log 5.367 - [\log 12.93 + \log 0.06321]$$

1	log 5.367	= 0.7298	
2	log 12.93	= 1.1116	
3	log 0.06321	= 8.8008 − 10	
4	(2 + 3)	= 9.9124 − 10	
5	(1 − 4)	= 0.8174	$N = 6.567$

Note: To subtract (4) from (1) we first change 0.7298 to 10.7298 − 10.

2. Find N if $N = \dfrac{(\sqrt[3]{0.9573})(3.21)^2}{98.32}$.

SOLUTION

$$\log N = \tfrac{1}{3} \log 0.9573 + 2 \log 3.21 - \log 98.32$$

1	$\tfrac{1}{3} \log 0.9573$ = $\tfrac{1}{3}[9.9810 - 10]$	=	$9.9937 - 10$
2	$2 \log 3.21$ = $2[0.5065]$	=	1.0130
3	$(1 + 2)$	=	$11.0067 - 10$
4	$\log 98.32$	=	1.9927
5	$(3 - 4)$	=	$9.0140 - 10$
	$\boxed{N = 0.1033}$		

Note: To find

$$\tfrac{1}{3}[9.9810 - 10] = \tfrac{1}{3}[29.9810 - 30] = 9.9937 - 10$$

always rearrange a negative characteristic so that *after* dividing the result will be x.xxxx minus an integer.

3. Find N if $N = \sqrt{\dfrac{(0.3592)^3}{673.5}}$.

SOLUTION

$$\log N = \tfrac{1}{2}[3 \log 0.3592 - \log 673.5]$$

1	$3 \log 0.3592$ = $3[9.5553 - 10]$	=	$28.6659 - 30$
2	$\log 673.5$	=	2.8284
3	$(1 - 2)$	=	$25.8375 - 30$
4	$\tfrac{1}{2}(3)$ = $\tfrac{1}{2}[15.8375 - 20]$	=	$7.9188 - 10$
	$\boxed{N = 0.008294}$		

Note: In going from (3) to (4) we made the change

$$25.8375 - 30 = 15.8375 - 20$$

This was done so that after dividing by 2 we would have x.xxxx − 10.

D. Natural Logarithms and e^x

The natural logarithms (logarithms to the base e) do not separate into a characteristic and a mantissa. The tables give the complete value of the logarithm. As an example ln 4.13 = 1.41828. (See page 167.)

The exponential function e^x may also be tabulated for values of x. These values become large for $x > 0$. If x is negative (see the table for e^{-x}), the values are small and in order to retain the same number of significant digits a new format is used.* For each entry there is printed an additional number with a sign. This number refers to the power of 10 which is to be multiplied into the tabular entry; for example,

$$2.08584 - 2 = 2.08584 \times 10^{-2} = 0.0208584$$

In practice, the additional number indicates the number of places that the decimal point is to be moved in the entry. If the number is *negative*, the decimal point is moved to the *left*; if *positive*, to the *right*. As an example, let us find e^{-x} for $x = 4.35$. Refer to the table on p. 171 where in the row 4.3 and the column headed by "5" the entry is 1.29068 − 2. Consequently, $e^{-4.35} = 0.0129068$.

---

*This format is becoming very popular for many tables since it adapts itself to direct printing from electronic calculators. It is based on the idea of the "floating" decimal point and the scientific notation of powers of 10.

Appendix C: List of Symbols

Symbol	Meaning or Translation
$p(x), q(x), r(x)$	open sentences
$\sim p$	negation: not p
$p \wedge q$	conjunction: p and q
$p \vee q$	disjunction: p or q or both
$\{x \mid p(x)\}$	the set of all x such that $p(x)$ is true
$\{\ \}, A, B, C$	sets
a, b, c	elements of a set
$\in$	is an element of
U, I	the universal set
$\phi, \{\ \}$	the empty (null) set
$A \subseteq B$	A is a subset of B
$A \subset B$	A is a proper subset of B
$A \equiv B$	A and B are equivalent sets
$n(A) = N$	the cardinal number of A is N
A'	the complement of A
$A - B$	the difference of A and B
$A \cap B$	the intersection of A and B
$A \cup B$	the union of A and B
$P(x)$	polynomial in x
$P(x, y)$	polynomial in x and y
$>$	greater than
$\geq$	greater than or equal to
$<$	less than
$\leq$	less than or equal to
$(x, y) \in R$	(x, y) is an element of the relation R
xRy	x is related to y
i	identity element; imaginary unit: $i^2 = -1$
a'	inverse element for a; successor of the natural number a
GCD	greatest common divisor
f, g, h	functions
$f(x)$	value of f at x
f^{-1}	inverse of f
$\sqrt{\ }$	positive square root; square root function
$|\ |$	absolute value; absolute value function
$\begin{pmatrix} 3 & 4 & 2 \\ 8 & 6 & 1 \end{pmatrix}$	2 by 3 matrix
$\delta(A)$	value of the determinant of matrix A
$\begin{vmatrix} 3 & 5 \\ 6 & 4 \end{vmatrix}$	second order determinant
e	an irrational number: $e \approx 2.71828$
$\log_b x$	logarithm of x, base b
$\log x$	common logarithm of x, base 10
$\ln x$	natural logarithm of x, base e
$a = \{a\}_{n=1}^{\infty}$	sequence
Σ	summation
$\sum_{n=1}^{\infty} a_n$	series
$\lim_{n \to \infty} a_n$	limit of a sequence as n becomes infinite
$\lim_{x \to a} f(x)$	limit of a function as x approaches a
$n!$	n factorial
$_nP_r$ or $P_{n,r}$	number of permutations of n things taken r at a time
$_nC_r, C_{n,r}$ or $\binom{n}{r}$	number of combinations of n things taken r at a time
PR$[p]$	probability of statement p
$m(P)$	probability measure of set P
PR$[p \mid q]$	probability of p given q

Four-Place Common Logarithms.

N	0	1	2	3	4	5	6	7	8	9
10	0000	0043	0086	0128	0170	0212	0253	0294	0334	0374
11	0414	0453	0492	0531	0569	0607	0645	0682	0719	0755
12	0792	0828	0864	0899	0934	0969	1004	1038	1072	1106
13	1139	1173	1206	1239	1271	1303	1335	1367	1399	1430
14	1461	1492	1523	1553	1584	1614	1644	1673	1703	1732
15	1761	1790	1818	1847	1875	1903	1931	1959	1987	2014
16	2041	2068	2095	2122	2148	2175	2201	2227	2253	2279
17	2304	2330	2355	2380	2405	2430	2455	2480	2504	2529
18	2553	2577	2601	2625	2648	2672	2695	2718	2742	2765
19	2788	2810	2833	2856	2878	2900	2923	2945	2967	2989
20	3010	3032	3054	3075	3096	3118	3139	3160	3181	3201
21	3222	3243	3263	3284	3304	3324	3345	3365	3385	3404
22	3424	3444	3464	3483	3502	3522	3541	3560	3579	3598
23	3617	3636	3655	3674	3692	3711	3729	3747	3766	3784
24	3802	3820	3838	3856	3874	3892	3909	3927	3945	3962
25	3979	3997	4014	4031	4048	4065	4082	4099	4116	4133
26	4150	4166	4183	4200	4216	4232	4249	4265	4281	4298
27	4314	4330	4346	4362	4378	4393	4409	4425	4440	4456
28	4472	4487	4502	4518	4533	4548	4564	4579	4594	4609
29	4624	4639	4654	4669	4683	4698	4713	4728	4742	4757
30	4771	4786	4800	4814	4829	4843	4857	4871	4886	4900
31	4914	4928	4942	4955	4969	4983	4997	5011	5024	5038
32	5051	5065	5079	5092	5105	5119	5132	5145	5159	5172
33	5185	5198	5211	5224	5237	5250	5263	5276	5289	5302
34	5315	5328	5340	5353	5366	5378	5391	5403	5416	5428
35	5441	5453	5465	5478	5490	5502	5514	5527	5539	5551
36	5563	5575	5587	5599	5611	5623	5635	5647	5658	5670
37	5682	5694	5705	5717	5729	5740	5752	5763	5775	5786
38	5798	5809	5821	5832	5843	5855	5866	5877	5888	5899
39	5911	5922	5933	5944	5955	5966	5977	5988	5999	6010
40	6021	6031	6042	6053	6064	6075	6085	6096	6107	6117
41	6128	6138	6149	6160	6170	6180	6191	6201	6212	6222
42	6232	6243	6253	6263	6274	6284	6294	6304	6314	6325
43	6335	6345	6355	6365	6375	6385	6395	6405	6415	6425
44	6435	6444	6454	6464	6474	6484	6493	6503	6513	6522
45	6532	6542	6551	6561	6571	6580	6590	6599	6609	6618
46	6628	6637	6646	6656	6665	6675	6684	6693	6702	6712
47	6721	6730	6739	6749	6758	6767	6776	6785	6794	6803
48	6812	6821	6830	6839	6848	6857	6866	6875	6884	6893
49	6902	6911	6920	6928	6937	6946	6955	6964	6972	6981
50	6990	6998	7007	7016	7024	7033	7042	7050	7059	7067
51	7076	7084	7093	7101	7110	7118	7126	7135	7143	7152
52	7160	7168	7177	7185	7193	7202	7210	7218	7226	7235
53	7243	7251	7259	7267	7275	7284	7292	7300	7308	7316
54	7324	7332	7340	7348	7356	7364	7372	7380	7388	7396

Continued.

N	0	1	2	3	4	5	6	7	8	9
55	7404	7412	7419	7427	7435	7443	7451	7459	7466	7474
56	7482	7490	7497	7505	7513	7520	7528	7536	7543	7551
57	7559	7566	7574	7582	7589	7597	7604	7612	7619	7627
58	7634	7642	7649	7657	7664	7672	7679	7686	7694	7701
59	7709	7716	7723	7731	7738	7745	7752	7760	7767	7774
60	7782	7789	7796	7803	7810	7818	7825	7832	7839	7846
61	7853	7860	7868	7875	7882	7889	7896	7903	7910	7917
62	7924	7931	7938	7945	7952	7959	7966	7973	7980	7987
63	7993	8000	8007	8014	8021	8028	8035	8041	8048	8055
64	8062	8069	8075	8082	8089	8096	8102	8109	8116	8122
65	8129	8136	8142	8149	8156	8162	8169	8176	8182	8189
66	8195	8202	8209	8215	8222	8228	8235	8241	8248	8254
67	8261	8267	8274	8280	8287	8293	8299	8306	8312	8319
68	8325	8331	8338	8344	8351	8357	8363	8370	8376	8382
69	8388	8395	8401	8407	8414	8420	8426	8432	8439	8445
70	8451	8457	8463	8470	8476	8482	8488	8494	8500	8506
71	8513	8519	8525	8531	8537	8543	8549	8555	8561	8567
72	8573	8579	8585	8591	8597	8603	8609	8615	8621	8627
73	8633	8639	8645	8651	8657	8663	8669	8675	8681	8686
74	8692	8698	8704	8710	8716	8722	8727	8733	8739	8745
75	8751	8756	8762	8768	8774	8779	8785	8791	8797	8802
76	8808	8814	8820	8825	8831	8837	8842	8848	8854	8859
77	8865	8871	8876	8882	8887	8893	8899	8904	8910	8915
78	8921	8927	8932	8938	8943	8949	8954	8960	8965	8971
79	8976	8982	8987	8993	8998	9004	9009	9015	9020	9025
80	9031	9036	9042	9047	9053	9058	9063	9069	9074	9079
81	9085	9090	9096	9101	9106	9112	9117	9122	9128	9133
82	9138	9143	9149	9154	9159	9165	9170	9175	9180	9186
83	9191	9196	9201	9206	9212	9217	9222	9227	9232	9233
84	9243	9248	9253	9258	9263	9269	9274	9279	9284	9289
85	9294	9299	9304	9309	9315	9320	9325	9330	9335	9340
86	9345	9350	9355	9360	9365	9370	9375	9380	9385	9390
87	9395	9400	9405	9410	9415	9420	9425	9430	9435	9440
88	9450	9450	9455	9460	9465	9469	9474	9479	9484	9489
89	9494	9499	9504	9509	9513	9518	9523	9528	9533	9538
90	9542	9547	9552	9557	9562	9566	9571	9576	9581	9586
91	9590	9595	9600	9605	9609	9614	9619	9624	9628	9633
92	9638	9643	9647	9652	9657	9661	9666	9671	9675	9680
93	9685	9689	9694	9699	9703	9708	9713	9717	9722	9727
94	9731	9736	9741	9745	9750	9754	9759	9763	9768	9773
95	9777	9782	9786	9791	9795	9800	9805	9809	9814	9818
96	9823	9827	9832	9836	9841	9845	9850	9854	9859	9863
97	9868	9872	9877	9881	9886	9890	9894	9899	9903	9908
98	9912	9917	9921	9926	9930	9934	9939	9943	9948	9952
99	9956	9961	9965	9969	9974	9978	9983	9987	9991	9996

Natural Logarithms, ln N

N	0	1	2	3	4
0.0	.....	-4.60517	-3.91202	-3.50656	-3.21888
0.1	-2.30258	-2.20727	-2.12026	-2.04022	-1.96611
0.2	-1.60943	-1.56065	-1.51413	-1.46968	-1.42712
0.3	-1.20397	-1.17118	-1.13943	-1.10866	-1.07881
0.4	-0.91629	-0.89160	-0.86750	-0.84397	-0.82098
0.5	-0.69314	-0.67334	-0.65393	-0.63488	-0.61619
0.6	-0.51082	-0.49430	-0.47804	-0.46204	-0.44629
0.7	-0.35667	-0.34249	-0.32850	-0.31471	-0.30111
0.8	-0.22314	-0.21072	-0.19845	-0.18633	-0.17435
0.9	-0.10536	-0.09431	-0.08338	-0.07257	-0.06188
1.0	0.00000	0.00995	0.01980	0.02956	0.03922
1.1	0.09531	0.10436	0.11333	0.12222	0.13103
1.2	0.18232	0.19062	0.19885	0.20701	0.21511
1.3	0.26236	0.27003	0.27763	0.28518	0.29267
1.4	0.33647	0.34359	0.35066	0.35767	0.36464
1.5	0.40547	0.41211	0.41871	0.42527	0.43178
1.6	0.47000	0.47623	0.48243	0.48858	0.49470
1.7	0.53063	0.53649	0.54232	0.54812	0.55389
1.8	0.58779	0.59333	0.59884	0.60432	0.60977
1.9	0.64185	0.64710	0.65233	0.65752	0.66269
2.0	0.69315	0.69813	0.70310	0.70804	0.71295
2.1	0.74194	0.74669	0.75142	0.75612	0.76081
2.2	0.78846	0.79299	0.79751	0.80200	0.80648
2.3	0.83291	0.83725	0.84157	0.84587	0.85015
2.4	0.87547	0.87963	0.88377	0.88789	0.89200
2.5	0.91629	0.92028	0.92426	0.92822	0.93216
2.6	0.95551	0.95935	0.96317	0.96698	0.97078
2.7	0.99325	0.99695	1.00063	1.00430	1.00796
2.8	1.02962	1.03318	1.03674	1.04028	1.04380
2.9	1.06471	1.06815	1.07158	1.07500	1.07841
3.0	1.09861	1.10194	1.10526	1.10856	1.11186
3.1	1.13140	1.13462	1.13783	1.14103	1.14422
3.2	1.16315	1.16627	1.16938	1.17248	1.17557
3.3	1.19392	1.19695	1.19996	1.20297	1.20597
3.4	1.22378	1.22671	1.22964	1.23256	1.23547
3.5	1.25276	1.25562	1.25846	1.26130	1.26413
3.6	1.28093	1.28371	1.28647	1.28923	1.29198
3.7	1.30833	1.31103	1.31372	1.31641	1.31909
3.8	1.33500	1.33763	1.34025	1.34286	1.34547
3.9	1.36098	1.36354	1.36609	1.36864	1.37118
4.0	1.38629	1.38879	1.39128	1.39377	1.39624
4.1	1.41099	1.41342	1.41585	1.41828	1.42070
4.2	1.43508	1.43746	1.43984	1.44220	1.44456
4.3	1.45861	1.46094	1.46326	1.46557	1.46787
4.4	1.48160	1.48387	1.48614	1.48840	1.49065
4.5	1.50408	1.50630	1.50851	1.51072	1.51293
4.6	1.52606	1.52823	1.53039	1.53256	1.53471
4.7	1.54756	1.54969	1.55181	1.55393	1.55604
4.8	1.56862	1.57070	1.57277	1.57485	1.57691
4.9	1.58924	1.59127	1.59331	1.59534	1.59737

Natural Logarithms, ln N

N	5	6	7	8	9
0.0	-2.99573	-2.81341	-2.65926	-2.52573	-2.40795
0.1	-1.89712	-1.83258	-1.77196	-1.71480	-1.66073
0.2	-1.38629	-1.34707	-1.30933	-1.27297	-1.23787
0.3	-1.04982	-1.02165	-0.99425	-0.96758	-0.94161
0.4	-0.79851	-0.77653	-0.75502	-0.73397	-0.71335
0.5	-0.59784	-0.57982	-0.56212	-0.54473	-0.52763
0.6	-0.43078	-0.41552	-0.40048	-0.38566	-0.37106
0.7	-0.28768	-0.27444	-0.26136	-0.24846	-0.23572
0.8	-0.16252	-0.15082	-0.13926	-0.12783	-0.11653
0.9	-0.05129	-0.04082	-0.03046	-0.02020	-0.01005
1.0	0.04879	0.05827	0.06766	0.07696	0.08618
1.1	0.13976	0.14842	0.15700	0.16551	0.17395
1.2	0.22314	0.23111	0.23902	0.24686	0.25464
1.3	0.30010	0.30748	0.31481	0.32208	0.32930
1.4	0.37156	0.37844	0.38526	0.39204	0.39878
1.5	0.43825	0.44469	0.45108	0.45742	0.46373
1.6	0.50078	0.50682	0.51282	0.51879	0.52473
1.7	0.55962	0.56531	0.57098	0.57661	0.58222
1.8	0.61519	0.62058	0.62594	0.63127	0.63658
1.9	0.66783	0.67294	0.67803	0.68310	0.68813
2.0	0.71784	0.72271	0.72755	0.73237	0.73716
2.1	0.76547	0.77011	0.77473	0.77932	0.78390
2.2	0.81093	0.81536	0.81978	0.82418	0.82855
2.3	0.85442	0.85866	0.86289	0.86710	0.87129
2.4	0.89609	0.90016	0.90422	0.90826	0.91228
2.5	0.93609	0.94001	0.94391	0.94779	0.95166
2.6	0.97456	0.97833	0.98208	0.98582	0.98954
2.7	1.01160	1.01523	1.01885	1.02245	1.02604
2.8	1.04732	1.05082	1.05431	1.05779	1.06126
2.9	1.08181	1.08519	1.08856	1.09192	1.09527
3.0	1.11514	1.11841	1.12168	1.12493	1.12817
3.1	1.14740	1.15057	1.15373	1.15688	1.16002
3.2	1.17865	1.18173	1.18479	1.18784	1.19089
3.3	1.20896	1.21194	1.21491	1.21788	1.22083
3.4	1.23837	1.24127	1.24415	1.24703	1.24990
3.5	1.26695	1.26976	1.27257	1.27536	1.27815
3.6	1.29473	1.29746	1.30019	1.30291	1.30563
3.7	1.32176	1.32442	1.32708	1.32972	1.33237
3.8	1.34807	1.35067	1.35325	1.35584	1.35841
3.9	1.37372	1.37624	1.37877	1.38128	1.38379
4.0	1.39872	1.40118	1.40364	1.40610	1.40854
4.1	1.42311	1.42552	1.42792	1.43031	1.43270
4.2	1.44692	1.44927	1.45161	1.45395	1.45629
4.3	1.47018	1.47247	1.47476	1.47705	1.47933
4.4	1.49290	1.49515	1.49739	1.49962	1.50185
4.5	1.51513	1.51732	1.51951	1.52170	1.52388
4.6	1.53687	1.53902	1.54116	1.54330	1.54543
4.7	1.55814	1.56025	1.56235	1.56444	1.56653
4.8	1.57898	1.58104	1.58309	1.58515	1.58719
4.9	1.59939	1.60141	1.60342	1.60543	1.60744

Natural Logarithms, ln N

N	0	1	2	3	4
5.0	1.60944	1.61144	1.61343	1.61542	1.61741
5.1	1.62924	1.63120	1.63315	1.63511	1.63705
5.2	1.64866	1.65058	1.65250	1.65441	1.65632
5.3	1.66771	1.66959	1.67147	1.67335	1.67523
5.4	1.68640	1.68825	1.69010	1.69194	1.69378
5.5	1.70475	1.70656	1.70838	1.71019	1.71199
5.6	1.72277	1.72455	1.72633	1.72811	1.72988
5.7	1.74047	1.74222	1.74397	1.74572	1.74746
5.8	1.75786	1.75958	1.76130	1.76302	1.76473
5.9	1.77495	1.77665	1.77834	1.78002	1.78171
6.0	1.79176	1.79342	1.79509	1.79675	1.79840
6.1	1.80829	1.80993	1.81156	1.81319	1.81482
6.2	1.82455	1.82616	1.82777	1.82938	1.83098
6.3	1.84055	1.84214	1.84372	1.84530	1.84688
6.4	1.85630	1.85786	1.85942	1.86097	1.86253
6.5	1.87180	1.87334	1.87487	1.87641	1.87794
6.6	1.88707	1.88858	1.89010	1.89160	1.89311
6.7	1.90211	1.90360	1.90509	1.90658	1.90806
6.8	1.91692	1.91839	1.91986	1.92132	1.92279
6.9	1.93152	1.93297	1.93442	1.93586	1.93730
7.0	1.94591	1.94734	1.94876	1.95019	1.95161
7.1	1.96009	1.96150	1.96291	1.96431	1.96571
7.2	1.97408	1.97547	1.97685	1.97824	1.97962
7.3	1.98787	1.98924	1.99061	1.99198	1.99334
7.4	2.00148	2.00283	2.00418	2.00553	2.00687
7.5	2.01490	2.01624	2.01757	2.01890	2.02022
7.6	2.02815	2.02946	2.03078	2.03209	2.03340
7.7	2.04122	2.04252	2.04381	2.04511	2.04640
7.8	2.05412	2.05540	2.05668	2.05796	2.05924
7.9	2.06686	2.06813	2.06939	2.07065	2.07191
8.0	2.07944	2.08069	2.08194	2.08318	2.08443
8.1	2.09186	2.09310	2.09433	2.09556	2.09679
8.2	2.10413	2.10535	2.10657	2.10779	2.10900
8.3	2.11626	2.11746	2.11866	2.11986	2.12106
8.4	2.12823	2.12942	2.13061	2.13180	2.13298
8.5	2.14007	2.14124	2.14242	2.14359	2.14476
8.6	2.15176	2.15292	2.15409	2.15524	2.15640
8.7	2.16332	2.16447	2.16562	2.16677	2.16791
8.8	2.17475	2.17589	2.17702	2.17816	2.17929
8.9	2.18605	2.18717	2.18830	2.18942	2.19054
9.0	2.19722	2.19834	2.19944	2.20055	2.20166
9.1	2.20827	2.20937	2.21047	2.21157	2.21266
9.2	2.21920	2.22029	2.22138	2.22246	2.22354
9.3	2.23001	2.23109	2.23216	2.23324	2.23431
9.4	2.24071	2.24177	2.24284	2.24390	2.24496
9.5	2.25129	2.25234	2.25339	2.25444	2.25549
9.6	2.26176	2.26280	2.26384	2.26488	2.26592
9.7	2.27213	2.27316	2.27419	2.27521	2.27624
9.8	2.28238	2.28340	2.28442	2.28544	2.28646
9.9	2.29253	2.29354	2.29455	2.29556	2.29657

Natural Logarithms, ln N

N	5	6	7	8	9
5.0	1.61939	1.62137	1.62334	1.62531	1.62728
5.1	1.63900	1.64094	1.64287	1.64481	1.64673
5.2	1.65823	1.66013	1.66203	1.66393	1.66582
5.3	1.67710	1.67896	1.68083	1.68269	1.68455
5.4	1.69562	1.69745	1.69928	1.70111	1.70293
5.5	1.71380	1.71560	1.71740	1.71919	1.72098
5.6	1.73166	1.73342	1.73519	1.73695	1.73871
5.7	1.74920	1.75094	1.75267	1.75440	1.75613
5.8	1.76644	1.76815	1.76985	1.77156	1.77326
5.9	1.78339	1.78507	1.78675	1.78842	1.79009
6.0	1.80006	1.80171	1.80336	1.80500	1.80665
6.1	1.81645	1.81808	1.81970	1.82132	1.82294
6.2	1.83258	1.83418	1.83578	1.83737	1.83896
6.3	1.84845	1.85003	1.85160	1.85317	1.85473
6.4	1.86408	1.86563	1.86718	1.86872	1.87026
6.5	1.87947	1.88099	1.88251	1.88403	1.88555
6.6	1.89462	1.89612	1.89762	1.89912	1.90061
6.7	1.90954	1.91102	1.91250	1.91398	1.91545
6.8	1.92425	1.92571	1.92716	1.92862	1.93007
6.9	1.93874	1.94018	1.94162	1.94305	1.94448
7.0	1.95303	1.95445	1.95586	1.95727	1.95869
7.1	1.96711	1.96851	1.96991	1.97130	1.97269
7.2	1.98100	1.98238	1.98376	1.98513	1.98650
7.3	1.99470	1.99606	1.99742	1.99877	2.00013
7.4	2.00821	2.00956	2.01089	2.01223	2.01357
7.5	2.02155	2.02287	2.02419	2.02551	2.02683
7.6	2.03471	2.03601	2.03732	2.03862	2.03992
7.7	2.04769	2.04898	2.05027	2.05156	2.05284
7.8	2.06051	2.06179	2.06306	2.06433	2.06560
7.9	2.07317	2.07443	2.07568	2.07694	2.07819
8.0	2.08567	2.08691	2.08815	2.08939	2.09063
8.1	2.09802	2.09924	2.10047	2.10169	2.10291
8.2	2.11021	2.11142	2.11263	2.11384	2.11505
8.3	2.12226	2.12346	2.12465	2.12585	2.12704
8.4	2.13417	2.13535	2.13653	2.13771	2.13889
8.5	2.14593	2.14710	2.14827	2.14943	2.15060
8.6	2.15756	2.15871	2.15987	2.16102	2.16217
8.7	2.16905	2.17020	2.17134	2.17248	2.17361
8.8	2.18042	2.18155	2.18267	2.18380	2.18493
8.9	2.19165	2.19277	2.19389	2.19500	2.19611
9.0	2.20276	2.20387	2.20497	2.20607	2.20717
9.1	2.21375	2.21485	2.21594	2.21703	2.21812
9.2	2.22462	2.22570	2.22678	2.22786	2.22894
9.3	2.23538	2.23645	2.23751	2.23858	2.23965
9.4	2.24601	2.24707	2.24813	2.24918	2.25024
9.5	2.25654	2.25759	2.25863	2.25968	2.26072
9.6	2.26696	2.26799	2.26903	2.27006	2.27109
9.7	2.27727	2.27829	2.27932	2.28034	2.28136
9.8	2.28747	2.28849	2.28950	2.29051	2.29152
9.9	2.29757	2.29858	2.29958	2.30058	2.30158

Exponential, e^x

x	0	1	2	3	4
0.0	1.00000	1.01005	1.02020	1.03045	1.04081
0.1	1.10517	1.11628	1.12750	1.13883	1.15027
0.2	1.22140	1.23368	1.24608	1.25860	1.27125
0.3	1.34986	1.36343	1.37713	1.39097	1.40495
0.4	1.49182	1.50682	1.52196	1.53726	1.55271
0.5	1.64872	1.66529	1.68203	1.69893	1.71601
0.6	1.82212	1.84043	1.85893	1.87761	1.89648
0.7	2.01375	2.03399	2.05443	2.07508	2.09594
0.8	2.22554	2.24791	2.27050	2.29332	2.31637
0.9	2.45960	2.48432	2.50929	2.53451	2.55998
1.0	2.71828	2.74560	2.77319	2.80107	2.82922
1.1	3.00417	3.03436	3.06485	3.09566	3.12677
1.2	3.32012	3.35348	3.38719	3.42123	3.45561
1.3	3.66930	3.70617	3.74342	3.78104	3.81904
1.4	4.05520	4.09596	4.13712	4.17870	4.22070
1.5	4.48169	4.52673	4.57223	4.61818	4.66459
1.6	4.95303	5.00281	5.05309	5.10387	5.15517
1.7	5.47395	5.52896	5.58453	5.64065	5.69734
1.8	6.04965	6.11045	6.17186	6.23389	6.29654
1.9	6.68589	6.75309	6.82096	6.88951	6.95875
2.0	7.38906	7.46332	7.53832	7.61409	7.69061
2.1	8.16617	8.24824	8.33114	8.41487	8.49944
2.2	9.02501	9.11572	9.20733	9.29987	9.39333
2.3	9.97418	10.0744	10.1757	10.2779	10.3812
2.4	11.0232	11.1340	11.2459	11.3589	11.4730
2.5	12.1825	12.3049	12.4286	12.5535	12.6797
2.6	13.4637	13.5991	13.7357	13.8738	14.0132
2.7	14.8797	15.0293	15.1803	15.3329	15.4870
2.8	16.4446	16.6099	16.7769	16.9455	17.1158
2.9	18.1741	18.3568	18.5413	18.7276	18.9158
3.0	20.0855	20.2874	20.4913	20.6972	20.9052
3.1	22.1980	22.4210	22.6464	22.8740	23.1039
3.2	24.5325	24.7791	25.0281	25.2797	25.5337
3.3	27.1126	27.3851	27.6604	27.9383	28.2191
3.4	29.9641	30.2652	30.5694	30.8766	31.1870
3.5	33.1155	33.4483	33.7844	34.1240	34.4669
3.6	36.5982	36.9661	37.3376	37.7128	38.0918
3.7	40.4473	40.8538	41.2644	41.6791	42.0980
3.8	44.7012	45.1504	45.6042	46.0625	46.5255
3.9	49.4024	49.8990	50.4004	50.9070	51.4186
4.0	54.5982	55.1469	55.7011	56.2609	56.8263
4.1	60.3403	60.9467	61.5592	62.1779	62.8028
4.2	66.6863	67.3565	68.0335	68.7172	69.4079
4.3	73.6998	74.4405	75.1886	75.9443	76.7075
4.4	81.4509	82.2695	83.0963	83.9314	84.7749
4.5	90.0171	90.9218	91.8356	92.7586	93.6908
4.6	99.4843	100.484	101.494	102.514	103.544
4.7	109.947	111.052	112.168	113.296	114.434
4.8	121.510	122.732	123.965	125.211	126.469
4.9	134.290	135.639	137.003	138.380	139.770
x	0	1	2	3	4

Exponential, e^x

x	5	6	7	8	9
0.0	1.05127	1.06184	1.07251	1.08329	1.09417
0.1	1.16183	1.17351	1.18530	1.19722	1.20925
0.2	1.28403	1.29693	1.30996	1.32313	1.33643
0.3	1.41907	1.43333	1.44773	1.46228	1.47698
0.4	1.56831	1.58407	1.59999	1.61607	1.63232
0.5	1.73325	1.75067	1.76827	1.78604	1.80399
0.6	1.91554	1.93479	1.95424	1.97388	1.99372
0.7	2.11700	2.13828	2.15977	2.18147	2.20340
0.8	2.33965	2.36316	2.38691	2.41090	2.43513
0.9	2.58571	2.61170	2.63794	2.66446	2.69123
1.0	2.85765	2.88637	2.91538	2.94468	2.97427
1.1	3.15819	3.18993	3.22199	3.25437	3.28708
1.2	3.49034	3.52542	3.56085	3.59664	3.63279
1.3	3.85743	3.89619	3.93535	3.97490	4.01485
1.4	4.26311	4.30596	4.34924	4.39295	4.43710
1.5	4.71147	4.75882	4.80665	4.85496	4.90375
1.6	5.20698	5.25931	5.31217	5.36556	5.41948
1.7	5.75460	5.81244	5.87085	5.92986	5.98945
1.8	6.35982	6.42374	6.48830	6.55350	6.61937
1.9	7.02869	7.09933	7.17068	7.24274	7.31553
2.0	7.76790	7.84597	7.92482	8.00447	8.08492
2.1	8.58486	8.67114	8.75828	8.84631	8.93521
2.2	9.48774	9.58309	9.67940	9.77668	9.87494
2.3	10.4856	10.5910	10.6974	10.8049	10.9135
2.4	11.5883	11.7048	11.8224	11.9413	12.0613
2.5	12.8071	12.9358	13.0658	13.1971	13.3298
2.6	14.1540	14.2963	14.4400	14.5851	14.7317
2.7	15.6426	15.7998	15.9586	16.1190	16.2810
2.8	17.2878	17.4615	17.6370	17.8143	17.9933
2.9	19.1060	19.2980	19.4919	19.6878	19.8857
3.0	21.1153	21.3276	21.5419	21.7584	21.9771
3.1	23.3361	23.5706	23.8075	24.0468	24.2884
3.2	25.7903	26.0495	26.3113	26.5758	26.8429
3.3	28.5027	28.7892	29.0785	29.3708	29.6660
3.4	31.5004	31.8170	32.1367	32.4597	32.7859
3.5	34.8133	35.1632	35.5166	35.8735	36.2341
3.6	38.4747	38.8613	39.2519	39.6464	40.0448
3.7	42.5211	42.9484	43.3801	43.8160	44.2564
3.8	46.9931	47.4654	47.9424	48.4242	48.9109
3.9	51.9354	52.4573	52.9845	53.5170	54.0549
4.0	57.3975	57.9743	58.5570	59.1455	59.7399
4.1	63.4340	64.0715	64.7155	65.3659	66.0228
4.2	70.1054	70.8100	71.5216	72.2404	72.9665
4.3	77.4785	78.2571	79.0436	79.8380	80.6404
4.4	85.6269	86.4875	87.3567	88.2347	89.1214
4.5	94.6324	95.5835	96.5441	97.5144	98.4944
4.6	104.585	105.636	106.698	107.770	108.853
4.7	115.584	116.746	117.919	119.104	120.301
4.8	127.740	129.024	130.321	131.631	132.954
4.9	141.175	142.594	144.027	145.474	146.936
x	5	6	7	8	9

Exponential, e^x

x	0	1	2	3	4
5.0	148.413	149.905	151.411	152.933	154.470
5.1	164.022	165.670	167.335	169.017	170.716
5.2	181.272	183.094	184.934	186.793	188.670
5.3	200.337	202.350	204.384	206.438	208.513
5.4	221.406	223.632	225.879	228.149	230.442
5.5	244.692	247.151	249.635	252.144	254.678
5.6	270.426	273.144	275.889	278.662	281.463
5.7	298.867	301.871	304.905	307.969	311.064
5.8	330.300	333.619	336.972	340.359	343.779
5.9	365.037	368.706	372.412	376.155	379.935
6.0	403.429	407.483	411.579	415.715	419.893
6.1	445.858	450.339	454.865	459.436	464.054
6.2	492.749	497.701	502.703	507.755	512.859
6.3	544.572	550.045	555.573	561.157	566.796
6.4	601.845	607.894	614.003	620.174	626.407
6.5	665.142	671.826	678.578	685.398	692.287
6.6	735.093	742.483	749.945	757.482	765.095
6.7	812.406	820.571	828.818	837.147	845.561
6.8	897.847	906.871	915.985	925.191	934.489
6.9	992.275	1002.25	1012.32	1022.49	1032.77
7.0	1096.63	1107.65	1118.79	1130.03	1141.39
7.1	1211.97	1224.15	1236.45	1248.88	1261.43
7.2	1339.43	1352.89	1366.49	1380.22	1394.09
7.3	1480.30	1495.18	1510.20	1525.38	1540.71
7.4	1635.98	1652.43	1669.03	1685.81	1702.75
7.5	1808.04	1826.21	1844.57	1863.11	1881.83
7.6	1998.20	2018.28	2038.56	2059.05	2079.74
7.7	2208.35	2230.54	2252.96	2275.60	2298.47
7.8	2440.60	2465.13	2489.52	2514.93	2540.20
7.9	2697.28	2724.39	2751.77	2779.43	2807.36
8.0	2980.96	3010.92	3041.18	3071.74	3102.61
8.1	3294.47	3327.58	3361.02	3394.80	3428.92
8.2	3640.95	3677.54	3714.50	3751.83	3789.54
8.3	4023.87	4064.31	4105.16	4146.42	4188.09
8.4	4447.07	4491.76	4536.90	4582.50	4628.56
8.5	4914.77	4964.16	5014.05	5064.45	5115.34
8.6	5431.66	5486.25	5541.39	5597.08	5653.33
8.7	6002.91	6063.24	6124.18	6185.73	6247.90
8.8	6634.24	6700.92	6768.26	6836.29	6904.99
8.9	7331.97	7405.66	7480.09	7555.27	7631.20
9.0	8103.08	8184.52	8266.78	8349.86	8433.78
9.1	8955.29	9045.29	9136.20	9228.02	9320.77
9.2	9897.13	9996.60	10097.1	10198.5	10301.0
9.3	10938.0	11047.9	11159.0	11271.1	11384.4
9.4	12088.4	12209.9	12332.6	12456.5	12581.7
9.5	13359.7	13494.0	13629.6	13766.6	13904.9
9.6	14764.8	14913.2	15063.1	15214.4	15367.3
9.7	16317.6	16481.6	16647.2	16814.6	16983.5
9.8	18033.7	18215.0	18398.1	18583.0	18769.7
9.9	19930.4	20130.7	20333.0	20537.3	20743.7

Exponential, e^x

x	5	6	7	8	9
5.0	156.022	157.591	159.174	160.774	162.390
5.1	172.431	174.164	175.915	177.683	179.469
5.2	190.566	192.481	194.416	196.370	198.343
5.3	210.608	212.725	214.863	217.022	219.203
5.4	232.758	235.097	237.460	239.847	242.257
5.5	257.238	259.823	262.434	265.072	267.736
5.6	284.291	287.149	290.035	292.949	295.894
5.7	314.191	317.348	320.538	323.759	327.013
5.8	347.234	350.724	354.249	357.809	361.405
5.9	383.753	387.610	391.506	395.440	399.415
6.0	424.113	428.375	432.681	437.029	441.421
6.1	468.717	473.428	478.186	482.992	487.846
6.2	518.013	523.219	528.477	533.789	539.153
6.3	572.493	578.246	584.058	589.928	595.857
6.4	632.702	639.061	645.484	651.971	658.523
6.5	699.244	706.272	713.370	720.539	727.781
6.6	772.784	780.551	788.396	796.319	804.322
6.7	854.059	862.642	871.312	880.069	888.914
6.8	943.881	953.367	962.949	972.626	982.401
6.9	1043.15	1053.63	1064.22	1074.92	1085.72
7.0	1152.86	1166.45	1176.15	1187.97	1199.91
7.1	1274.11	1286.91	1299.84	1312.91	1326.10
7.2	1408.10	1422.26	1436.55	1450.99	1465.57
7.3	1556.20	1571.84	1587.63	1603.59	1619.71
7.4	1719.86	1737.15	1754.61	1772.24	1790.05
7.5	1900.74	1919.85	1939.14	1958.63	1978.31
7.6	2100.65	2121.76	2143.08	2164.62	2186.37
7.7	2321.57	2344.90	2368.47	2392.27	2416.32
7.8	2565.73	2591.52	2617.57	2643.87	2670.44
7.9	2835.58	2864.07	2892.86	2921.93	2951.30
8.0	3133.80	3165.29	3197.10	3229.23	3261.69
8.1	3463.38	3498.19	3533.34	3568.85	3604.72
8.2	3827.63	3866.09	3904.95	3944.19	3983.83
8.3	4230.18	4272.69	4315.64	4359.01	4402.82
8.4	4675.07	4722.06	4769.52	4817.45	4865.87
8.5	5166.75	5218.68	5271.13	5324.11	5377.61
8.6	5710.15	5767.33	5825.50	5884.05	5943.18
8.7	6310.69	6374.11	6438.17	6502.88	6568.23
8.8	6974.39	7044.48	7115.28	7186.79	7259.02
8.9	7707.89	7785.36	7863.60	7942.63	8022.46
9.0	8518.54	8604.15	8690.62	8777.97	8866.19
9.1	9414.44	9509.06	9604.62	9701.15	9798.65
9.2	10404.6	10509.1	10614.8	10721.4	10829.2
9.3	11498.8	11614.4	11731.1	11849.0	11968.1
9.4	12708.2	12835.9	12964.9	13095.2	13226.8
9.5	14044.7	14185.8	14328.4	14472.4	14617.9
9.6	15521.8	15677.8	15835.3	15994.5	16155.2
9.7	17154.2	17326.6	17500.8	17676.7	17854.3
9.8	18958.4	19148.9	19341.3	19535.7	19732.1
9.9	20952.2	21162.8	21375.5	21590.3	21807.3

Exponential, e^{-x}

x	0	1	2	3	4
0.0	1.00000	9.90050 −1	9.80199 −1	9.70446 −1	9.60789 −1
0.1	9.04837 −1	8.95834 −1	8.86920 −1	8.78095 −1	8.69358 −1
0.2	8.18731 −1	8.10584 −1	8.02519 −1	7.94534 −1	7.86628 −1
0.3	7.40818 −1	7.33447 −1	7.26149 −1	7.18924 −1	7.11770 −1
0.4	6.70320 −1	6.63650 −1	6.57047 −1	6.50509 −1	6.44036 −1
0.5	6.06531 −1	6.00496 −1	5.94521 −1	5.88605 −1	5.82748 −1
0.6	5.48812 −1	5.43351 −1	5.37944 −1	5.32592 −1	5.27292 −1
0.7	4.96585 −1	4.91644 −1	4.86752 −1	4.81909 −1	4.77114 −1
0.8	4.49329 −1	4.44858 −1	4.40432 −1	4.36049 −1	4.31711 −1
0.9	4.06570 −1	4.02524 −1	3.98519 −1	3.94554 −1	3.90628 −1
1.0	3.67879 −1	3.64219 −1	3.60595 −1	3.57007 −1	3.53455 −1
1.1	3.32871 −1	3.29559 −1	3.26280 −1	3.23033 −1	3.19819 −1
1.2	3.01194 −1	2.98197 −1	2.95230 −1	2.92293 −1	2.89384 −1
1.3	2.72532 −1	2.69820 −1	2.67135 −1	2.64477 −1	2.61846 −1
1.4	2.46597 −1	2.44143 −1	2.41714 −1	2.39309 −1	2.36928 −1
1.5	2.23130 −1	2.20910 −1	2.18712 −1	2.16536 −1	2.14381 −1
1.6	2.01897 −1	1.99888 −1	1.97899 −1	1.95930 −1	1.93980 −1
1.7	1.82684 −1	1.80866 −1	1.79066 −1	1.77284 −1	1.75520 −1
1.8	1.65299 −1	1.63654 −1	1.62026 −1	1.60414 −1	1.58817 −1
1.9	1.49569 −1	1.48080 −1	1.46607 −1	1.45148 −1	1.43704 −1
2.0	1.35335 −1	1.33989 −1	1.32655 −1	1.31336 −1	1.30029 −1
2.1	1.22456 −1	1.21238 −1	1.20032 −1	1.18837 −1	1.17655 −1
2.2	1.10803 −1	1.09701 −1	1.08609 −1	1.07528 −1	1.06459 −1
2.3	1.00259 −1	9.92613 −2	9.82736 −2	9.72958 −2	9.63276 −2
2.4	9.07180 −2	8.98153 −2	8.89216 −2	8.80368 −2	8.71609 −2
2.5	8.20850 −2	8.12682 −2	8.04596 −2	7.96590 −2	7.88664 −2
2.6	7.42736 −2	7.35345 −2	7.28029 −2	7.20785 −2	7.13613 −2
2.7	6.72055 −2	6.65368 −2	6.58748 −2	6.52193 −2	6.45703 −2
2.8	6.08101 −2	6.02050 −2	5.96059 −2	5.90129 −2	5.84257 −2
2.9	5.50232 −2	5.44757 −2	5.39337 −2	5.33970 −2	5.28657 −2
3.0	4.97871 −2	4.92917 −2	4.88012 −2	4.83156 −2	4.78349 −2
3.1	4.50492 −2	4.46010 −2	4.41572 −2	4.37178 −2	4.32828 −2
3.2	4.07622 −2	4.03566 −2	3.99551 −2	3.95575 −2	3.91639 −2
3.3	3.68832 −2	3.65162 −2	3.61528 −2	3.57931 −2	3.54370 −2
3.4	3.33733 −2	3.30412 −2	3.27124 −2	3.23869 −2	3.20647 −2
3.5	3.01974 −2	2.98969 −2	2.95994 −2	2.93049 −2	2.90133 −2
3.6	2.73237 −2	2.70518 −2	2.67827 −2	2.65162 −2	2.62523 −2
3.7	2.47235 −2	2.44771 −2	2.42330 −2	2.39928 −2	2.37501 −2
3.8	2.23708 −2	2.21482 −2	2.19278 −2	2.17096 −2	2.14936 −2
3.9	2.02419 −2	2.00405 −2	1.98411 −2	1.96437 −2	1.94482 −2
4.0	1.83156 −2	1.81334 −2	1.79530 −2	1.77743 −2	1.75975 −2
4.1	1.65727 −2	1.64078 −2	1.62445 −2	1.60829 −2	1.59229 −2
4.2	1.49956 −2	1.48464 −2	1.46986 −2	1.45524 −2	1.44076 −2
4.3	1.35686 −2	1.34336 −2	1.32999 −2	1.31675 −2	1.30365 −2
4.4	1.22773 −2	1.21552 −2	1.20342 −2	1.19145 −2	1.17959 −2
4.5	1.11090 −2	1.09985 −2	1.08890 −2	1.07807 −2	1.06734 −2
4.6	1.00518 −2	9.95182 −3	9.85280 −3	9.75476 −3	9.65770 −3
4.7	9.09528 −3	9.00478 −3	8.91518 −3	8.82647 −3	8.73865 −3
4.8	8.22975 −3	8.14786 −3	8.06679 −3	7.98652 −3	7.90705 −3
4.9	7.44658 −3	7.37249 −3	7.29913 −3	7.22650 −3	7.15460 −3
x	0	1	2	3	4

Exponential, e^{-x}

x	5	6	7	8	9
0.0	9.51229 −1	9.41765 −1	9.32394 −1	9.23116 −1	9.13931 −1
0.1	8.60708 −1	8.52144 −1	8.43665 −1	8.35270 −1	8.26959 −1
0.2	7.78801 −1	7.71052 −1	7.63380 −1	7.55784 −1	7.48264 −1
0.3	7.04688 −1	6.97676 −1	6.90734 −1	6.83861 −1	6.77057 −1
0.4	6.37628 −1	6.31284 −1	6.25002 −1	6.18783 −1	6.12626 −1
0.5	5.76950 −1	5.71209 −1	5.65525 −1	5.59898 −1	5.54327 −1
0.6	5.22046 −1	5.16851 −1	5.11709 −1	5.06617 −1	5.01576 −1
0.7	4.72367 −1	4.67666 −1	4.63013 −1	4.58406 −1	4.53845 −1
0.8	4.27415 −1	4.23162 −1	4.18952 −1	4.14783 −1	4.10656 −1
0.9	3.86741 −1	3.82893 −1	3.79083 −1	3.75311 −1	3.71577 −1
1.0	3.49938 −1	3.46456 −1	3.43009 −1	3.39596 −1	3.36216 −1
1.1	3.16637 −1	3.13486 −1	3.10367 −1	3.07279 −1	3.04221 −1
1.2	2.86505 −1	2.83654 −1	2.80832 −1	2.78037 −1	2.75271 −1
1.3	2.59240 −1	2.56661 −1	2.54107 −1	2.51579 −1	2.49075 −1
1.4	2.34570 −1	2.32236 −1	2.29925 −1	2.27638 −1	2.25373 −1
1.5	2.12248 −1	2.10136 −1	2.08045 −1	2.05975 −1	2.03926 −1
1.6	1.92050 −1	1.90139 −1	1.88247 −1	1.86374 −1	1.84520 −1
1.7	1.73774 −1	1.72045 −1	1.70333 −1	1.68638 −1	1.66960 −1
1.8	1.57237 −1	1.55673 −1	1.54124 −1	1.52590 −1	1.51072 −1
1.9	1.42274 −1	1.40858 −1	1.39457 −1	1.38069 −1	1.36695 −1
2.0	1.28735 −1	1.27454 −1	1.26186 −1	1.24930 −1	1.23687 −1
2.1	1.16484 −1	1.15325 −1	1.14178 −1	1.13042 −1	1.11917 −1
2.2	1.05399 −1	1.04350 −1	1.03312 −1	1.02284 −1	1.01266 −1
2.3	9.53962 −2	9.44202 −2	9.34807 −2	9.25506 −2	9.16297 −2
2.4	8.62936 −2	8.54350 −2	8.45849 −2	8.37432 −2	8.29100 −2
2.5	7.80817 −2	7.73047 −2	7.65355 −2	7.57740 −2	7.50200 −2
2.6	7.06512 −2	6.99482 −2	6.92522 −2	6.85632 −2	6.78809 −2
2.7	6.39279 −2	6.32918 −2	6.26620 −2	6.20385 −2	6.14212 −2
2.8	5.78443 −2	5.72688 −2	5.66989 −2	5.61348 −2	5.55762 −2
2.9	5.23397 −2	5.18189 −2	5.13033 −2	5.07928 −2	5.02874 −2
3.0	4.73589 −2	4.68877 −2	4.64212 −2	4.59593 −2	4.55020 −2
3.1	4.28521 −2	4.24257 −2	4.20036 −2	4.15857 −2	4.11719 −2
3.2	3.87742 −2	3.83884 −2	3.80064 −2	3.76283 −2	3.72538 −2
3.3	3.50844 −2	3.47353 −2	3.43896 −2	3.40475 −2	3.37087 −2
3.4	3.17456 −2	3.14298 −2	3.11170 −2	3.08074 −2	3.05009 −2
3.5	2.87246 −2	2.84388 −2	2.81559 −2	2.78757 −2	2.75983 −2
3.6	2.59911 −2	2.57325 −2	2.54765 −2	2.52230 −2	2.49720 −2
3.7	2.35177 −2	2.32827 −2	2.30521 −2	2.28227 −2	2.25956 −2
3.8	2.12797 −2	2.10680 −2	2.08584 −2	2.06508 −2	2.04453 −2
3.9	1.92547 −2	1.90631 −2	1.88734 −2	1.86856 −2	1.84997 −2
4.0	1.74224 −2	1.72490 −2	1.70774 −2	1.69075 −2	1.67392 −2
4.1	1.57644 −2	1.56076 −2	1.54523 −2	1.52985 −2	1.51463 −2
4.2	1.42642 −2	1.41223 −2	1.39818 −2	1.38427 −2	1.37049 −2
4.3	1.29068 −2	1.27784 −2	1.26512 −2	1.25254 −2	1.24007 −2
4.4	1.16786 −2	1.15624 −2	1.14473 −2	1.13334 −2	1.12206 −2
4.5	1.05672 −2	1.04621 −2	1.03580 −2	1.02549 −2	1.01529 −2
4.6	9.56160 −3	9.46646 −3	9.37227 −3	9.27901 −3	9.18669 −3
4.7	8.65170 −3	8.56561 −3	8.48038 −3	8.39600 −3	8.31246 −3
4.8	7.82838 −3	7.75048 −3	7.67337 −3	7.59701 −3	7.52142 −3
4.9	7.08341 −3	7.01293 −3	6.94315 −3	6.87406 −3	6.80566 −3
x	5	6	7	8	9

171

Exponential, e^{-x}

x	0	1	2	3	4
5.0	6.73795 −3	6.67090 −3	6.60453 −3	6.53881 −3	6.47375 −3
5.1	6.09675 −3	6.03608 −3	5.97602 −3	5.91656 −3	5.85769 −3
5.2	5.51656 −3	5.46167 −3	5.40733 −3	5.35353 −3	5.30026 −3
5.3	4.99159 −3	4.94193 −3	4.89275 −3	4.84407 −3	4.79587 −3
5.4	4.51658 −3	4.47164 −3	4.42715 −3	4.38310 −3	4.33948 −3
5.5	4.08677 −3	4.04611 −3	4.00585 −3	3.96599 −3	3.92653 −3
5.6	3.69786 −3	3.66107 −3	3.62464 −3	3.58858 −3	3.55287 −3
5.7	3.34597 −3	3.31267 −3	3.27971 −3	3.24708 −3	3.21477 −3
5.8	3.02755 −3	2.99743 −3	2.96761 −3	2.93808 −3	2.90884 −3
5.9	2.73944 −3	2.71219 −3	2.68520 −3	2.65848 −3	2.63203 −3
6.0	2.47875 −3	2.45409 −3	2.42967 −3	2.40549 −3	2.38156 −3
6.1	2.24287 −3	2.22055 −3	2.19846 −3	2.17658 −3	2.15492 −3
6.2	2.02943 −3	2.00924 −3	1.98925 −3	1.96945 −3	1.94986 −3
6.3	1.83630 −3	1.81803 −3	1.79994 −3	1.78203 −3	1.76430 −3
6.4	1.66156 −3	1.64502 −3	1.62866 −3	1.61245 −3	1.59641 −3
6.5	1.50344 −3	1.48848 −3	1.47367 −3	1.45901 −3	1.44449 −3
6.6	1.36037 −3	1.34683 −3	1.33343 −3	1.32016 −3	1.30703 −3
6.7	1.23091 −3	1.21866 −3	1.20654 −3	1.19453 −3	1.18265 −3
6.8	1.11378 −3	1.10269 −3	1.09172 −3	1.08086 −3	1.07010 −3
6.9	1.00779 −3	9.97758 −4	9.87801 −4	9.78001 −4	9.68270 −4
7.0	9.11882 −4	9.02809 −4	8.93825 −4	8.84932 −4	8.76127 −4
7.1	8.25105 −4	8.16895 −4	8.08767 −4	8.00719 −4	7.92752 −4
7.2	7.46586 −4	7.39157 −4	7.31802 −4	7.24521 −4	7.17312 −4
7.3	6.75539 −4	6.68817 −4	6.62162 −4	6.55574 −4	6.49051 −4
7.4	6.11253 −4	6.05171 −4	5.99149 −4	5.93188 −4	5.87285 −4
7.5	5.53084 −4	5.47581 −4	5.42133 −4	5.36738 −4	5.31398 −4
7.6	5.00451 −4	4.95472 −4	4.90542 −4	4.85661 −4	4.80828 −4
7.7	4.52827 −4	4.48321 −4	4.43861 −4	4.39444 −4	4.35072 −4
7.8	4.09735 −4	4.05658 −4	4.01622 −4	3.97625 −4	3.93669 −4
7.9	3.70744 −4	3.67055 −4	3.63402 −4	3.59786 −4	3.56206 −4
8.0	3.35463 −4	3.32125 −4	3.28820 −4	3.25548 −4	3.22309 −4
8.1	3.03539 −4	3.00519 −4	2.97529 −4	2.94568 −4	2.91637 −4
8.2	2.74654 −4	2.71921 −4	2.69215 −4	2.66536 −4	2.63884 −4
8.3	2.48517 −4	2.46044 −4	2.43596 −4	2.41172 −4	2.38772 −4
8.4	2.24867 −4	2.22630 −4	2.20415 −4	2.18221 −4	2.16050 −4
8.5	2.03468 −4	2.01444 −4	1.99439 −4	1.97455 −4	1.95490 −4
8.6	1.84106 −4	1.82274 −4	1.80460 −4	1.78665 −4	1.76887 −4
8.7	1.66586 −4	1.64928 −4	1.63287 −4	1.61662 −4	1.60054 −4
8.8	1.50733 −4	1.49233 −4	1.47748 −4	1.46278 −4	1.44823 −4
8.9	1.36389 −4	1.35032 −4	1.33688 −4	1.32358 −4	1.31041 −4
9.0	1.23410 −4	1.22182 −4	1.20966 −4	1.19762 −4	1.18571 −4
9.1	1.11666 −4	1.10555 −4	1.09455 −4	1.08366 −4	1.07287 −4
9.2	1.01039 −4	1.00034 −4	9.90387 −5	9.80532 −5	9.70776 −5
9.3	9.14242 −5	9.05145 −5	8.96139 −5	8.87222 −5	8.78394 −5
9.4	8.27241 −5	8.19009 −5	8.10860 −5	8.02792 −5	7.94804 −5
9.5	7.48518 −5	7.41070 −5	7.33697 −5	7.26396 −5	7.19168 −5
9.6	6.77287 −5	6.70348 −5	6.63876 −5	6.57271 −5	6.50731 −5
9.7	6.12835 −5	6.06737 −5	6.00700 −5	5.94723 −5	5.88805 −5
9.8	5.54516 −5	5.48998 −5	5.43536 −5	5.38128 −5	5.32773 −5
9.9	5.01747 −5	4.96754 −5	4.91812 −5	4.86918 −5	4.82073 −5
x	0	1	2	3	4

Exponential, e^{-x}

x	5	6	7	8	9
5.0	6.40933 −3	6.34556 −3	6.28242 −3	6.21991 −3	6.15802 −3
5.1	5.79940 −3	5.74170 −3	5.68457 −3	5.62801 −3	5.57201 −3
5.2	5.24752 −3	5.19530 −3	5.14361 −3	5.09243 −3	5.04176 −3
5.3	4.74815 −3	4.70091 −3	4.65413 −3	4.60782 −3	4.56197 −3
5.4	4.29630 −3	4.25356 −3	4.21123 −3	4.16933 −3	4.12784 −3
5.5	3.88746 −3	3.84878 −3	3.81048 −3	3.77257 −3	3.73503 −3
5.6	3.51752 −3	3.48252 −3	3.44787 −3	3.41356 −3	3.37959 −3
5.7	3.18278 −3	3.15111 −3	3.11976 −3	3.08872 −3	3.05798 −3
5.8	2.87990 −3	2.85124 −3	2.82287 −3	2.79479 −3	2.76698 −3
5.9	2.60584 −3	2.57991 −3	2.55424 −3	2.52883 −3	2.50366 −3
6.0	2.35786 −3	2.33440 −3	2.31117 −3	2.28818 −3	2.26541 −3
6.1	2.13348 −3	2.11225 −3	2.09124 −3	2.07043 −3	2.04983 −3
6.2	1.93045 −3	1.91125 −3	1.89223 −3	1.87340 −3	1.85476 −3
6.3	1.74675 −3	1.72937 −3	1.71216 −3	1.69512 −3	1.67826 −3
6.4	1.58052 −3	1.56480 −3	1.54923 −3	1.53381 −3	1.51855 −3
6.5	1.43012 −3	1.41589 −3	1.40180 −3	1.38785 −3	1.37404 −3
6.6	1.29402 −3	1.28115 −3	1.26840 −3	1.25578 −3	1.24328 −3
6.7	1.17088 −3	1.15923 −3	1.14769 −3	1.13627 −3	1.12497 −3
6.8	1.05946 −3	1.04891 −3	1.03848 −3	1.02814 −3	1.01791 −3
6.9	9.58635 −4	9.49097 −4	9.39653 −4	9.30303 −4	9.21047 −4
7.0	8.67409 −4	8.58778 −4	8.50233 −4	8.41773 −4	8.33397 −4
7.1	7.84864 −4	7.77055 −4	7.69323 −4	7.61668 −4	7.54089 −4
7.2	7.10174 −4	7.03108 −4	6.96112 −4	6.89186 −4	6.82328 −4
7.3	6.42592 −4	6.36198 −4	6.29868 −4	6.23601 −4	6.17396 −4
7.4	5.81442 −4	5.75656 −4	5.69928 −4	5.64257 −4	5.58643 −4
7.5	5.26110 −4	5.20875 −4	5.15692 −4	5.10561 −4	5.05481 −4
7.6	4.76044 −4	4.71307 −4	4.66618 −4	4.61975 −4	4.57378 −4
7.7	4.30743 −4	4.26457 −4	4.22213 −4	4.18012 −4	4.13853 −4
7.8	3.89752 −4	3.85874 −4	3.82034 −4	3.78233 −4	3.74470 −4
7.9	3.52662 −4	3.49153 −4	3.45679 −4	3.42239 −4	3.38834 −4
8.0	3.19102 −4	3.15927 −4	3.12783 −4	3.09671 −4	3.06590 −4
8.1	2.88735 −4	2.85862 −4	2.83018 −4	2.80202 −4	2.77414 −4
8.2	2.61259 −4	2.58659 −4	2.56085 −4	2.53537 −4	2.51014 −4
8.3	2.36397 −4	2.34044 −4	2.31716 −4	2.29410 −4	2.27127 −4
8.4	2.13900 −4	2.11772 −4	2.09665 −4	2.07579 −4	2.05513 −4
8.5	1.93545 −4	1.91619 −4	1.89713 −4	1.87825 −4	1.85956 −4
8.6	1.75127 −4	1.73384 −4	1.71659 −4	1.69951 −4	1.68260 −4
8.7	1.58461 −4	1.56885 −4	1.55324 −4	1.53778 −4	1.52248 −4
8.8	1.43382 −4	1.41955 −4	1.40543 −4	1.39144 −4	1.37760 −4
8.9	1.29737 −4	1.28446 −4	1.27168 −4	1.25903 −4	1.24650 −4
9.0	1.17391 −4	1.16223 −4	1.15067 −4	1.13922 −4	1.12788 −4
9.1	1.06220 −4	1.05163 −4	1.04117 −4	1.03081 −4	1.02055 −4
9.2	9.61117 −5	9.51553 −5	9.42085 −5	9.32711 −5	9.23431 −5
9.3	8.69654 −5	8.61001 −5	8.52434 −5	8.43952 −5	8.35555 −5
9.4	7.86896 −5	7.79066 −5	7.71314 −5	7.63639 −5	7.56041 −5
9.5	7.12013 −5	7.04928 −5	6.97914 −5	6.90969 −5	6.84094 −5
9.6	6.44256 −5	6.37845 −5	6.31499 −5	6.25215 −5	6.18994 −5
9.7	5.82947 −5	5.77146 −5	5.71404 −5	5.65718 −5	5.60089 −5
9.8	5.27472 −5	5.22224 −5	5.17027 −5	5.11883 −5	5.06789 −5
9.9	4.77276 −5	4.72527 −5	4.67826 −5	4.63171 −5	4.58562 −5
x	5	6	7	8	9

Powers & Roots

n	n²	n³	$\sqrt{n}$	$\sqrt{10n}$	$\sqrt[3]{n}$
1	1	1	1.00000	3.16228	1.00000
2	4	8	1.41421	4.47214	1.25992
3	9	27	1.73205	5.47723	1.44225
4	16	64	2.00000	6.32456	1.58740
5	25	125	2.23607	7.07107	1.70998
6	36	216	2.44949	7.74597	1.81712
7	49	343	2.64575	8.36660	1.91293
8	64	512	2.82843	8.94427	2.00000
9	81	729	3.00000	9.48683	2.08008
10	100	1000	3.16228	10.00000	2.15444
11	121	1331	3.31662	10.48809	2.22398
12	144	1728	3.46410	10.95445	2.28943
13	169	2197	3.60555	11.40175	2.35134
14	196	2744	3.74166	11.83216	2.41014
15	225	3375	3.87298	12.24745	2.46621
16	256	4096	4.00000	12.64911	2.51984
17	289	4913	4.12311	13.03840	2.57128
18	324	5832	4.24264	13.41641	2.62074
19	361	6859	4.35890	13.78405	2.66840
20	400	8000	4.47214	14.14214	2.71442
21	441	9261	4.58258	14.49138	2.75892
22	484	10648	4.69042	14.83240	2.80204
23	529	12167	4.79583	15.16575	2.84387
24	576	13824	4.89898	15.49193	2.88450
25	625	15625	5.00000	15.81139	2.92402
26	676	17576	5.09902	16.12452	2.96250
27	729	19683	5.19615	16.43168	3.00000
28	784	21952	5.29150	16.73320	3.03659
29	841	24389	5.38516	17.02939	3.07232
30	900	27000	5.47723	17.32051	3.10723
31	961	29791	5.56776	17.60682	3.14138
32	1024	32768	5.65685	17.88854	3.17480
33	1089	35937	5.74456	18.16590	3.20753
34	1156	39304	5.83095	18.43909	3.23961
35	1225	42875	5.91608	18.70829	3.27107
36	1296	46656	6.00000	18.97367	3.30193
37	1369	50653	6.08276	19.23538	3.33222
38	1444	54872	6.16441	19.49359	3.36198
39	1521	59319	6.24500	19.74842	3.39121
40	1600	64000	6.32456	20.00000	3.41995
41	1681	68921	6.40312	20.24846	3.44822
42	1764	74088	6.48074	20.49390	3.47603
43	1849	79507	6.55744	20.73644	3.50340
44	1936	85184	6.63325	20.97618	3.53035
45	2025	91125	6.70820	21.21320	3.55689
46	2116	97336	6.78233	21.44761	3.58305
47	2209	103823	6.85566	21.67948	3.60883
48	2304	110592	6.92820	21.90890	3.63424
49	2401	117649	7.00000	22.13594	3.65931
50	2500	125000	7.07107	22.36068	3.68403
51	2601	132651	7.14143	22.56318	3.70843
52	2704	140608	7.21110	22.80351	3.73251
53	2809	148877	7.28011	23.02173	3.75629
54	2916	157464	7.34847	23.23790	3.77976
55	3025	166375	7.41620	23.45208	3.80295
56	3136	175616	7.48332	23.66432	3.82586
57	3249	185193	7.54983	23.87467	3.84850
58	3364	195112	7.61577	24.08319	3.87088
59	3481	205379	7.68115	24.28992	3.89300
60	3600	216000	7.74597	24.49490	3.91487
61	3721	226981	7.81025	24.69818	3.93650
62	3844	238328	7.87401	24.89980	3.95789
63	3969	250047	7.93725	25.09980	3.97906
64	4096	262144	8.00000	25.29822	4.00000
65	4225	274625	8.06226	25.49510	4.02073
66	4356	287496	8.12404	25.69047	4.04124
67	4489	300763	8.18535	25.88436	4.06155
68	4624	314432	8.24621	26.07681	4.08166
69	4761	328509	8.30662	26.26785	4.10157
70	4900	343000	8.36660	26.45751	4.12128
71	5041	357911	8.42615	26.64583	4.14082
72	5184	373248	8.48528	26.83282	4.16017
73	5329	389017	8.54400	27.01851	4.17934
74	5476	405224	8.60233	27.20294	4.19834
75	5625	421875	8.66025	27.38613	4.21716
76	5776	438976	8.71780	27.56810	4.23582
77	5929	456533	8.77496	27.74887	4.25432
78	6084	474552	8.83176	27.92848	4.27266
79	6241	493039	8.88819	28.10694	4.29084
80	6400	512000	8.94427	28.28427	4.30887
81	6561	531441	9.00000	28.46050	4.32675
82	6724	551368	9.05539	28.63564	4.34448
83	6889	571787	9.11043	28.80972	4.36207
84	7056	592704	9.16515	28.98275	4.37952
85	7225	614125	9.21954	29.15476	4.39683
86	7396	636056	9.27362	29.32576	4.41400
87	7569	658503	9.32738	29.49576	4.43105
88	7744	681472	9.38083	29.66479	4.44796
89	7921	704969	9.43398	29.83287	4.46474
90	8100	729000	9.48683	30.00000	4.48140
91	8281	753571	9.53939	30.16621	4.49794
92	8464	778688	9.59166	30.33150	4.51436
93	8649	804357	9.64365	30.49590	4.53066
94	8836	830584	9.69536	30.65942	4.54684
95	9025	857375	9.74679	30.82207	4.56290
96	9216	884736	9.79796	30.98387	4.57886
97	9409	912673	9.84886	31.14482	4.59470
98	9604	941192	9.89949	31.30495	4.61044
99	9801	970299	9.94987	31.46427	4.62606
100	10000	1000000	10.00000	31.62278	4.64159

1 radian = 57.29577 95130 82321 degrees

RADIANS TO DEGREES

Radians	Tenths	Hundredths	Thousandths	Ten-thousandths
1 57°17'44".8	5°43'46".5	0°34'22".6	0° 3'26".3	0° 0'20".6
2 114°35'29".6	11°27'33".0	1° 8'45".3	0° 6'52".5	0° 0'41".3
3 171°53'14".4	17°11'19".4	1°43'07".9	0°10'18".8	0° 1'01".9
4 229°10'59".2	22°55'05".9	2°17'30".6	0°13'45".1	0° 1'22".5
5 286°28'44".0	28°38'52".4	2°51'53".2	0°17'11".3	0° 1'43".1
6 343°46'28".8	34°22'38".9	3°26'15".9	0°20'37".6	0° 2'03".8
7 401° 4'13".6	40° 6'25".4	4° 0'38".5	0°24'03".9	0° 2'24".4
8 458°21'58".4	45°50'11".8	4°35'01".2	0°27'30".1	0° 2'45".0
9 515°39'43".3	51°33'58".3	5° 9'23".8	0°30'56".4	0° 3'05".6

FACTORIALS

n	n!	n	n!
0	1	8	40,320
1	1	9	362,880
2	2	10	3,628,800
3	6	11	39,916,800
4	24	12	479,001,600
5	120	13	6,227,020,800
6	720	14	87,178,291,200
7	5040	15	1,307,674,368,000

DEGREES TO RADIANS

1 degree = 0.01745 32925 19943 radians

°	Degrees	°	Degrees	°	Degrees	Minutes	Seconds
0	0.00000 00	60	1.04719 76	120	2.09439 51	0 0.00000 00	0 0.00000 00
1	0.01745 33	61	1.06465 08	121	2.11184 84	1 0.00029 09	1 0.00000 48
2	0.03490 66	62	1.08210 41	122	2.12930 17	2 0.00058 18	2 0.00000 97
3	0.05235 99	63	1.09955 74	123	2.14675 50	3 0.00087 27	3 0.00001 45
4	0.06981 32	64	1.11701 07	124	2.16420 83	4 0.00116 36	4 0.00001 94
5	0.08726 65	65	1.13446 40	125	2.18166 16	5 0.00145 44	5 0.00002 42
6	0.10471 98	66	1.15191 73	126	2.19911 49	6 0.00174 53	6 0.00002 91
7	0.12217 30	67	1.16937 06	127	2.21656 82	7 0.00203 62	7 0.00003 39
8	0.13962 63	68	1.18682 39	128	2.23402 14	8 0.00232 71	8 0.00003 88
9	0.15707 96	69	1.20427 72	129	2.25147 47	9 0.00261 80	9 0.00004 36
10	0.17453 29	70	1.22173 05	130	2.26892 80	10 0.00290 89	10 0.00004 85
11	0.19198 62	71	1.23918 38	131	2.28638 13	11 0.00319 98	11 0.00005 33
12	0.20943 95	72	1.25663 71	132	2.30383 46	12 0.00349 07	12 0.00005 82
13	0.22689 28	73	1.27409 04	133	2.32128 79	13 0.00378 15	13 0.00006 30
14	0.24434 61	74	1.29154 36	134	2.33874 12	14 0.00407 24	14 0.00006 79
15	0.26179 94	75	1.30899 69	135	2.35619 45	15 0.00436 33	15 0.00007 27
16	0.27925 27	76	1.32645 02	136	2.37364 78	16 0.00465 42	16 0.00007 76
17	0.29670 60	77	1.34390 35	137	2.39110 11	17 0.00494 51	17 0.00008 24
18	0.31415 93	78	1.36135 68	138	2.40855 44	18 0.00523 60	18 0.00008 73
19	0.33161 26	79	1.37881 01	139	2.42600 77	19 0.00552 69	19 0.00009 21
20	0.34906 59	80	1.39626 34	140	2.44346 10	20 0.00581 78	20 0.00009 70
21	0.36651 91	81	1.41371 67	141	2.46091 42	21 0.00610 87	21 0.00010 18
22	0.38397 24	82	1.43117 00	142	2.47836 75	22 0.00639 95	22 0.00010 67
23	0.40142 57	83	1.44862 33	143	2.49582 08	23 0.00669 04	23 0.00011 15
24	0.41887 90	84	1.46607 66	144	2.51327 41	24 0.00698 13	24 0.00011 64
25	0.43633 23	85	1.48352 99	145	2.53072 74	25 0.00727 22	25 0.00012 12
26	0.45378 56	86	1.50098 32	146	2.54818 07	26 0.00756 31	26 0.00012 61
27	0.47123 89	87	1.51843 64	147	2.56563 40	27 0.00785 40	27 0.00013 09
28	0.48869 22	88	1.53588 97	148	2.58308 73	28 0.00814 49	28 0.00013 57
29	0.50614 55	89	1.55334 30	149	2.60054 06	29 0.00843 58	29 0.00014 06
30	0.52359 88	90	1.57079 63	150	2.61799 39	30 0.00872 66	30 0.00014 54
31	0.54105 21	91	1.58824 96	151	2.63544 72	31 0.00901 75	31 0.00015 03
32	0.55850 54	92	1.60570 29	152	2.65290 05	32 0.00930 84	32 0.00015 51
33	0.57595 87	93	1.62315 62	153	2.67035 38	33 0.00959 93	33 0.00016 00
34	0.59341 19	94	1.64060 95	154	2.68780 70	34 0.00989 02	34 0.00016 48
35	0.61086 52	95	1.65806 28	155	2.70526 03	35 0.01018 11	35 0.00016 97
36	0.62831 85	96	1.67551 61	156	2.72271 36	36 0.01047 20	36 0.00017 45
37	0.64577 18	97	1.69296 94	157	2.74016 69	37 0.01076 29	37 0.00017 94
38	0.66322 51	98	1.71042 27	158	2.75762 02	38 0.01105 38	38 0.00018 42
39	0.68067 84	99	1.72787 60	159	2.77507 35	39 0.01134 46	39 0.00018 91
40	0.69813 17	100	1.74532 93	160	2.79252 68	40 0.01163 55	40 0.00019 39
41	0.71558 50	101	1.76278 25	161	2.80998 01	41 0.01192 64	41 0.00019 88
42	0.73303 83	102	1.78023 58	162	2.82743 34	42 0.01221 73	42 0.00020 36
43	0.75049 16	103	1.79768 91	163	2.84488 67	43 0.01250 82	43 0.00020 85
44	0.76794 49	104	1.81514 24	164	2.86234 00	44 0.01279 91	44 0.00021 33
45	0.78539 82	105	1.83259 57	165	2.87979 33	45 0.01309 00	45 0.00021 82
46	0.80285 15	106	1.85004 90	166	2.89724 66	46 0.01338 09	46 0.00022 30
47	0.82030 47	107	1.86750 23	167	2.91469 99	47 0.01367 17	47 0.00022 79
48	0.83775 80	108	1.88495 56	168	2.93215 31	48 0.01396 26	48 0.00023 27
49	0.85521 13	109	1.90240 89	169	2.94960 64	49 0.01425 35	49 0.00023 76
50	0.87266 46	110	1.91986 22	170	2.96705 97	50 0.01454 44	50 0.00024 24
51	0.89011 79	111	1.93731 55	171	2.98451 30	51 0.01483 53	51 0.00024 73
52	0.90757 12	112	1.95476 88	172	3.00196 63	52 0.01512 62	52 0.00025 21
53	0.92502 45	113	1.97222 21	173	3.01941 96	53 0.01541 71	53 0.00025 70
54	0.94247 78	114	1.98967 53	174	3.03687 29	54 0.01570 80	54 0.00026 18
55	0.95993 11	115	2.00712 86	175	3.05432 62	55 0.01599 89	55 0.00026 66
56	.097738 44	116	2.02458 19	176	3.07177 95	56 0.01628 97	56 0.00027 15
57	0.99483 77	117	2.04203 52	177	3.08923 28	57 0.01658 06	57 0.00027 63
58	1.01229 10	118	2.05948 85	178	3.10668 61	58 0.01687 15	58 0.00028 12
59	1.02974 43	119	2.07694 18	179	3.12413 94	59 0.01716 24	59 0.00028 60
60	1.04719 76	120	2.09439 51	180	3.14159 27	60 0.01745 33	60 0.00029 09

DICTIONARY-INDEX

Abscissa: horizontal coordinate, x, of a point, **11**
Absolute value: the numerical value of a number, $|a| = a$, if $a > 0$; $|a| = -a$, if $a < 0$, **41**
Addition principle: 136
Algebra, fundamental theorem of: every rational integral equation $f(x) = 0$ has at least one root, **112**
Algebraic expression: any expression formed by a finite number of additions, subtractions, multiplications, divisions, or extractions of roots on a finite set of variables, **52**
    binomial: an algebraic expression consisting of two terms, **52**
    fractional: a rational algebraic expression with variables in denominators, **52, 58**
    irrational: an algebraic expression which cannot be written without variables under radical signs, **52, 59**
    monomial: an algebraic expression consisting of one term, **52**
    polynomial: an algebraic expression generated by additions, subtractions, and multiplications, **52**
    radical: an algebraic expression denoting the root of a variable, **59**
    rational: an algebraic expression which can be written without variables under radical signs, **52**
    trinomial: an algebraic expression consisting of three terms, **52**
Amplitude of a complex number: *see* **Complex number, amplitude of a**
Angle: *see* **Point sets, angle**
Associative law: *see* **Law of operations, associative**
Asymptote: a line is said to be an asymptote to a curve if the line and curve get closer and closer together but never touch each other, **107**
Axes: the perpendicular lines OX and OY in the rectangular coordinate system, **11**
Axiom: an unproven statement in an axiomatic system, **22**
    of mathematical induction: *see* **Mathematical induction, axiom of**
Axiomatic systems: 22–23
    definition of: logical systems having primitive terms and postulates, **22**

Binary operation: *see* **Operation on a set, binary**
Binomial expansion, general terms of: the $(r + 1)$st term of the expansion of $(a + b)^n$ is $\binom{n}{r} a^{n-r} b^r$, **137**
Binomial theorem: $(a + b)^n = \sum_{r=0}^{n} \binom{n}{r} a^{n-r} b^r$, for any positive integer n, **137, 143**

Cancellation law: *see* **Law of operations, cancellation**
Cardinal number of a set: *see* **Set, cardinal number of a**
Cartesian coordinate system: a rectangular coordinate system used for the graphic representation of ordered pairs of real numbers, **11**

Cartesian product of two sets: *see* **Sets, Cartesian product of two**
Circle: the locus of a point that moves at a constant distance from a fixed point, **106, 107**
    center of a: the fixed point $C(h, k)$ in the general equation $(x - h)^2 + (y - k)^2 = r^2$, **106, 107**
    radius of a: the distance from the center to the circumference, **106, 107**
Closed interval: *see* **Interval of real numbers, closed**
Closure law: *see* **Laws of operations, closure**
Coefficient: the numerical factor of an algebraic term, **52**
Cofactor: *see* **Determinant, cofactor of an element of a**
Combination: any selection of objects taken without regard to order, **133–138**
Combined variations: if three or more variables are so related that one varies directly, jointly, and/or inversely as the remaining two, then we have a combined variation; e.g., $z = kxy$, **65**
Commutative law: *see* **Laws of operations, commutative**
Complement of a set: *see* **Set, complement of a**
Completing the square: the process for changing $ax^2 + bx + c$ to the form $a\left(x + \frac{b}{2a}\right)^2 + c - \frac{b}{4a}$, **94**
Complex number: $a + bi = r(\cos \theta + i \sin \theta)$, $i^2 = -1$, **46**
    amplitude of a: in the trigonometric form of a complex number $a + bi = r(\cos \theta + i \sin \theta)$, the angle θ is called the amplitude, **47**
    conjugate of a: the same complex number with the sign of the imaginary part changed, **46**
    graphical representation of a: Argand diagram, **46**
    imaginary part of a: the part bi of the number $a - bi$, **46**
    roots of a: $[r(\cos \theta + i \sin \theta)]^{\frac{1}{n}}$
$= \sqrt[n]{r} \cos \frac{\theta + k360°}{n} + i \sin \frac{\theta + k360°}{n}$, **47**
    trigonometric form of a:
$x + iy = r(\cos \theta + i \sin \theta)$, **47**
Composite number: a positive integer which is neither 1 nor prime, **40**
Conditional probability of p given q: *see* **Probability, conditional**
Conic section: a curve whose equation is of the second degree; the path of a point which moves so that the ratio of its distance from a fixed point to its distance from a fixed line is constant, **106**
    degenerate: a conic section whose discriminant is zero; its locus may be parallel lines, intersecting lines, or a point, **107**
    proper: a conic section whose discriminant is not zero, **107**
Conjugate of a binomial: the binomial with the sign of the second term changed, **59**
Conjunction: the truth-functional operation 'and,' **17**
Consistent postulate set: a postulate set which implies no contradictory statements, **22**
Constant: a symbol which, throughout a discussion, represents a fixed number; a symbol that varies over a set which consists of only one element, **64**

Continuity of a function: 131

    at a point: f is continuous at point a if $\lim_{x \to a} f(x) = f(a)$, **131**

    in an interval: f is continuous in an interval if f is continuous at every point in that interval, **131**

Coordinates: the real numbers which locate a point in a plane or space; the abscissa, x, and the ordinate, y, of a point, $P(x, y)$ in a place, **11**

Counting numbers, set of: $\{1, 2, 3, \ldots\}$, **4, 28**, *see also* **Natural numbers**

Cramer's rule: a method for solving systems of linear equations by determinants, **83**

De Moivre's theorem: if n is any positive integer, then $(\cos \theta + i \sin \theta)^n = \cos n\theta + i \sin n\theta$, **47**

Descartes' rule of signs: the number of positive roots of an equation $f(x) = 0$ with real coefficients cannot exceed the number of variations in sign in the polynomial $f(x)$; the number of negative roots cannot exceed the number of variations in sign in $f(-x)$, **112**

Determinant: 79–84

    cofactor of an element of a: the product of the minor of that element and $(-1)^{i+j}$, where i is the number of the row and j is the number of the element, **82**

    definition of a: a real number associated with a matrix, **82**

    element of a: the single entry in a specific row and column, **82**

    evaluation of a: 82

    minor of an element of a: the determinant obtained by deleting the row and column of that element, **82**

    third-order: a determinant with three rows and three columns, **83**

Difference of two real numbers: $a - b = a + (-b)$, **41**

Difference of two sets: *see* **Sets, difference of two**

Directrix: the fixed line in the definition of a conic section, **106, 107**

Discriminant of a quadratic equation: the expression $b^2 - 4ac$ for the equation $ax^2 + bx + c = 0$, **94**

Disjoint sets: *see* **Sets, disjoint**

Disjunction: the truth-functional operation 'or,' **17**

Distance from a point to a line: $P_1(x_1, y_1)$ to the line $ax + by + c = 0$, $d = \left| \dfrac{ax_1 + by_1 + c}{\sqrt{a^2 + b^2}} \right|$, **77**

Distributive law: *see* **Laws of operations, distributive**

Distributive property of multiplication: the product of an expression of two or more terms by a single factor is equal to the sum of the products of each term of the expression by the single factor, $(a + b - c)d = ad + bd - cd$, **46**

Division: the inverse operation of multiplication, **40**

    synthetic: the rapid process for dividing a polynomial by a linear factor, **112**

Divisor: an integer $a \neq 0$ is a divisor of an integer b if and only if there exists an integer x such that $b = ax$, **40**

    common: an integer a is a common divisor of integers b and c if a is a divisor of both b and c, **40**

    greatest common: the largest of the common divisors, **40**

Domain: *see* **Function, domain of a**

Domain of a variable: *see* **Variable, domain of a**

Dot product: a scalar equal to the sum of the products of the corresponding vector elements, **149**

Duality, principle of: In any true statement involving only union, intersection, or complementation, if $\cup$ and $\cap$ are interchanged and U and $\emptyset$ are interchanged, a true statement results, **16**

e: $e \sim 2.71828$, **124**

Eccentricity: the constant ratio of the distance to a fixed point and a fixed line in the definition of a conic section, **106**

Ellipse: the locus of a point which moves so that the sum of its distances from two fixed points is a constant; the conic section for which the eccentricity is less than one, **106**

    vertices of an: the points of intersection of the ellipse and its major axis, **106**

Empty set: *see* **Set, empty**

Equality: *see* **Identity**

    conditional: a statement of equality which is true for some values of the variables for which the expression is defined, **53, 58**

Equal sets: *see* **Sets, equal**

Equations: statements of equality, **70, 124–125**

    conditional: a statement of equality which is true for only specific values of the variable, **70**

    consistent linear: two linear equations that have one and only one point in common, **89**

    cubic: equations in which the highest power of the variable is three; $ax^3 + bx^2 + cx + d = 0$, **113**

    dependent: two linear equations in x and y whose graphs coincide, **89**

    derived: equations whose form has been changed by algebraic operations, **70**

    exponential: equations in which the unknown occurs in an exponent, **125**

    homogeneous: equations in which all terms are of the same degree, **89**

    inconsistent: two linear equations which define lines that are parallel; they have no common solution, **89**

    in matrix form: $AX = C$, **89**

    linear: statements of equality connecting constants and variables and in which any term involving the variables is of the first degree, **71, 76**

    logarithmic: equations in which the unknown occurs in an expression whose logarithm is indicated, **125**

    null: equations which have no solutions; e.g., $x = x + 1$, **70**

    quadratic: $ax^2 + bx + c = 0$, $a \neq 0$, **94**

    simultaneous linear: two (or more) linear equations in the same two (or more) unknowns, **88**

    solutions of a system of: the values of the variables which satisfy all of the equations simultaneously, **100**

    systems of: two or more equations in two or more variables when considered simultaneously, **88–89, 100–101**

Equivalence relation: *see* **Relation, equivalence**

Equivalent sets: *see* **Sets, equivalent**

Event: any subset of a sample space, **142**

    mutually exclusive: disjoint events, **142**

    simple: elements of a sample space, **142**

Experiment: 142

Exponential equations: *see* **Equations, exponential**

Exponential functions: 124–125

    definition of: $f : x \to b^x$, $b > 0$, **124**

    graphs of: 124

    inverse of: $f : x \to \log_b x$, **125**

    properties of: 124

Extraneous roots: *see* **Roots, extraneous**

Factor: an integer $a \neq 0$ is a factor of an integer b if and only if a is a divisor of b, **40**
Factorial: $n! = n(n-1)(n-2) \cdots 3 \cdot 2 \cdot 1$, **137**
Factorization: the process of finding two or more expressions whose product is a given expression; the increase of multiplication resulting in exact division, **40, 53**
   of polynomials, 53
   of real numbers, 40
Factor theorem: if r is a root of the equation $f(x) = 0$, then $(x - r)$ is a factor of $f(x)$, **112**
Fibonacci sequence: 132
Field properties of real numbers: 41
Finite set: *see* Set, finite
Focus: the fixed point in the definition of a conic section, **106, 107**
Fraction: a part of a unit; the indicated quotient of two algebraic expressions, **58**
   complex: a fraction whose numerator or denominator, or both, is also a fraction, **59**
Function: a set of ordered pairs such that no two ordered pairs have the same element; the relation between x and y so that for an assigned value of x a unique value of y is determined, **64**
   composite: a function obtained by replacing the independent variable of f by a function g, **65**
   continuity of a: *see* Continuity of a function
   determinant: *see* Determinant
   domain of a: the set of first elements of an ordered pair; the interval of values for the independent variable in a functional relation, **64**
   even: a function such that $f(-x) = f(x)$, **64**
   exponential: *see* Exponential functions
   hyperbolic: 124
   inverse: the function obtained by interchanging the role of the domain and the range of a given function, **64, 65**
   limit of a: the limit of a function f at a is the number F if, for every positive number ϵ, there exists a positive number δ such that $|f(x) - F| < \epsilon$ for all x having the property that $0 < |x - a| < \delta$, **131**
   logarithmic: *see* Logarithmic functions
   odd: a function such that $f(-x) = -f(x)$, **64**
   range of a: the set of second elements of an ordered pair; the interval of values for the dependent variable in a functional relation, **64**
   sequence: *see* Sequence
   zero of a: the values of a variable for which the function equals zero, **70, 94**

Geometric means: 132
Goldbach's conjecture: 27
Graphs: 64, 65
   of exponential functions: 124
   of functions: 64, 65
   of inverse relations: 65
   of logarithmic functions: 125
   of ordered pairs: 11
   of real numbers: 10
Group: 22
Grouping symbols: 52

Half-line: *see* Point sets, half-line
Half-plane: *see* Point sets, half-plane
Horner's method: a procedure for finding the approximate irrational roots of an equation of degree higher than 2, **112**

Hyperbola: the locus of a point which moves so that the difference of its distance from the fixed points is a constant; the conic section for which $e > 1$, **106, 107**
   vertices of a: the points of intersection of the hyperbola and its transverse axis, **107**
Hyperbolic functions: 124

Identity: a statement of equality which is true for all values of the variable(s) for which the expression is defined, **53, 58, 70**
Identity element: if $a * i = i * a = a$ for all $a \in S$, i is the identity in S for $*$, **23**
Identity matrix: *see* Matrix, identity for an n by n
Imaginary numbers: numbers of the form $a + bi$, $b \neq 0$, **46**
   see also Complex number
   operations on: 46
   pure: numbers of the form bi, $b \neq 0$, **46**
Imaginary unit: $i = \sqrt{-1}$, **46**
Inconsistent equations: *see* Equations, inconsistent
Independent statements: p and q are independent statements if $\text{Pr}[p \wedge q] = \text{Pr}[p] \cdot \text{Pr}[q]$, **142**
Independent trials: if a sequence of experiments consists of the repetition of a single experiment in such a manner that the result of any one experiment in no way affects the result of any other experiment, each experiment is an independent trial, **143**
Inequality: a statement that one quantity is greater than or less than another, **118**
   conditional: a statement of inequality which is true for only specific values of the variables, **119**
Infinite set: *see* Set, infinite
Integers, set of: $\{\ldots -2, -1, 0, 1, 2, \ldots\}$, **4**
   properties of the: 41
Intercept: the value of the variable for which a curve crosses the coordinate axis, **65, 76**
   x-: the value of x for which a curve crosses the X-axis, **65, 76**
   y-: the value of y for which a curve crosses the Y-axis, **65, 76**
Intercept form of a straight line: $\frac{x}{a} + \frac{y}{b} = 1$, **76**
Intersection of two sets: *see* Sets, intersection of two
Interval of real numbers: 5
   closed: $[a, b] = \{x(\text{real}) \mid a \leq x \leq b\}$, **5**
   half-closed, half-open: $[a, b) = \{x(\text{real}) \mid a \leq x < b\}$, **5**
   half-open, half-closed: $(a, b] = \{x(\text{real}) \mid a < x \leq b\}$, **5**
   open: $(a, b) = \{x(\text{real}) \mid a < x < b\}$, **5**
Inverse element: if i is the identity for $*$ and $a * a' = a' * a = i$, than a' is the inverse of a, **23**
Irrational numbers, set of: 40

Latus rectum: the line segment between the points of intersection of a line through the focus parallel to the directrix and the conic section, **106, 107**
Laws of operations: 23
   associative: $(a * b) * c = a * (b * c)$, **23**
   cancellation: if $a * b = a * c$, then $b = c$, **23**
   closure: if the result of operating on any two elements of a set is always an element of the set, the set is closed with respect to that operation, **23, 29, 94**
   commutative: $a * b = b * a$, **23, 29**
   distributive: if $a * (b \circ c) = (a * b) \circ (a * c)$, then $*$ is distributive over $\circ$, **23**

Limit: 127–132
    of a function: *see* **Function, limit of a**
    of a sequence: *see* **Sequence, limit of a**
    theorems for evaluating the: 130, 131
Line: a set of points along a straight path determined by any two points of Euclidean space, 5, 76
Line segment: the portion of a line between two specified points, 5, 76
    division of a: if the point $P(x, y)$ divides the line segment from $P_1(x_1, y_1)$ to $P_2(x_2, y_2)$ so that P_1P is a fraction of P_1P_2, then $x = x_1 + k(x_2 - x_1)$ and $y = y_1 + k(y_2 - y_1)$, 64, 65
    midpoint of a: the coordinates of the midpoint of $\overline{P_1P_2}$ are $x = \frac{1}{2}(x_1 + x_2)$, $y = \frac{1}{2}(y_1 + y_2)$, 77
Locus: the totality of all points, and only those points, whose coordinates satisfy the given equation, 64, 106
Logarithmic equation: *see* **Equations, logarithmic**
Logarithmic functions: 121–126
    common: logarithms to the base 10, 124
    definition of: $f: x \to \log_b x$, $b \neq 1$, 125
    graphs of: 125
    inverse of: $f: x \to b^x$, $b > 0$, 125
    natural: logarithms to the base e, 124
    properties of: 125
    tables of: 166
Logic related to set theory: 17

Mathematical induction: 28, 29
    axiom of: if S is a set of natural numbers with the properties that $1 \in S$ and that $k \in S \to k + 1 \in S$, then all natural numbers belong to S, 28
    problems: 31–36
    proof by: 29
Matrices: 79–84
    definition of: rectangular arrays of numbers, 82
    equal: matrices which have the same dimensions and equal corresponding elements, 82
    m by n: matrices which have m rows and n columns, 82
    product of two: the product AB is an m by n matrix whose elements are such that the entry in any row i and any column j is the sum of the products formed by multiplying each element in row i of A by the corresponding element in column j of B, 82
    row-equivalent: those which can be obtained from each other by row operations, 83
    square: matrices with n rows and n columns, 82
    sum of two: the sum is a matrix whose elements are the sums of the corresponding elements of the given matrices, 82
Matrix: 79–84
    identity for an n by n: an n by n matrix whose elements in the principal diagonal are all ones and whose other elements are all zeros, 82
    inverse of an n by n: an n by n matrix which satisfies the equations $A^{-1}A = I$ and $AA^{-1} = I$, where A is the given matrix, I is the identity for A, and A is the inverse of A, 82
    negative of a: any matrix whose elements are the negatives of the corresponding elements of the given matrix, 82
    scalar product of a real number k and a: a matrix whose elements are k times the corresponding elements of the given matrix, 82

    zero: an m by n matrix each of whose elements is zero, 82
Midpoint of a line segment: *see* **Line segment, midpoint of a**
Minor: *see* **Determinant, minor of an element of a**
Multiplication principle: 136
Mutually exclusive events: X_1 and X_2 are mutually exclusive events if $X_1 \cap X_2 = \emptyset$, 142
Mutually exclusive tasks: tasks that cannot occur simultaneously, 136

Natural numbers: 25–30
    axioms for: 28
    operations on: 29
    properties of: 29
    set of: $\{1, 2, 3, \ldots\}$, 28
Negation: the truth-functional operation 'not,' 17
Null equation: *see* **Equations, null**
Null set: *see* **Set, null**
Number line: 10
Numeral: a symbol for a number, 28

One-to-one correspondence between sets: *see* **Sets, one-to-one correspondence between**
Open interval: *see* **Interval of real numbers, open**
Open sentence: a statement containing variables which is neither true nor false, 17
Operation on a set: a rule whose result is a unique element of the set, 23
    binary: an operation performed on two elements, 23
    tertiary: an operation performed on three elements, 23
    unary: an operation performed on one element, 23
Ordered pairs: sets containing two elements and having the property that the left element and the right element can be distinguished from each other by virtue of the order in which they appear, 23
    graphs of: 11
Ordinate: the vertical coordinate, y, of a point $P(x, y)$ in a plane, 11
Origin: the point of intersection of the coordinate axes, 11
Overlapping sets: *see* **Sets, overlapping**

Parabola: the locus of points which are equidistant from a fixed point and a fixed line; the conic section for which $e = 1$, 106
    vertex of a: the point of intersection of the parabola and its axis, 106
Parallel lines: *see* **Point sets, parallel**
Pascal's rule: used to evaluate $\binom{n}{r}$, 136
Pascal's triangle: gives values of $\binom{n}{r}$, 137
Peano's axioms: 28
Permutation: any arrangement of elements in a definite order, 133–138
Perpendicular lines: *see* **Point sets, perpendicular**
Plane: *see* **Point sets, plane**
Point sets: sets of points of Euclidean space, 5
    angle: point sets which are the union of two rays such that the intersection of those rays is the end-point of both, 5, 17
    half-line: point sets which include all points on a line on one side of a given point on the line, but not the given point, 5

half-plane: point sets which contain all points on a plane on one side of a given line in the plane, but not the given line, **5**
line: 5
  parallel: sets of two lines which are disjoint point sets, **76**
  perpendicular: sets of two lines which overlap in one and only one point forming four congruent angles, **77**
  plane: point sets which extend indefinitely in all directions in a flat region of space, **5**
  polygon: point sets which are the union of line segments such that each segment intersects two others and each such intersection is an endpoint of exactly two of the segments, **17**
  ray: the points on a line including an endpoint and all the points on one side of the endpoint, **5**
Polygon: *see* **Point sets, polygon**
Polynomial equation in one variable: $a_0 x^n + a_1 x^{n-1} + \cdots + a_n = 0$, where n is a positive integer and a_i are constants, **71**
Polynomial expressions: 49–54
Postulates: unproven statements containing only primitive terms, **22**
Prime factorization: 40
Prime numbers: positive integers greater than 1 which have no divisors other than themselves and 1, **40**
  relatively: two positive integers whose only common divisor is 1, **40**
Primitive terms: undefined technical terms including a set of elements, relations which can hold among the elements, and operations which can be performed on the elements, **22**
Probability: 139–144
  conditional: $\text{PR}[p \mid q] = \dfrac{\text{PR}[p \wedge q]}{\text{PR}[q]}$, **142**
  measure of a set: the sum of the weights of the elements of the set, **142**
  of a statement: the probability measure of its truth set, **142**
Proper conic: *see* **Conic, proper**
Proper subset: *see* **Subset, proper**

Quadrants: the four parts into which a plane is divided by the coordinate axes, **11**
Quadratic equation: 91–102
  in x: $ax^2 + bx + c = 0$, $a \neq 0$, **94**
  in x **and** y: $ax^2 + bxy + cy^2 + dx + ey + f = 0$, **100**
  irrational roots of a: the roots of $ax^2 + bx + c = 0$ if $b^2 - 4ac$ is not a perfect square, **94**
  product of the roots of a: for a quardatic equation in x, $r_1 r_2 = \dfrac{c}{a}$, **95**
  rational roots of a: the roots of $ax^2 + bx + c = 0$ if $b^2 - 4ac$ is a perfect square, **94**
  sum of the roots of a: for a quadratic equation in x, $r_1 + r_2 = -\dfrac{b}{a}$, **94**
Quadratic formula: $x = \dfrac{-b \pm \sqrt{b^2 - 4ac}}{2a}$, **94**
Quotient of two real numbers: $\dfrac{a}{b} = a \times \dfrac{1}{b}$, **41**

Radius of a circle: *see* **Circle, radius of a**
Radius vector: the line segment from the origin to a point P in a plane, **11**

Range: *see* **Function, range of a**
Rational numbers: 40–41
  properties of: 41
  set of: $\left\{ \dfrac{a}{b} \;\middle|\; a \text{ and } b \text{ are integers}, b \neq 0 \right\}$, **40**
Ray: *see* **Point sets, ray**
Real numbers: 37–42
  graphs of: 10, 11
  intervals of: *see* **Interval of real numbers**
  properties of: 41
  set of: the set consisting of all rational and irrational numbers, **5, 41**
Relation: an association among two or more things, **22**
  dyadic: associates two elements, **22**
  equivalence: a relation which is reflexive, symmetric, and transitive, **22**
  reflexive: aRa, **23**
  symmetric: $aRb \rightarrow bRa$, **23**
  transitive: $(aRb \text{ and } bRc) \rightarrow aRc$, **23**
  triadic: associates three elements, **22**
Relatively prime: *see* **Prime numbers, relatively**
Remainder theorem: if a polynomial $f(x)$ is divided by $(x - r)$ until a remainder independent of x is obtained, this remainder is equal to $f(r)$, **112**
Repeating decimals: decimal fractions in which a series of digits repeat; e.g., $0.333\ldots = 0.\overline{3}$, **40**
Roots: the roots of an equation are the solution set formed when the equation is equated to zero, **94, 112**
  extraneous: roots of a derived redundant equation that are not roots of the original equation, **70**
  irrational: solutions of equations that are not rational numbers, **94, 112**
  of a complex number: *see* **Complex number, roots of a**
  of a quadratic: 94
  rational: in the general equation, $a_0 x^n + a_1 x^{n-1} + \cdots + a_{n-1} x + a_n = 0$, a number $\dfrac{b}{c}$ where b is a factor of a_n and c is a factor of a_0, **112**
Row operations: 82–83

Sample space: a set of logical outcomes of an experiment, **142**
Scalar: a number as distinguished from a matrix or vector, **82, 148**
Sequence: 127–132
  arithmetic: consecutive terms have a constant difference, **130**
  convergent: a sequence which has a limit, **130**
  definition of a: $a = a_1, a_2, a_3, \ldots, a_n, \ldots$, **130**
  divergent: a sequence which has no limit, **130**
  general term of a: 130
  geometric: consecutive terms have a constant ratio, **130**
  limit of a: the limit of a sequence is the number A if, given any positive number ϵ, there exists a real number N such that $|a_n - A| < \epsilon$ for all $n > N$, **130**
Series: 127–132
  arithmetic: the indicated sum of an arithmetic sequence, **130**
  definition of a: the indicated sum of a sequence, **130**
  geometric: the indicated sum of a geometric sequence, **130**

Set: 1–18
    cardinal number of a: the number of elements in the set, **11**
    complement of a: $A' = \{x \mid x \notin A\}$, **16**
    definition of a: any collection of objects, **4**
    element of a: 4
    empty: the set with no elements, **22**
    finite: a set which contains a countable number of elements, **4**
    infinite: a set which contains more elements than can be counted, **4**
    null: the empty set, **4, 22**
    replacement: the domain of a variable, **4**
    truth: the values of x for which p_x is true, **4, 142**
    universal: the set which includes all elements under consideration in a given discourse, **4**

Sets: 1–18
    cartesian product of two: $A \times B = \{(a,b) \mid a \in A \text{ and } b \in B\}$, **11**
    difference of two: $A - B = \{x \mid x \in A \text{ and } x \notin B\}$, **16**
    disjoint: A and B if $A \cap B = \emptyset$, **10**
    equal: two sets with the same elements, **10**
    equivalent: two sets with the same number of elements, **11**
    intersection of two: $A \cap B = \{x \mid x \in A \text{ and } x \in B\}$, **16**
    methods of describing: 5
    one-to-one correspondence between two: a pairing of the elements of the first set with the elements of the second set such that each element of the first set corresponds to only one element of the second set, and each element of the second set corresponds to only one element of the first set, **10**
    operations on: complementation, difference, intersection, and union, **13–18**
    overlapping: A and B if $A \cap B \neq \emptyset$, **10**
    point: *see* **Point sets**
    union of two: $A \cup B = \{x \mid x \in A \text{ or } x \in B\}$, **17**

Slope: the tangent of the angle of inclination (the acute angle between a line and the positive X-axis), **76**
Square matrix: *see* **Matrix, square**
Subset: $A \subseteq B$ if and only if $a \in A \rightarrow a \in B$, **10**
    proper: $A \subset B$ if and only if $A \subseteq B$ and $A \neq B$, **10**
Symmetric relation: *see* **Relation, symmetric**
Synthetic division: *see* **Division, synthetic**

Tertiary operation: *see* **Operation on a set, tertiary**
Theorem: any non-primitive statement of a system, **22**
Transfinite number: 11

Transitive relation: *see* **Relation, transitive**
Tree diagram: 136, 143
Tree measure: 143
Trichotomy property: for each real number a, exactly one of the following is true: a is positive, a is negative, or a is zero, **41**
Truth-functional operations: conjunction, disjunction, and negation, **17**
Truth set: *see* **Set, truth**

Unary operation: *see* **Operation on a set, unary**
Union of two sets: *see* **Sets, union of two**
Universal set: *see* **Set, universal**
Unknown: the variable in a statement of comparison, **70**

Variable: a symbol which, throughout a discussion, may assume different values, **4, 64**
    dependent: the range of a function, **64**
    domain of a: the set of meaningful replacements for the variable, **4**
    independent: the domain of a function, **64**
Variation: the effect on one variable by a change in another, **64, 65**
    direct: $y = kx$, **64**
    inverse: $y = \dfrac{k}{x}$, **65**
    joint: $y = kxz$, **65**
Vector: an ordered n-tuple of n real numbers, **148–149**
    norm of: square root of the sum of the squares of the n real numbers, **149**
    length of: the norm of the vector, **149**
    unit vector: a vector divided by its norm, **149**
Venn diagram: a pictorial representation of sets, **10, 16, 17**
Vertex, of a parabola: *see* **Parabola, vertex of a**
Vertices, of an ellipse: *see* **Ellipse, vertices of an**
Vertices, of a hyperbola: *see* **Hyperbola, vertices of a**

Whole numbers, set of: $\{0, 1, 2, 3, \ldots\}$, **38**

x-intercept: *see* **Intercept, x-**

y-intercept: *see* **Intercept, y-**

Zeros of a function: *see* **Roots**
Zero matrix: *see* **Matrix, zero**
Zero of a function: *see* **Function, zero of a**

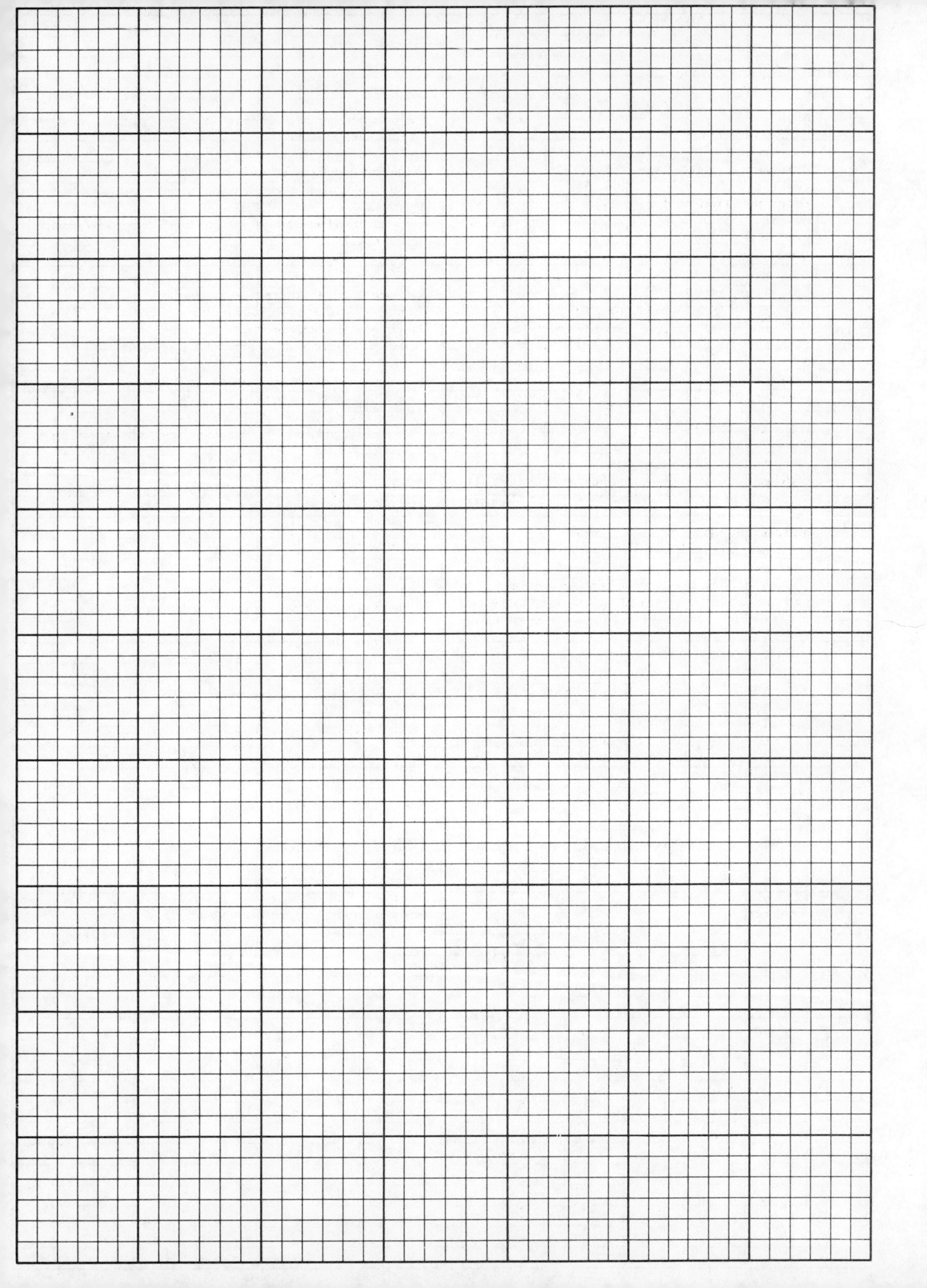

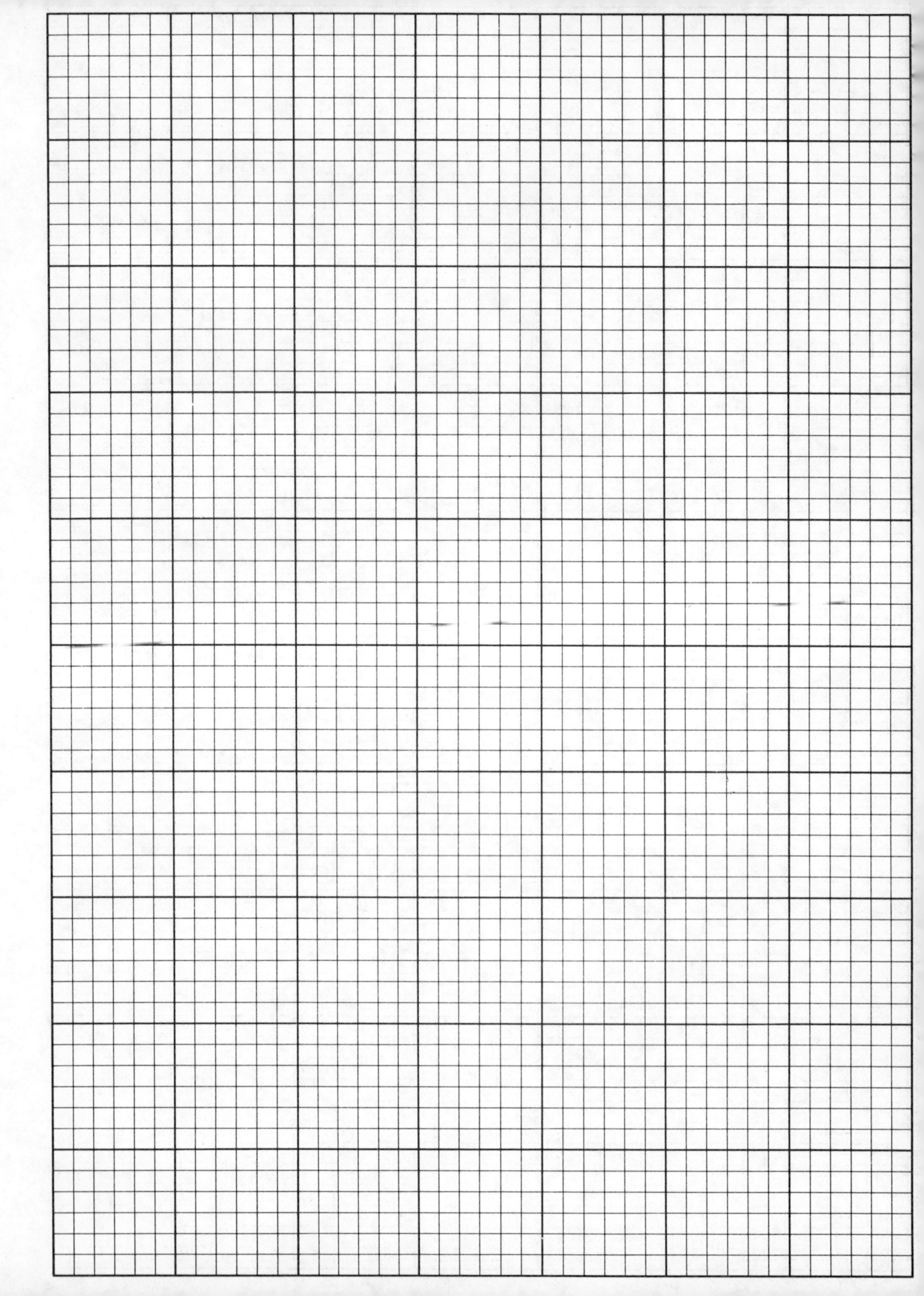

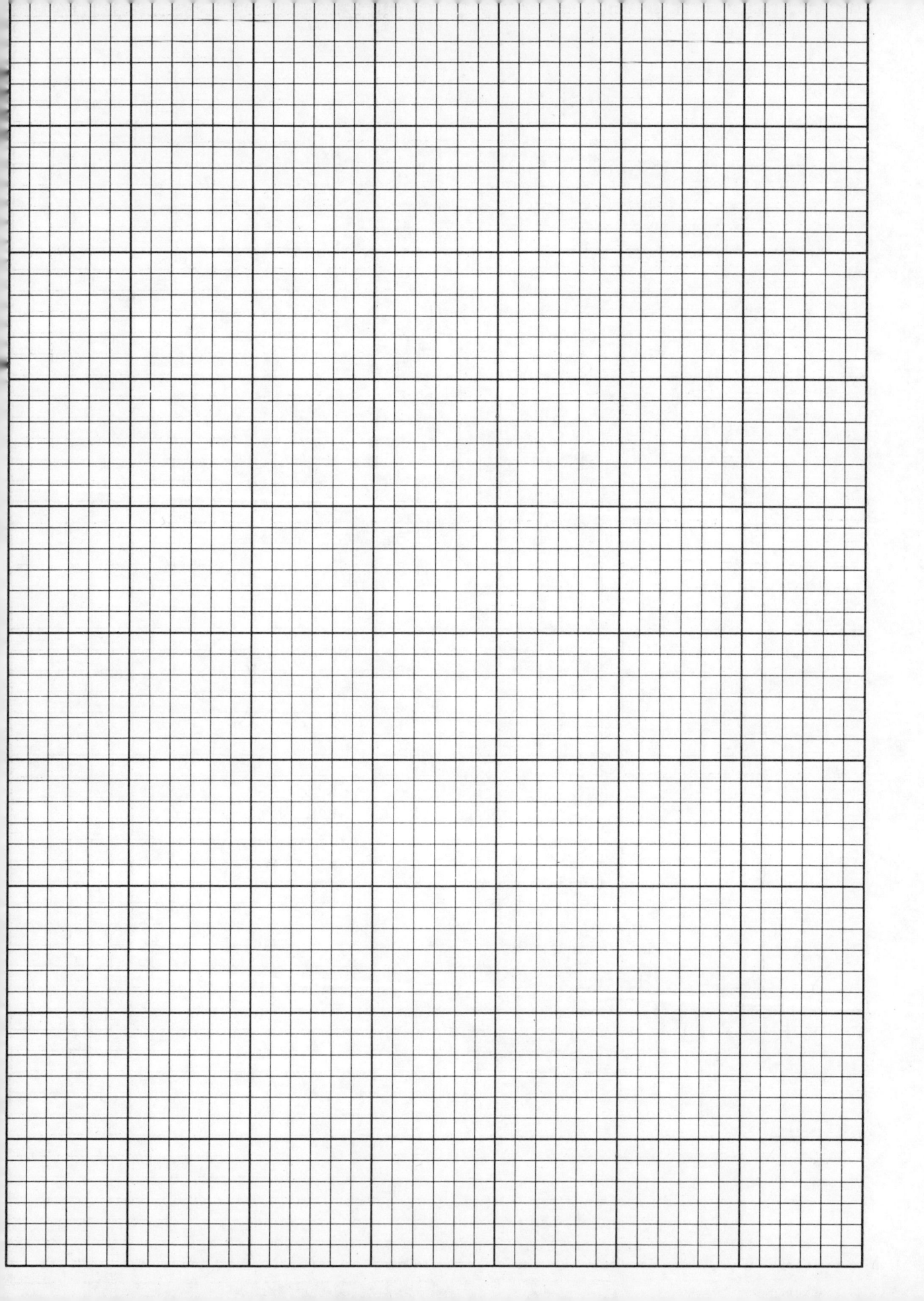

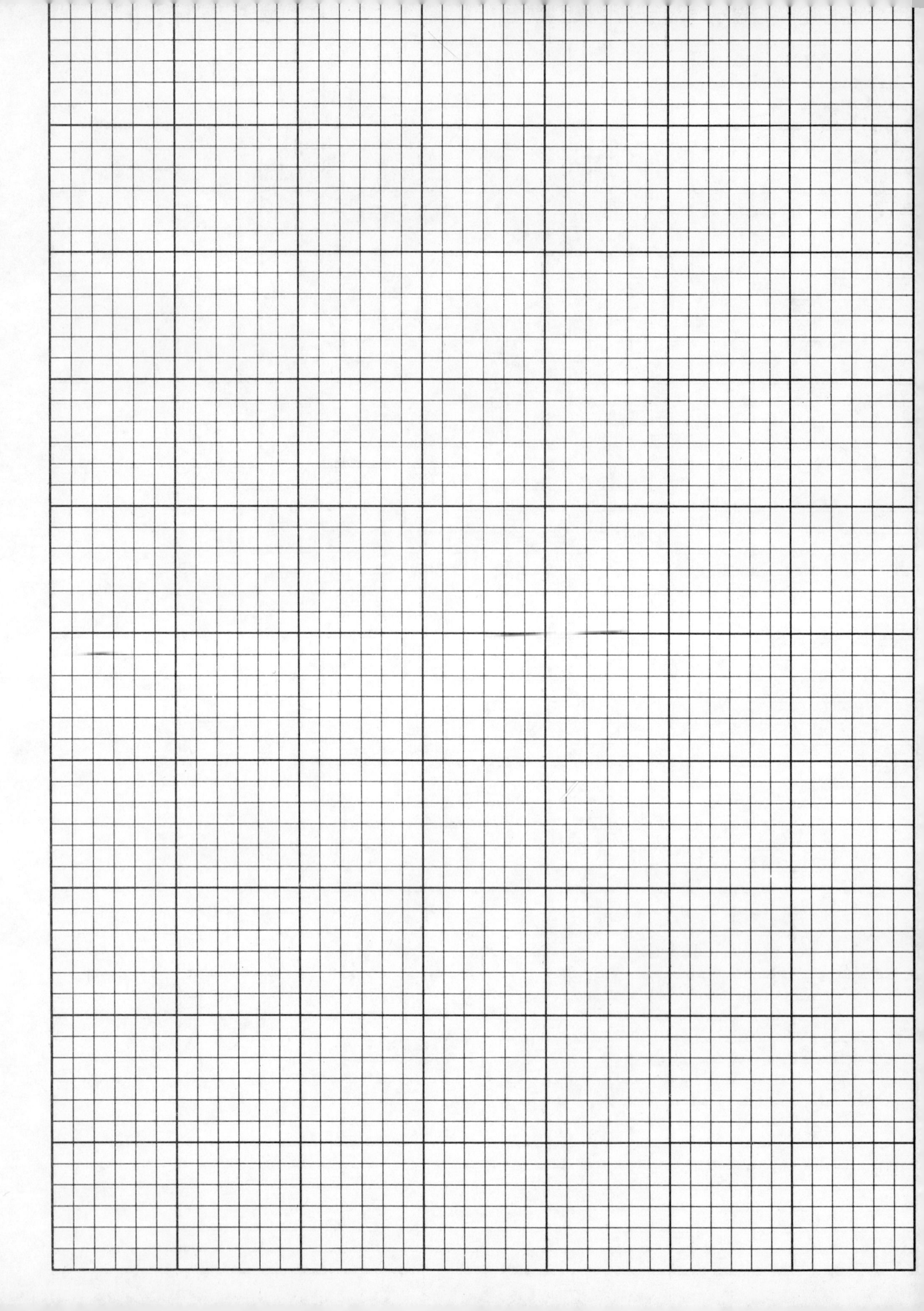

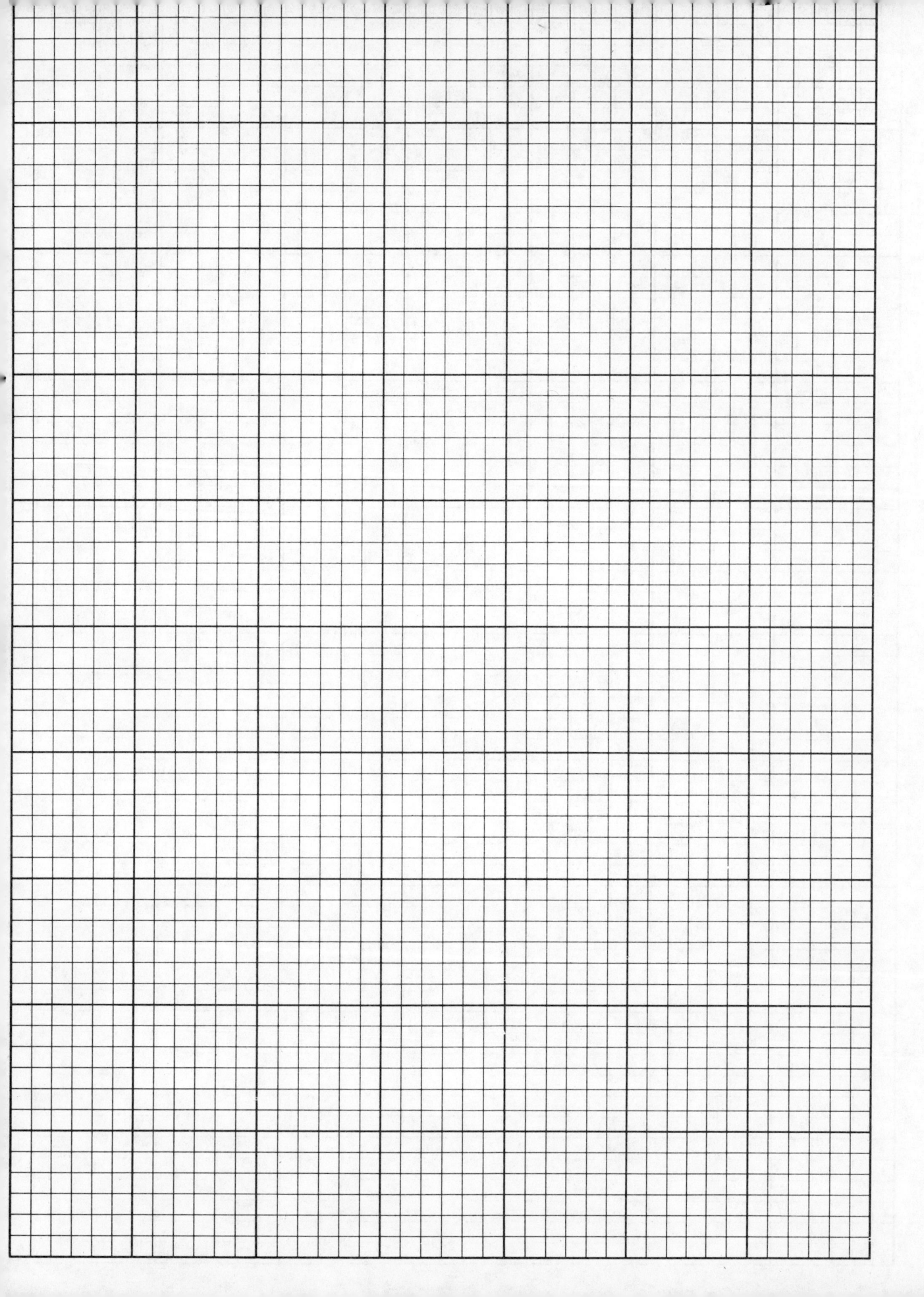

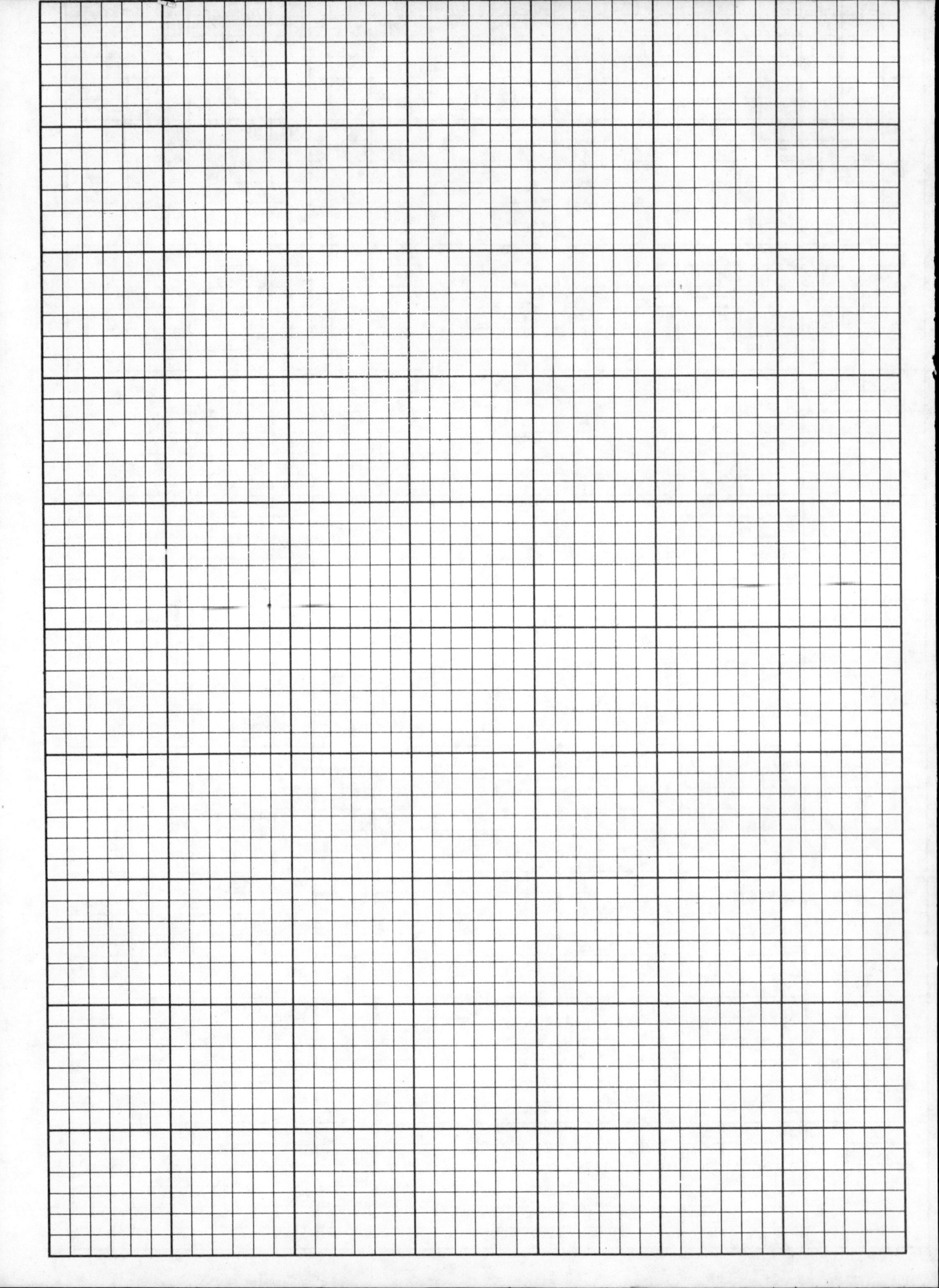